THE SOIL CHEMISTRY

OF HAZARDOUS MATERIALS

THE SOIL CHEMISTRY OF HAZARDOUS MATERIALS

James Dragun, Ph.D.

Hazardous Materials Control Research Institute
Silver Spring, Maryland

Contents

Acknowledgments

The author thanks the following individuals who spent many hours of their personal time reviewing this document and discussing their suggestions on revisions with the author:

Dr. Abdul Abdul, Hydrogeologist, General Motors Corporation

Dr. Bruce Baughman, Soil Chemist, American Petroleum Institute

Mr. Joseph Daniels, Environmental Engineer, Stalwart Environmental Sciences & Services

Dr. Walter Grube, Soil Scientist, U.S. Environmental Protection Agency

Ms. Sharon Mason, Hydrogeologist, Ford Motor Company

Dr. Randall Ransom, Environmental Specialist, Dow Corning Corporation

Dr. Michael Roulier, Soil Scientist, U.S. Environmental Protection Agency

Mr. Ronald Wilhelm, Soil Scientist, U.S. Environmental Protection Agency

Mr. Bruce Yare, Hydrogeologist, Monsanto Chemical Company

The author especially thanks Dr. Dale E. Baker, Professor of Soil Chemistry, Penn State University for the years he devoted to mentoring and counseling the author.

The author thanks Dr. Stanley Kirschner, Professor of Chemistry, Wayne State University for his guidance and encouragement during the author's undergraduate years.

The author also thanks Dr. Ed Calabrese, Professor of Toxicology, University of Massachusetts for his encouragement, insight and guidance as the author wrote this book.

To:

The Lord God Almighty
The Creator of Earth

PREFACE

Soil chemistry involves the study of the chemical principles governing the behavior of soil and the behavior of chemicals in soil. Although several soil chemistry texts have been published, this text is unique for several reasons.

First, this text focuses primarily on those soil chemistry principles which are pertinent to the behavior of hazardous materials in soil. The term hazardous material encompasses a wide variety of toxic, corrosive, ignitable, flammable, and/or radioactive products and wastes. These can be solids, liquids, gases and etiologic agents. Although the scope of this text was limited primarily to introducing and discussing basic principles, case studies involving a wide variety of hazardous materials have been incorporated into the text, wherever feasible.

Second, this text was written primarily for those hydrologists, geologists, civil engineers, environmental engineers, geotechnical engineers, biologists, chemists, and environmental scientists who must identify and determine the significance of chemicals in soil. Since soil chemistry and its related disciplines are usually taught only at the graduate level in relatively few universities, this manuscript was written with the understanding that most scientists and engineers have no background in soil chemistry and have only been exposed to college freshman-level chemistry. The author strongly suggests that those individuals who are unfamiliar with soil chemistry should read the text in its entirety, beginning with Chapter One. Then, the reader will be more capable of finding answers to questions about how a chemical may behave in a particular situation because the reader will have a better understanding of where information is found in this book. For example, after reading the entire text, one would know that in order to ascertain how to assess the significance of a copper acid plating bath spill, information from Chapters Five (acids in soil), Three (fixation), and Four (adsorption) would be most useful.

A thorough understanding of the basic principles presented in this manuscript is essential for those involved in site remediation, in siting or assessing waste treatment, storage, and disposal activities and facilities, and in analyzing chemical releases from and the cleanup and closure of industrial facilities. These principles govern the proper estimation of the migration and degradation potential of chemicals in topsoil, the unsaturated zone, and the saturated zone. Mathematical models must take into account these principles or else they fail. The determination of "how clean is clean" depends upon a thorough understanding and application of these principles. All soil and groundwater in-situ treatment technologies must work in unison with these principles, or else the technologies will fail. These principles apply to chemical behavior in solid and hazardous waste treatment and disposal units employing or impacting soil, and other geologic materials. These units include landfarms, landfills, deepwell injection systems, compost piles, and sites affected by leaks, spills, or other types of deliberate or accidental chemical releases.

Third, an important objective of this text is to dispel commonly-held misconceptions and to correct misapplication of soil chemistry principles. During the past several years, the author has worked in association with numerous professionals from local, state, and federal agencies, from private industry, and from engineering-consulting firms. During this time, the author has identified over 100 common misconceptions and misapplication of basic principles. These misconceptions and misapplications have been addressed in this text.

Fourth, this text addresses the behavior of organic as well as inorganic hazardous materials, bulk hazardous materials as well as hazardous materials present in dilute concentrations in aqueous and nonaqueous solvents.

Fifth, this text stresses the importance of chemical structure and its effect on the behavior of chemicals in soil. This text shows in many ways that chemicals possessing similar structure will react in a similar manner.

Sixth, in addition to introducing and discussing basic principles, this text focuses on the practical application of these principles. This text has incorporated information from other soil science and environmental disciplines, whenever necessary or appropriate, in order to achieve this goal.

Seventh, this text presents several simple and relatively uncomplicated models that give general information on chemical migration and degradation in soil and groundwater. Because models need (as input parameters) data on soil physical/chemical properties and hazardous material physical/chemical properties, this text also contains data which the reader can use as heshe experiments with these models.

This text is based on the professional experience and opinions of the author as well as on the technical information available to the author at the time the text was written. The text reflects the state of the science: in some areas, the science is relatively well-developed, while in other areas the science is in its infancy. In any event, the author hopes that the proper understanding of the information presented in this text, when utilized in conjunction with laboratory and field data on a site-by-site basis, will allow a more prompt and meaningful assessment of the relative magnitude of hazards, will save time, and will assist in the cost-effective allocation of funds and other resources.

1
OVERVIEW:
How Soil Governs Water Quality

INTRODUCTION

The term "soil' has different shades of meaning and connotation, depending on the professional field in which it is being considered. For example, a civil engineer traditionally views soil as a geologic body possessing physical properties which makes it suitable as construction material or as a body upon which structures and facilities are built. The geologist views soil primarily as a natural body, derived over long periods of time, which transports water at rates dependent upon its physical composition. The geologist is primarily concerned with obtaining (a) an estimate of the texture (i.e. sand, silt, and clay composition) of this natural body, and (b) general information concerning the age and genesis of the body. The hydrologist views soil as a naturally occurring porous body that affects the rate and direction of water flow, due to the size of interconnecting pores.

Within the vast field of soil science, soil is viewed differently by the agronomist, the soil classifier, the soil surveyor, and the soil chemist. The agronomist views soil as a natural medium for the growth of land plants. According to this viewpoint, the growth response of a crop is related to the soil's nutrient supplying capacity, its ability to provide a suitable medium for root development, and its responsiveness to management practices. The soil classifier or pedologist views soil as a natural body comprised of soil layers (horizons) residing near the soil surface, each with distinct, unique morphology. The soil surveyor studies the areal distribution of soil classes. The objectives of soil mapping is the delineation of areal units onto soil maps, the study and rationalization of the distribution patterns, and the appraisal of the suitability of soil units for various land uses.

The soil chemist's perception of soil is significantly different than the perception of the geologist and the hydrologist. The soil chemist views soil as a naturally occurring, unconsolidated material residing above bedrock which supports numerous physical, chemical, and microbial reactions. In order to characterize soil and to determine how soil governs the concentration of hazardous materials in water, the soil chemist draws information from numerous fields: colloid chemistry, surface chemistry, physical chemistry, organic chemistry, inorganic chemistry, geochemistry, electrochemistry, organometallic chemistry, radiochemistry, analytical chemistry, geology, clay and soil mineralogy, soil genesis, soil fertility, soil microbiology, and soil physics. This chapter will provide a general overview on the soil chemist's perception of soil with regard to hazardous materials. The remaining chapters of this book will discuss the physical, chemical, and microbial reactions in greater detail. Also, these chapters will show how the basic principles of the many disciplines listed above affect the behavior of hazardous materials in soil.

SOIL REACTIONS WITH HAZARDOUS MATERIALS

Soil governs the concentrations of hazardous materials in water found in both unsaturated and saturated zone soil. In this text, the term groundwater will refer to water existing beneath the water table in the soil's saturated zone. The term soil water will refer to water existing above the water table in the soil's unsaturated zone (i.e. the vadose zone). The term water will specifically refer to both soil water and ground water. Figure 1.1 illustrates how soil governs water quality. When a water sample from a monitoring well is retrieved and analyzed, the sample may contain some concentration of a hazardous material. This concentration is the net result of the reactions and interactions illustrated in Figure 1.1. An analysis of Figure 1.1 will reveal several important facts regarding soil reactions with hazardous materials.

First, soil is a heterogeneous, reactive mass of solids and water. These solids possess distinct physical and chemical properties and chemical reactivities which exert a profound effect on the concentration of hazardous materials in water. These soil solids are comprised of (a) primary, secondary, and amorphous minerals and (b) organic matter. These minerals and organic matter do not exist in an independent, non-interacting state. In fact, a number of processes involving the transport, transformation, and interactions of soil solids are continually occurring. Twenty-four of these processes which occur over geologic time to form soil horizons are listed in Table 1.1.

Second, soil possesses solids with significant surface areas that mediate various physical and chemical surface reactions (see Table 1.2). The primary physical reaction occurring at soil surfaces, which governs the concentration of hazardous materials in water, are hydrolysis, oxidation, reduction, bound residue formation, and various fixation reactions. It is most important to note that water is in intimate contact with soil surfaces at all times. As a result, hazardous materials dissolved in water are not isolated from these surfaces but are in intimate contact with them. Even desert soils contain a few percent of water on a volume basis.

Third, soil contains a significant amount of water in which aqueous chemical reactions may occur. Water may comprise from 33 to 50 percent of the total volume of a saturated soil or from 107,000 to 163,000 gal/acre-foot of soil.

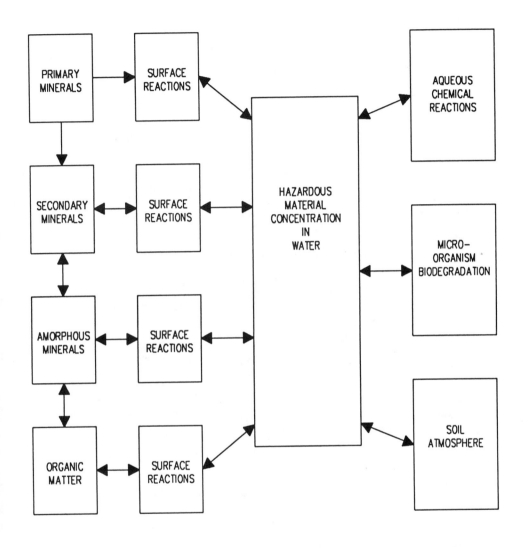

FIGURE 1.1 A diagram illustrating the many soil phases and reactions that govern the concentration of a hazardous material in water.

TABLE 1.1 Soil Processes Involving the Transport, Transformations, and Interactions of Soil Solids.[a]

Process	Description
Alkalization	Accumulation of sodium ions and salts in a soil horizon.
Braunification	Release of iron from primary minerals.
Calcification	Accumulation of calcium carbonate in a soil horizon.
Cumulization	Addition of minerals at soil surfaces due to aeolian and hydrologic depositions.
Dealkalization	Migration of sodium ions and salts from natric horizons.
Decalcification	Removal of calcium carbonate from a soil horizon.
Decomposition	The breakdown of minerals and organic matter.
Desalinization	Removal of soluble salts from salic soil horizons.
Eluviation	Movement of solids and salts out of a soil.
Enrichment	Addition of solids and salts to a soil.
Erosion	Removal of solids and salts from a soil surface.
Gleization	The reduction of iron under anaerobic environments. Results in development of bluish to greenish gray soil colors, and possibly (a) yellowish brown, brown, and black mottles, and (b) iron and manganese-iron concretions.
Humification	Transformation of soil organic matter into humus.
Laterization	Movement of silica out of soil, resulting in a soil whose mineralogy is dominated by sesquioxides and possibly laterite, plinthite, and concretions.
Lessivage	Mechanical migration of small mineral particles from topsoil into subsoil, producing a relatively enriched clay horizon.
Leucinization	Lightening or paling of soil color due to the disappearance of organic matter.
Littering	Accumulation of organic matter and humus on the surface of mineral soil.
Melanization	Darkening of soil by addition of organic matter.
Mineralization	The release of mineral oxides through organic matter decomposition.
Pedoturbation	Churning and cycling of soil via biological and physical processes (e.g. freeze-thaw cycles, wet-dry cycles).
Podzolization	The migration of aluminum, iron, and/or organic matter, resulting in higher silica concentrations in the eluviated horizon.
Ripening	Alterations in the physical, chemical, and biological properties of organic soil due to aeration.
Salinization	Accumulation of soluble salts such as calcium, magnesium, sodium, and potassium chlorides and sulfates.
Synthesis	The formation of new minerals and organic matter.

[a] Modified From Data Presented in Ref. 1.

Water is usually the solvent in a soil system that is responsible for moving dilute concentrations of hazardous materials. In some cases, however, organic solvents may be present, sometimes to the exclusion of water. Hazardous materials will react differently in different soil-solvent systems. For exam-

TABLE 1.2   Surface Areas of Soils and Soil Minerals[1,4,5,6,9]

	Surface Area (m²/gram)
Soil	
Sandy Loam	10-40
Loam	50-100
Silt Loam	50-150
Silty Clay Loam	120-200
Clay	150-250
Soil Clay Minerals (crystalline)	
Kaolinite	5-39
Halloysite	21-43
Chlorite	25-40
Illite	100-200
Montmorillonite (smectite)	700-800
Vermiculite	750-800
Saponite	700-800
Soil Clay Minerals (amorphous)	
Allophane	145-900
Imogolite	900-1100
Iron Oxides (crystalline)	
Goethite	14-90
Hematite	34-45
Quartz	
Platy	1-3
Spherical	2
Opal	1-200
Diatomaceous	89-123

ple, polychlorinated biphenyls (PCBs) and dioxin are usually adsorbed extensively onto the surface of finer particles in soil-water systems and are immobile for all practical purposes. However, PCBs and dioxin dissolved in oils are usually not adsorbed significantly and can migrate with any free-flowing oil. The effects of solvents on hazardous material migration and degradation in soil will be addressed in this text, when appropriate.

Fourth, soil contains a large mass of microorganisms possessing thousands of enzymes that can catalyze the transformations and degradation of many organic and inorganic molecules. For example, the number of bactera in top-soil under optimum conditions may reach 10 billion cells/gram soil. Concentrations of bacteria, measured in ground water obtained from monitoring wells, commonly exceed 1000 to 10,000 cells/ml.

Fifth, although 50 to 67 percent of the soil bulk volume is occupied by solids, 33 to 50 percent is comprised of an interconnected network of pores. In the unsaturated zone, these pores are occupied by soil water and the soil atmosphere. The soil atmosphere represents a significant transport pathway for many volatile hazardous materials.

Sixth, several reactions are simultaneously occurring in the soil system illustrated in Figure 1.1. For many hazardous materials, soil acts as a complicated buffer system which buffers the hazardous material's concentration in water. Physical, chemical, and microbial reactions will first operate to depress the hazardous material's concentration in water. If the hazardous material is held on soil exchange sites or adsorption sites on minerals and organic matter, these sites will release the hazards material (i.e. desorption) to attempt to re-establish its concentration in water, or, the depletion of some hazardous materials in water may cause some crystalline or amorphous materials to dissolve very slowly. This process is an attempt to resupply partially depleted soil exchange sites or adsorption sites and to eventually replenish the depleted water.

It is most important to note that each physical, chemical, and microbial reaction which may occur in one soil for one hazardous material may occur at a different rate and to a different extent in another soil. For example, PCBs applied to the surface of a sandy soil should eventually migrate to groundwater. On the other hand, in a topsoil with a significantly high amount of montmorillonite, and where water is the only solvent present, PCBs should be so extensively adsorbed that the migration rate to groundwater should be negligible. It is not uncommon to find one reaction such as adsorption which is so dominating that it will completely overshadow all others. Hazardous materials in many unsaturated zone soils do not migrate to groundwater due to the rates and dominance of one or more of the reactions discussed in this text.

Seventh, it is most important to note that data from the chemical analyses of groundwater gives the concentration of a chemical at that sampling site at the time the sample was collected. It represents a "snapshot" photograph that captures action at one moment in time. This concentration has already been reduced by physical, chemical, and microbial reactions occurring in the soil system. In many cases, information on past action and on future action can be estimated from these data. When appropriate, guidance will be given on estimating future action and on reconstructing past action.

Eighth, it is important to note that information on the historical development of soil over geologic time is not represented in Figure 1.1 and that it is generally not discussed in this book. How a soil developed over geologic time is not an important consideration in most cases in determining how the soil governs the concentration of hazardous materials in water. The physical, chemical, and biological properties of the soil of concern are of primary importance.

SOIL HETEROGENEITY AND SOIL REACTIONS
WITH HAZARDOUS MATERIALS

The term soil is a collective noun that refers to a natural resource which resides at the earth's surface as a continuum of many different types of soils. These different soil types result from the action of climate on a parent material which has been modified by the influence of vegetation, organisms, and relief over some period of geologic time.[1] These different soil types have a large spectrum of physical, chemical, and biological properties and reactivity.

An undue amount of emphasis and significance has been placed on the overall heterogeneity of soil properties among soil types. In assessing how a soil reacts with a hazardous material, what is important is not the heterogeneity of all properties of a soil type, but how heterogeneity of a property within the soil of concern affects the reaction of concern. For example, if we are studying the hydrolysis of chloromethane, it matters little if a particular soil's organic matter content varies significantly. Why? Because pH and not organic matter content is the soil property primarily affecting the rate of hydrolysis reactions. Furthermore, the soil of concern may have an A horizon pH of 3.5, a B horizon pH of 4.5, and a C horizon pH of 5.5. The pH of this soil is inherently variable. But, the hydrolysis rate for chloromethane is constant below pH 7 and shifts dramatically at higher pHs. Therefore, with regard to the hydrolysis reaction, the variability of pH in the soil of concern is not an issue; however, it would be an important issue in a soil whose pH varied from 6 to 8.

Although the soil system and its analysis is complex, the example cited in the paragraph above shows how the study of soil can be simplified via an analysis of individual phases. A phase is defined as a relatively homogeneous region within a soil system having some spatially uniform property or properties. In the absolute sense, no soil is homogeneous. However, for the purposes of determining hazardous material reactions in soil, the heterogeneity found in many soil phases is not great and, therefore, the phase in question can be treated as if it were homogeneous.

Figure 1.1 presented a general overview of several soil phases and reactions which govern water quality. The remaining chapters of this book will deal primarily with a more detailed identification and analysis of those soil phases and reactions governing water quality and an analysis of the components which serve as the fundamental building blocks of soil phases.

SOIL HETEROGENEITY AND
SOIL CLASSIFICATION

Although soil properties vary considerably from one major region of the

U.S.A. to another, soil variation is usually less pronounced within regions. In order to better define the variation of soil properties between regions and within regions, several classification schemes have been developed. These schemes attempt to classify soils on the basis of soil properties, soil use, or both. Two classification schemes are in widespread use in the U.S.A.: The Unified Soil Classification System (the USCS) and the U.S. Comprehensive Soil Classification System, which was developed by the Soil Staff of the U.S. Department of Agriculture (the USDA system).

The USCS provides a means of classifying soils in accordance with their engineering properties. Initially, this system was utilized to judge a soil's suitability as subgrade for roads and airfields. Today it is used for most engineering applications for soil.

The USCS is based on soil texture, gradation, and liquid limit. It may provide useful particle size information if a more detailed breakdown of particle size fraction commonly referred to as "fines" is measured. It does not address important soil properties and parameters such as pH, organic matter content, clay type, and cation exchange capacity. As a result, the USCS has several limitations restricting its usefulness in assessing soil reactions with hazardous materials.

The USDA system classifies soil on the basis of soil physical, chemical, and biological properties. Since most soil and groundwater quality investigations address subsurface chemical migration, it is most important to note that this system addresses both surface and subsurface soil. The majority of the chemical and physical properties which affect the reactions of hazardous materials in soil are utilized in this system as general classification criteria. For example, the system utilizes soil texture, soil structure, soil mineralogy, pH, salt content, and organic matter content.

The USDA system classifies soil in the continental U.S.A. into nine orders (see Figure 1.2). Table 1.3 lists general physical, chemical, and mineralogical properties of soil which are characteristics of each soil order. Detailed descriptions of this system and of soil survey reports are provided in other texts[5,6,8] and will not be discussed in this chapter. A thorough familiarization of the information provided in Figure 1.2 and Table 1.3 will be most helpful in assessing the relationships between soil phases and the reactivity of hazardous materials presented in the following chapters.

It is most important to remember that soil classification schemes do not necessarily utilize the same definitions, descriptive units, and ranges for all soil properties (see Figure 1.3 and 1.4). For example, boring logs for geologic investigations utilize the USCS for particle size analysis. However, most published literature on chemical migration and degradation is keyed into the USDA system. These two schemes classify soil texture using different textural

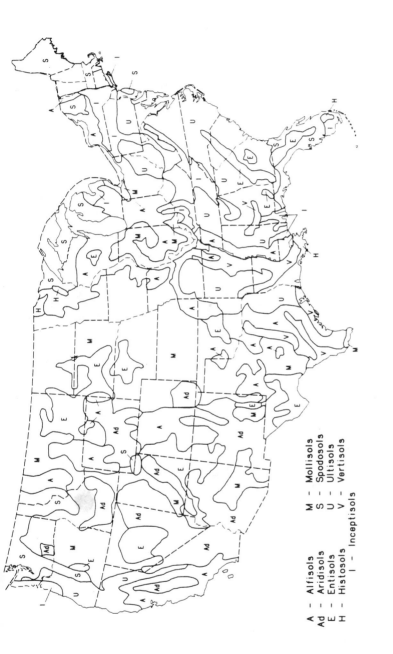

FIGURE 1.2 National locations of the 9 soil orders as listed. (From USDA, Soil Survey Staff, Soil Series in the United States, Their Taxonomic Classification, USDA, Soil Conservation Service, Washington, D.C., 1968, 433.)

A – Alfisols   M – Mollisols
Ad – Aridisols   S – Spodosols
E – Entisols   U – Ultisols
H – Histosols   V – Vertisols
        I – Inceptisols

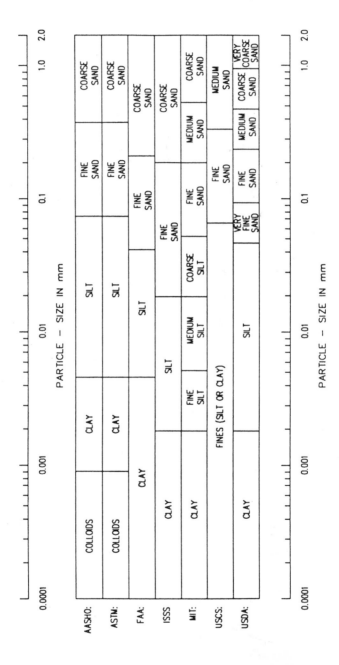

FIGURE 1.3 Ranges of particle sizes for several soil classification schemes.

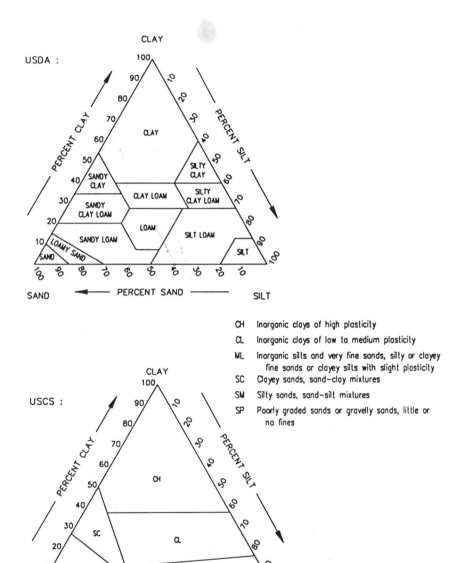

FIGURE 1.4 Textural Triangles for the USDA and the
USCS Soil Classification Systems.

TABLE 1.3

Soil Order	Physical/Chemical/Mineralogical Properties
Alfisol	• Acid to basic pH.
	• Some possess relatively high exchangeable Na.
	• Clay fraction predominantly mixtures of smectite with kaolinite, illite, vermiculite.
	• Sand and silt fractions predominantly mixed; siliceous minerals sometimes present.
	• A few alfisols may contain carbonates, oxides, and serpentine minerals.
Aridisol	• Neutral to very alkaline pH.
	• Very low accumulation of organic matter in surface horizons.
	• Clay fraction predominantly smectite with some kaolinite and illite.
	• Sand and silt fractions predominantly mixed minerals.
	• Possess relatively high amounts of carbonate, sulfate, and gypsic minerals and soluble salts.
Entisol	• Acid, neutral, or alkaline pH depending upon climate and parent material.
	• Highly variable amounts of organic matter in surface horizons.
	• Clay fraction predominantly smectite with some kaolinite and illite.
	• Sand and silt fractions predominantly mixed; siliceous with some micaceous minerals may be present.
	• May possess carbonates and some gypsic and oxide minerals.
Histosol	• Possess at least 20% organic matter.
	• Predominantly water saturated.
	• Possess low bulk density.
Inceptisol	• Acid, neutral, or alkaline pH depending upon climate and parent material.
	• Highly variable amounts of organic matter in surface horizons.
	• Clay fraction predominantly mixture of smectite, illite, kaolinite, and possibly halloysite.
	• Sand and silt fractions predominantly mixed; some siliceous and micaceous minerals present.
	• May possess carbonate and some oxide and serpentine minerals.
Mollisol	• Neutral to alkaline pH.
	• Relatively high native organic matter accumulation in surface horizons.
	• Clay fraction predominantly smectite with very minor amounts of kaolinite, illite, vermiculite, chlorite, and halloysite.
	• Sand and silt fractions predominantly mixed; minor amounts of siliceous and micaceous minerals may be present.
	• May possess carbonate and some serpentine minerals.
Oxisol[b]	• Very acid to acid pH.
	• Clay fraction predominantly kaolinite.
	• Possesses relatively high amounts of ferritic and oxide minerals.
	• Some quartz is present. However, Order possesses relatively low amounts of Si.
Spodosol	• Very acid to acid pH.
	• Possesses relatively little clay. Clay fraction predominantly mixed minerals.

TABLE 1.3 (cont.)

Soil Order	Physical/Chemical/Mineralogical Properties
Spodosol (cont.)	• Sand and silt fractions predominantly mixed; some siliceous minerals present.
	• Many possess oxide minerals and allophane.
Ultisol	• Very acid to acid pH.
	• Clay fraction predominantly kaolinite with some illite.
	• Sand and silt fractions primarily siliceous minerals with some mixed, micaceous minerals.
	• May contain oxide and possibly ferritic and glauconitic minerals.
Vertisol	• Acid to neutral pH.
	• Very high clay content.
	• Clay fraction predominantly expanding smectite with some kaolinite; causes excessive soil shrinking, cracking, and shearing.

a Compiled from data presented in Ref. 5, 6, and 7.

b Although Oxisols are a soil order, soils within this order are not found in the continental U.S.

classes and different ranges for their respective classes. Conversion of USCS data to the USDA system is not possible if the particle grain size diameter below 0.074 mm was reported as "fines"; the USDA system requires a determination of silt (0.002 to 0.05 mm) and clay (<0.002 mm) in order to classify soil texture.

Soil particle size data, which is utilized to assess hazardous material reactions in soil, must be derived from standard laboratory methods. Data derived from the professional judgment of a driller or on site geologist during a subsurface investigation is for the most part unsuitable for use with the information provided in this text on chemical reactions in soil.

WHY SOIL REACTS WITH HAZARDOUS MATERIALS

Chemicals possess certain structures and undergo various reactions because they seek to attain the state of lowest potential energy. The goal of all naturally-occurring environmental processes is maximum energy release. Soil is a mixture of numerous naturally-occurring inorganic and organic chemicals. The rules of chemical reactivity which apply to "pure" chemicals also apply to soil as well. Soil, like all systems, possesses potential energy and seeks to maintain a state of minimum potential energy. This process is analogous to a rock at the top of a hill. This rock is at the top of a gravitational potential energy hill and will be in the position of lower potential energy when it reaches the bottom of the hill.

A hazardous material entering a soil system is at the top of a potential energy hill, and the spontaneous trend is for the hazardous material to roll down this hill to a position of lower potential energy. Physical, chemical, and microbial reactions occur because the hazardous material seeks to exist in that state (e.g. the adsorbed state, the vapor state, the precipitated state, the state of transformed structure, etc.) which allows it to possess the lowest possible level of potential energy.

Another way of envisioning why soil reacts with hazardous materials is through LeChatelier's principle. According to this principle, when a stress is applied to a system at equilibrium, the system reacts to relieve the stress.[2] The addition of a hazardous material to a soil system represents the addition of a stress, in the form of a chemical reactant, in the system. Stress is relieved through a process known as mixing, in which added energy is mixed or spread over the entire system.[3] There are two distinct types of mixing. The first is the spreading of chemicals over positions in space. For example, as molecules of a chlorinated aliphatic solvent in soil air migrate from a zone of high concentration to a zone of lower concentration, chemicals are spread over positions in space. This results in an increase in the entropy or the randomness of the system. The second is the sharing or spreading of available energy between chemicals within the system. For example, the migration of a metal cation toward a negatively charged soil clay surface results in the conversion of potential energy into work and heat.

This text in essence is a discourse on the many ways soil relieves stress in the form of new potential energy due to the entrance of a hazardous material into soil. Stress and the release of stress is described throughout the text in terms of bond energies, ionization potentials, electronegativities, enthalpy, entropy, vaporization, electron affinities and in terms of the formation of bonds such as ionic, covalent, Van der Waals, hydrogen, and coordinate covalent bonds.

CHEMICAL STRUCTURE, CHEMICAL PROPERTIES, AND REACTIVITY IN SOIL

The term ''hazardous material'' encompasses a wide variety of inorganic and organic chemicals. These chemicals include toxic, corrosive, ignitable, flammable, and/or radioactive products and wastes that can be solids, liquids, or gases. There are literally hundreds of thousands of these products and wastes that possess different, unique chemical structures. Each chemical structure possesses a different arrangement of atoms which imparts to the chemical its own characteristic set of chemical and physical properties.

Structural theory is the framework of ideas about how atoms are combined to form molecules. It deals with the order in which atoms are attached to each other and with the electrons that hold them together. It deals with the shapes and sizes of the molecules these atoms form and with the way that electrons are distributed over these molecules.

Any chemical structure, when interpreted in terms of structural theory, tells us a great deal about the chemical. First, it foretells the magnitude of its physical properties. These include melting point, boiling point, octanol/water partition coefficient, dipole moment, vapor pressure, water solubility, extent of adsorption onto soil, specific gravity, solubility in organic solvents, color, diffusion coefficients in air and water, and others. Second, it foretells the chemical reactivity. This includes the kind of reagents the chemical will react with, the type of reaction (e.g. aqueous or soil surface catalyzed hydrolysis, oxidation, and reduction), the products that may be formed, the rate of reaction, the chemical's acidity or basicity, and its bio-degradation potential. Third, it foretells what arrangement in space—its conformation and configuration—the sequence of atoms in a molecule will display.

Structural theory is the basis of the study of the behavior of any chemical, not only in soil, but also in all environmental compartments. An analysis of the structure of the chemical(s) comprising the hazardous material is of paramount importance in assessing how the material will react in soil and water. If it is ignored, serious errors involving conclusions about the magnitude, importance, and remedial approach of soil and water contamination problems will result. Therefore, this manuscript will discuss, when appropriate, the structural theory of inorganic and organic chemicals in relation to their potential reactivity in soil.

REFERENCES

1. Greenland, D. J., and Hayes, M.H.B. (eds). The Chemistry of Soil Processes. New York: John Wiley & Sons (1981).
2. Snyder, Milton K. Chemistry: Structure and Reactions. New York: Holt, Rinehart, and Winston, Inc. (1966).
3. Denbigh, Kenneth. The Principles of Chemical Equilibrium. London: Cambridge at the University Press (1966).
4. Kohnke, H. Soil Physics. New York: McGraw-Hill (1968).
5. Dixon, J. B., and Weed, S. B. (eds.) Minerals in Soil Environments. Madison, WI: Soil Science Society of America (1977).
6. Buckman, Harry D., and Brady, Niles C. The Nature and Properties of Soils. London: The Macmillan Company, 6th printing (1972).
7. Fuller, Wallace H. and Warrick, Arthur W. Soils in Waste Treatment and Utilization. Volume 1. Boca Raton, FL: CRC Press, Inc. (1985).
8. U.S. Department of Agriculture. Soil Survey Manual. Handbook No. 18. Washington D.C.: U.S. Government Printing Office (1962).
9. Sethi, R. K., and Chopra, S. L. Adsorption and the Behavior of Pesticides in Soils. Pesticides **11**:15-25 (1977).

2

The Migration of Water and Bulk Hydrocarbons in Soil

INTRODUCTION

Water is the primary solvent in soil. Water is generally responsible for the transport of chemicals in soil systems. Therefore, a basic understanding of the principles governing the flow of water is essential in comprehending the migration of chemicals in soil.

It is most important to recognize that several factors governing water movement in the saturated zone are significantly different from those governing water movement in the unsaturated zone. As a result, chemical migration in the saturated and unsaturated zones will exhibit some similarities as well as some striking differences. This chapter will present a brief overview of the fundamentals of groundwater flow and a relatively more detailed discussion of soil water flow, based on excellent and detailed presentations addressing water flow in saturated zone soil[1,2,3,4,5] and in unsaturated zone soil[6,7,8,9,10,11].

In some situations, bulk hydrocarbons are the primary solvents in the soil system. Because the physical/chemical properties of bulk hydrocarbons are generally different than those of water, these chemicals will generally exhibit differences in flow relative to water. Many chemicals, which are insoluble and immobile in groundwater, are soluble in bulk hydrocarbons and will migrate along with the bulk hydrocarbon. This chapter will also present an overview of the fundamentals of bulk hydrocarbon flow in soil.

WHY WATER MIGRATES IN THE SATURATED AND UNSATURATED ZONES

When water moves from one location in a soil system to another, work is performed. The energy associated with water's ability to do work at any point in a soil system is called "potential energy" or "potential." Differences in potential energy create water movement in the direction of decreasing potential. Water movement will continue until the water potential difference between the two points is zero, and an equilibrium condition is attained.

The water potential E_p is the sum of several forces acting on water:

$$E_p = E_m + E_s + E_g + E_a + E_{hs} \qquad (2.1)$$

The matric potential E_m is caused by the attraction of soil surfaces (i.e. the matrix) for water. The solute potential E_s is due to the attraction of solutes for water. The gravitational potential E_g is the energy attributed to the downward pull of water by gravity. The pneumatic potential E_a is due to air or pneumatic pressures of the atmosphere on soil water. The hydrostatic potential E_{hs} is the energy attributed to hydrostatic or liquid water pressure (i.e. pressure head).

In the saturated zone, E_g and E_{hs} are the dominant potentials while E_m, E_s, and E_a are negligible in magnitude. The Darcy equation, which geologists and hydrologists routinely use, reflects the dominance of E_{hs} and E_g by relating water velocity to differences in hydrostatic potential. One form of this equation is:

$$V = -K \, (dh/dl) \, (p)^{-1} \qquad\qquad (2.2)$$

where V = average interstitial velocity of groundwater
           K = hydraulic conductivity or coefficient of permeability
               (see Figure 2.1)
           h = total head
           l = length along the total head
           p = porosity (see Figure 2.2)

The term dh/dl is commonly known as the hydraulic gradient, the unit loss in total head due to flow between two points a distance dl apart. The hydraulic conductivity and the hydraulic gradient are routinely obtained during hydrogeological investigations. Groundwater velocities may vary in the extreme from several feet per hour to less than a foot per year. The normal range of velocity is probably between 5 ft/yr and 5 ft/day.

Unlike the saturated zone, soil pores in the unsaturated zone soil are not completely filled with water. However, all unsaturated zone soil contains a measurable amount of water. The potentials governing water migration in the saturated zone, as described in Equation 2.1, also govern water migration in the unsaturated zone. However the relative importance of the potentials will change significantly as the volume of soil water in the unsaturated zone increases. The relative importance of these potentials will be described by illustrating how water infiltrates and percolates through soil in the unsaturated zone.

The process in which water penetrates a soil surface is known as infiltration. Figure 2.3 illustrates the general change in soil infiltration rate that occurs over time. The infiltration capacity is the maximum rate water can penetrate soil at a given time. At time t_o, a dry soil will have a relatively high infiltration capacity. As time progresses, however, the infiltration capacity decreases toward a steady-state capacity. This decrease is due to a shift in the relative contributions of the potentials comprising E_p in Equa-

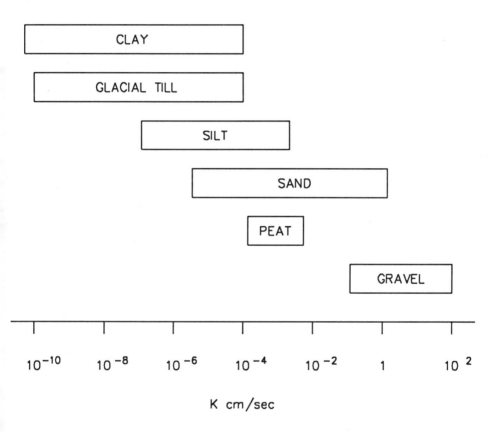

FIGURE 2.1 Ranges of hydraulic conductivity K for
subsurface soils.[4, 5, 8, 12]

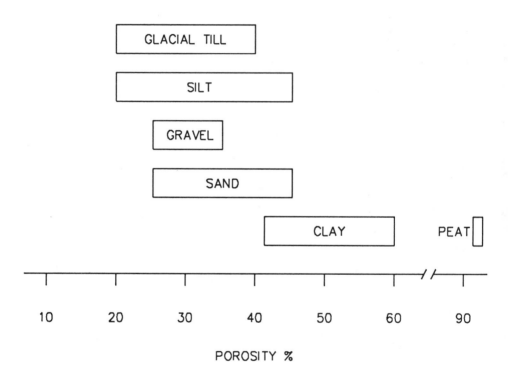

FIGURE 2.2   Ranges of porosity for subsurface soils.[3, 5]

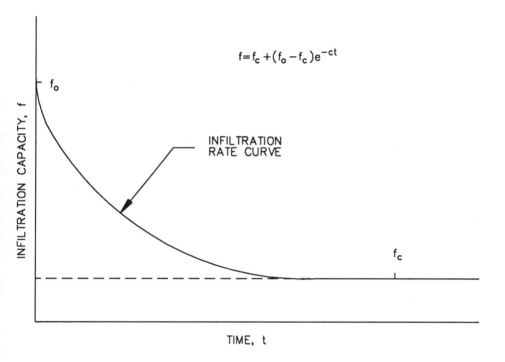

FIGURE 2.3   A general infiltration rate function. [13]

tion 2.1 and to a decrease in the effective porosity over time.

Table 2.1 lists general infiltration rates for soils with different textures. Also, infiltration rates can be estimated through use of infiltration equations[16,17,18].

TABLE 2.1   Ranges of Infiltration Rates for Soils With Different Textures[5,14,15]

Textural Class	Infiltration Rate (in./hr.)
Sand	0.5 – > 2.0
Loamy Sand	0.4 – 2.0
Sandy Loam	0.3 – 1.0
Loam, Silt Loam	0.2 – 1.0
Clay Loam	0.1 – 0.4
Sandy Clay, Silty Clay, Clay	< 0.1 – 0.2

Figure 2.4 can be utilized to illustrate how the relative importance of the potentials change and how the effective size of a soil pore changes as water is added to a dry soil. Dry soils contain a small increment of water which exists in the adsorbed state immediately adjacent to the soil particle surface. This class of water is known as "hygroscopic water" and exists as extremely thin films surrounding soil particles. The matric potential, E_m, is the dominant potential responsible for holding hygroscopic water next to soil particle surfaces. This class of water is held under tensions varying from 31 to 10,000 atm. As a result, hygroscopic water is not mobile, even though it may occupy from 3 to over 16 percent of the volume of soils.

If soil which contains only hygroscopic water had a small increment of water added to it, the added increment would exist as "capillary water" (Fig. 2.4). This class of water exists as films of varying thickness around and between soil particles with tensions ranging from approximately 0.1 to 31 atm. Capillary water moves slowly from thicker to thinner films; movement can occur in any direction. Capillary water can occupy from 4 to 18 percent of the volume of soils. As a result of the addition of capillary water, less air-filled pore space is present in soil, and the next increment of water added to this soil has a smaller pore through which it can migrate. As a result, its relative flux will decrease. Also, successive increments of water also encounter a small matric potential gradient near the soil surface as the wetting front migrates downward; this causes a lower infiltration rate.

When additional water is added to a soil containing hygroscopic and capillary water, the additional water can be removed from the unsaturated zone soil by gravity (i.e. the gravity potential, E_g). This class of water is known as "free water" (Fig. 2.4) and is the class of water primarily responsible for the downward migration of chemicals in soil pores. The process of free water

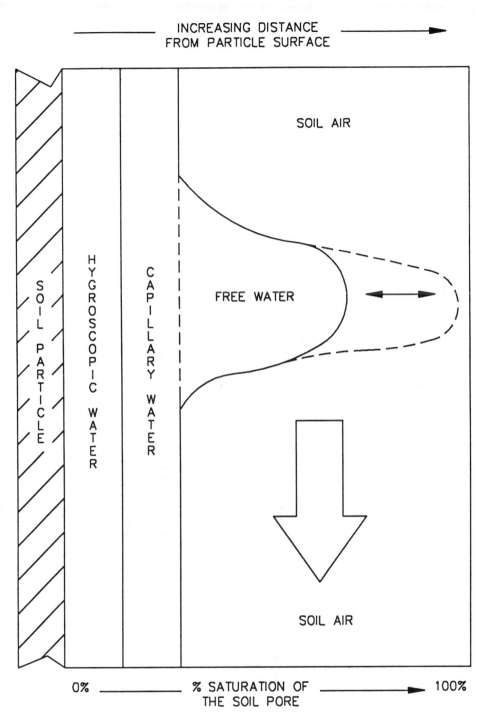

FIGURE 2.4   The classes of soil water.

migration is commonly known as mass flow or convection. An analysis of Fig. 2.4 will reveal that the soil pores need not be 100 percent water saturated in order for free water to migrate. Free water is held under tensions varying from less than 0.1 to 0.5 atm and can occupy from 25 to greater than 35 percent of the volume of soil.

Unsaturated zone soils generally do not accumulate free water for any appreciable period of time. However, unsaturated zone soils can hold very large volumes of water. For example, one acre-foot of saturated topsoil with a bulk density of 1.33 g/cc can hold approximately 163,000 gallons of water or about 3000 drums of water. Even extremely dry soil can hold as much as 10,500 gallons of hygroscopic water per acre/foot of soil. As a result, chemicals entering soil will encounter water.

In general, when water is gradually added to unsaturated zone soil with a relatively low soil water content, a distinct wetting front comprised primarily of free water and capillary water, which is really distributed throughout the soil, moves downward[6,19] (see Figure 2.5). If a sufficient amount of water is added to this soil, then the percolating water will eventually migrate to and recharge groundwater.

On the other hand, if the amount of percolating water added to this soil is limited in magnitude, which is usually the case under typical field conditions, the free water component of the percolating water will soon be transformed into capillary water (see Fig. 2.6). The rate of downward movement will decrease and will eventually cease. Then, slow movement of capillary water from regions of high to low film tension, the commonly operating mechanism of soil water movement or redistribution, will commence[6].

HOW SOIL STRUCTURE, MACROPORES, AND MICROPORES AFFECT WATER MIGRATION

Soil structure is the organization of individual sand, silt, clay, and organic matter particles at or near the soil surface into a relatively stable, aggregated unit commonly referred to as an aggregate or ped[14]. Figure 2.7 illustrates several common types of soil peds.

Soil structure is one primary reason why surface soils possess greater pore space (32–60%) and lower bulk density (approximately 1.0 – 1.3 g/cm^3), and more rapid water movement than subsurface soils[11,14]. Each ped possesses substantial numbers of capillary pores (micropores). Also, peds are separated from each other by voids (macropores). If water is rapidly added into a nearly saturated soil via heavy rains or during heavy irrigation, over 90% of the total water flow may occur through macropores (see Figure 2.8)[21]. Macropores are usually visible at the soil surface and may be continuous for distances of at least several feet in both vertical and lateral

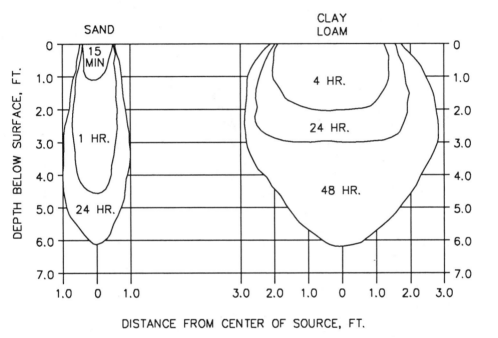

FIGURE 2.5   Rate of percolation of soil water from a continuous source in two soils. [20]

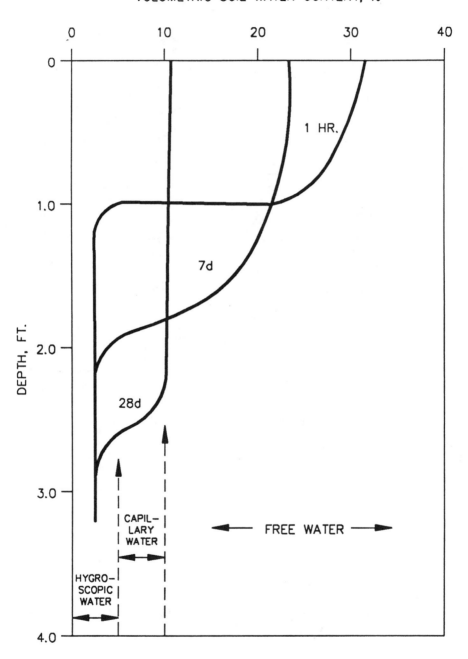

VOLUMETRIC SOIL WATER CONTENT, %

DEPTH, FT.

1 HR.

7d

28d

CAPIL-
LARY
WATER

FREE WATER

HYGRO-
SCOPIC
WATER

FIGURE 2.6   Redistribution of soil water in a sandy soil
after 1 hour of surface infiltration.

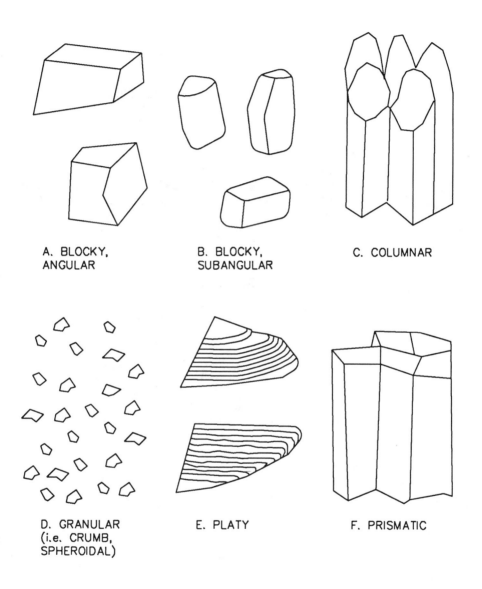

A. BLOCKY,
ANGULAR

B. BLOCKY,
SUBANGULAR

C. COLUMNAR

D. GRANULAR
(i.e. CRUMB,
SPHEROIDAL)

E. PLATY

F. PRISMATIC

FIGURE 2.7   Several common types of soil structure.[6,11,15]

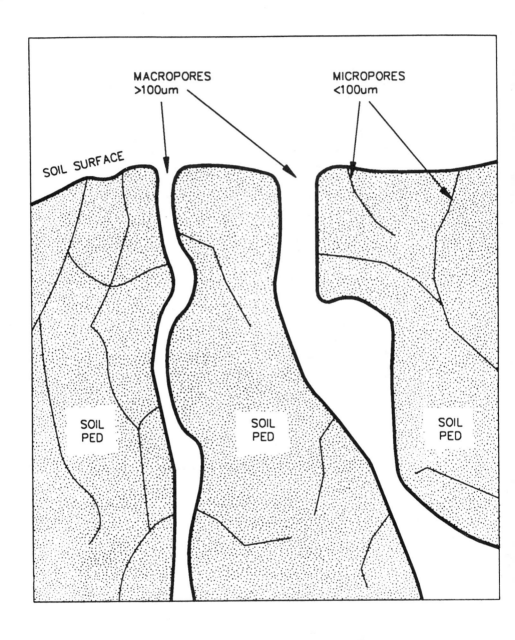

FIGURE 2.8 Relative sizes of macropores and micropores in soil.

directions[22]. Water in macropores can move into or below the root zone of soils in a matter of minutes after the addition of water to the soil surface[23]. However, flow through macropores appears to last no more than a few minutes or, in unusual cases, no more than a few hours after cessation of irrigation or heavy rainfall[19].

Interped voids are not the only source of macropores. Macropores which are tubular in shape are formed by soil fauna and plant roots[22]. Macropores which are cracks and fissures are formed either by (a) shrinkage resulting from the drying of clay soil, (b) chemical weathering of bedrock material, (c) freeze-thaw cycles, and (d) cultivation techniques[22]. Also, natural soil "pipes" may form due to the erosive action of subsurface flows if the forces imposed on individual soil particles exceed the structural competence of the soil. This condition generally occurs in highly permeable, noncohesive soils that are subjected to high hydraulic gradients.

Macropores usually comprise only a small percentage of the total soil porosity[22]. Also, macropores are generally not permanent structures. One macropore may last for several years, or it can be destroyed within one rainstorm by the inwashing of soil particles detached by rain splash[22]. The most long-lived macropores will occur in relatively undisturbed soils such as forest soils, while short-lived macropores will occur wherever land use is intensive[22,23].

If rain or irrigation water moves primarily through macropores, then relatively little mixing and interaction between chemicals in the rain or irrigation water with soil water occurs[19,24]; the macropores channel the rain or irrigation water away from the soil water. Also, since relatively little displacement of the initial soil water occurs[23], chemicals in soil should not contact clean, percolating water which is channeled away via macropores. If some interaction did occur, equilibrium distribution of a chemical between "clean" percolating water and adsorbed chemical could require several days[25], a period of time significantly greater than the average residence time of soil macropore water.

ESTIMATING AND ALTERING THE AMOUNT OF PERCOLATING SOIL WATER

The annual amount of soil water migrating downward (i.e. percolating) through the unsaturated zone can be grossly estimated using the water balance equation[26]:

$$Q = P - R - ET = dW \qquad (2.3)$$

where Q = amount of soil water migrating downward

P = amount of annual rainfall
R = amount of surface runoff
ET = amount of annual evapotranspiration
dW = change in the amount of stored water in the unsaturated
     zone.

Assuming R and dW are negligible, Equation 2.2 becomes:

$$Q = P - ET \qquad\qquad\qquad (2.4)$$

Figure 2.9 gives general estimates of Q for the continental U.S. An analysis of Figure 2.9 will reveal that mean annual precipitation exceeds potential evapotranspiration by 20 in. or more in only a few sections of the U.S. In many sections, potential evapotranspiration significantly exceeds mean annual precipitation. As a result, Q, the amount of water available to move chemicals through the unsaturated zone, may be only a few in. of water per year. This represents a very small amount of water relative to the amount of groundwater that is usually available to move chemicals through the saturated zone. Therefore, chemical migration through the unsaturated zone can be slow relative to chemical migration through the saturated zone.

At some hazardous waste sites, Q may still be relatively too high in order to prevent or mitigate unsaturated zone migration of chemicals to groundwater. Q can be decreased to negligible values by altering the permeability of the soil surface. This is usually accomplished by the use of soil caps or covers. A wide variety of soil and other materials are utilized to prevent or mitigate unsaturated zone migration of chemicals (see Table 2.2).

At some hazardous waste sites, Q may be too low in order to flush a chemical into groundwater where it can be removed via a groundwater extraction system and treated. Soil flushing or forced leaching refers to the surface application of water which percolates through the unsaturated zone to enhance the rate of vertical migration of a chemical. Soil flushing using water has been utilized at the BT-Kemi dumpsite in Sweden to enhance the migration of phenoxy acids, phenol derivatives, and DNBP[40]. Soil flushing using treated groundwater has been utilized at the Sydney Mine waste disposal site in Hillsborough County, Florida to enhance the migration of several volatile organic chlorinated solvents, chlorinated phenols, 2,4–D, and 2,4,5–TP[41].

THE CAPILLARY FRINGE AND CHEMICAL MIGRATION

The zone of water that is in direct contact with the water table and is held immediately above the water table by capillary forces acting against the force of gravity is known as the "capillary fringe." The "capillary rise" is the

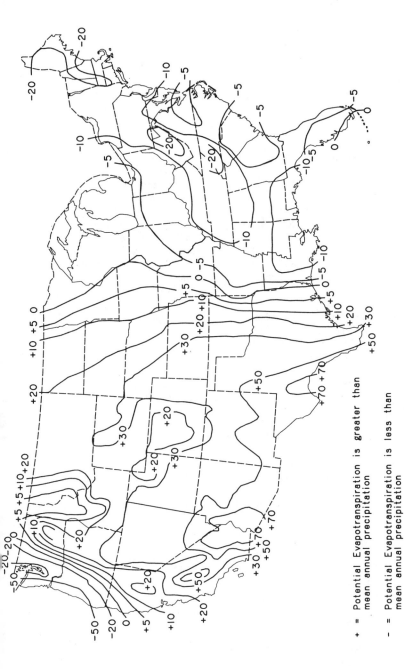

+ = Potential Evapotranspiration is greater than
    mean annual precipitation

- = Potential Evapotranspiration is less than
    mean annual precipitation

FIGURE 2.9   Potential evaporation versus
             mean annual precipitation.

(inches)

TABLE 2.2   Types of Soil Covers Utilized to Mitigate Water Flow Through Waste and Soil in the Unsaturated Zone.

Waste	Waste Generating Activity	Location[1]	Cover Material[2]	Ref.
Boiler ash, clinker, iron oxides, lime residue	Coal	United Kingdom	Topsoil over London Clay	27
Domestic, industrial wastes	Landfill	Girard, PA	Soil over geomembrane PVC over clay	28
		Middlesex Co., NJ	Soil over sand over synthetic membrane over compacted clay	29
		Nashua, NH	Soil over geomembrane over soil	29
		Pitman, NJ	Synthetic membrane	30
		Seffner, FL	Subsoil-montmorillonite mix	30
		Windham, CT	20 mil PVC	31 & 32
Metals in soil: As,Cd,Cr,Pb,Zn	Chromate & sulphuric acid works & Zn smelter	United Kingdom	Topsoil and sand	33
As,Cd,Hg,Pb,Zn	Refuse Dump	United Kingdom	Topsoil over sand over flint gravel	33
As,Cu,Pb,Zn	As & Cu smelting	United Kingdom	Acid sandy clay subsoil	33
	Sulphuric acid plant	United Kingdom	Silty clay over railway ash	33
Cd,Cu,Pb,Zn	Cu smelting	United Kingdom	Clay loam topsoil over limestone gravel	33
	Refuse dump	United Kingdom	Clay gravel subsoil	33
Cd,Pb,Zn	Pb & Zn Mining	United Kingdom	Sandy topsoil	33
	Pb & Zn Mining	United Kingdom	Stony subsoil over waste rock	33
	Pb & Zn Mining	United Kingdom	Coarse shale quarry waste	33
	Pb & Zn Mining	United Kingdom	Topsoil, burnt & unburnt colliery spoil	33
Cr	Chromate Works	United Kingdom	Coarse sandy soil	33
Cu,Pb	Pb smelting	United Kingdom	Clay topsoil over clay subsoil	33

TABLE 2.2 Types of Soil Covers Utilized to Mitigate Water Flow Through Waste and Soil in the Unsaturated Zone. (cont.)

Waste	Waste Generating Activity	Location[1]	Cover Material[2]	Ref.
Ni,Pb,Sb, Sulfuric acid	Secondary Pb smelting	Tampa, FL	Soil, asphalt	30
Zn	Sewage Sludge Disposal	United Kingdom	Topsoil over sand	33
	Steel Works	United Kingdom	Clay subsoil	33
Mill Tailings	Metal Mining	United Kingdom	Quarry overburden & crushed limestone	27
	Uranium Processing	Canonsburg, PA	Soil montmorillonite soil cover	34, 35
		Salt Lake City, UT	Unspecified soil	34
		Shiprock Navajoe Indian Reservation	Rock over sandy silt soil over silt soil	34
Oil, paint waste, PCBs, solvents in lagoon	—	East Central, NY	Unspecified material	36
PCBs in Soil	—	New Bedford, MA	Hydraulic asphalt concrete	37
	Transformer Service Shop	Oakland, CA	Gravel over soil/ montmorillonite mix	38
Sludge	Petroleum Refining	Salt Lake City, UT	Clay	30
	Steel Processing	Western, PA	Unspecific soil, PVC, geotextile	39
Waste Tip	Copper Refining	United Kingdom	Topsoil over sodium carbonate & sodium chloride	27

[1] Many sites are now being utilized as grazing land, parks, sports fields, gardens.
[2] Sites do not contain bottom liners.

distance between the water table and the top of the capillary fringe. The amount of water held in the capillary fringe depends upon the size of the particles comprising the soil.

If a soil pore could be idealized to represent a clean glass capillary tube, the capillary rise (h) for pure water would approximate:

$$h = 0.15/r \qquad (2.5)$$

Where r is the tube radius[3]. Figure 2.10 illustrates the values of capillary rise for various soils. Numerical and analytical models are available to estimate

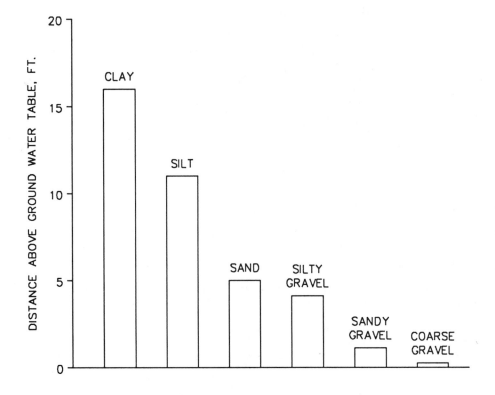

FIGURE 2.10   Typical values for capillary rise in various soils. [5,7]

capillary rise[42].

The capillary fringe does not exert a significant effect, for the most part, on the movement of chemicals in soils. Even with a fluctuating groundwater table, the movement of the capillary fringe should not exert a significant effect on the movement of chemicals in unsaturated zone soil. The reason can be exemplified by the following conceptual model. In a relatively uniform soil, a rise in the groundwater table toward the soil surface would cause a corresponding upward movement of the capillary fringe. This upward movement would result in the addition of a small amount of water from the capillary fringe to existing water, some quantity of hygroscopic and capillary water already within unsaturated zone soil pores. However, when the groundwater table drops, the capillary fringe moves downward. As a result, the net flux of a relatively small amount of water in the capillary fringe passing through a fixed point in the soil is approximately zero. Since the leaching of an organic or inorganic chemical in soil depends on water flux, chemical movement should be negligible because the net flux of water is zero.

There is one condition in which the upward flow of water from the ground water table will affect the transport of a chemical in the unsaturated zone. When the water table is five to seven feet below the soil surface, evapotranspiration causes the upward movement of dissolved salts and chemicals in saline soils[43,44,45,46,47]. This process is responsible for the salinization of agricultural soils in the western U.S.A. Under this circumstance, the net migration of a chemical in the unsaturated zone would be upward.

MIGRATION OF SOIL PARTICLES, BACTERIA, AND VIRUSES IN SOIL

The previous discussion on soil macropores revealed that macropores can be destroyed by the inwashing of soil particles detached by rain splash. In other words, soil particles can migrate into and through soil pores, under certain conditions. Also, bacteria and viruses can migrate for considerable distances through soil pores, under certain conditions. For example, coliform bacteria were transported in a loamy sand aquifer for more than 2500 ft[48].

The published literature identifies primarily three factors affecting soil particle, bacteria, and virus migration. First, the diameter of the migrating particle must be significantly smaller than the diameter of the pore. A general rule on the migration of bentonite grout in soil[49] can also be utilized to estimate the migration potential of a soil particle. A bentonite grout will penetrate soil pores if the ratio, R, where:

$$R = D_{15}/D_{85} \qquad (2.6)$$

and

D_{15} = Diameter of the particles comprising the soil, where 15 percent of the soil mass is finer.

D_{85} = Diameter of the migrating bentonite (or soil) particle, where 85 percent of the particles are finer.

is at least 29 and preferably greater than 24. This ratio can be utilized not only for identifying the migration potential of a soil particle, but apparently for bacterium and virus particles as well. Bacteria and viruses have diameters similar to that of clay (see Figure 2.11). An analysis of Figure 2.11 will reveal that D_{85} for bacteria is approximately 1.2u. For bacteria to migrate, the D_{15} of the soil must be 30.0u, based on Eq. 2.6 and with R equal to 25. A further analysis of Figure 2.13 will reveal that 85 percent of the soil texture must be comprised of coarser silt, sand, and gravel. The soil classes corresponding to this textural range are sandy loams, loamy sands and sands, based on an analysis of Fig. 1.4. Also note that bacteria should not migrate in silty and clay soils, based on this analysis. An analysis of published data on bacteria and virus migration in soil [48,50,51,52] revealed that these conclusions are generally valid. However, it is important to remember that bacteria and viruses can migrate through macropores in silty and clay soils.

Second, the migration of soil particles requires that the soil particle remain dispersed [53,54]. In other words, because soil particles generally possess a net negative surface charge, the migrating soil particles should be repelled from non-migrating soil particles. Electrostatic repulsion will occur, provided the soil possesses a low electrolyte content [53]. The presence of relative high amounts of dissolved cations in water will decrease the repulsive forces. Because many bacteria and viruses possess negative surface charges, the existence of electrostatic repulsion between bacterium/virus and soil will favor bacterium/virus migration. The presence of positively charged soil particles such as iron oxides will cause soil particle flocculation [53] as well as poliovirus adsorption to soil [55]. The adsorption of bacteria may be substantial in some soils and is utilized as the primary input parameter for a simple model assessing bacteria migration through unsaturated zone soil [56].

Third, bacteria and virus generally migrate negligible distances under unsaturated flow conditions, but move considerable distances under saturated flow [57]. Also, as soil water velocity increases, the number of transported organisms also increases [11]. The same effect should occur for migrating soil particles as well.

THE MIGRATION OF BULK HYDROCARBONS IN SOIL

In general, water is the primary solvent in soil. However, bulk hydrocar-

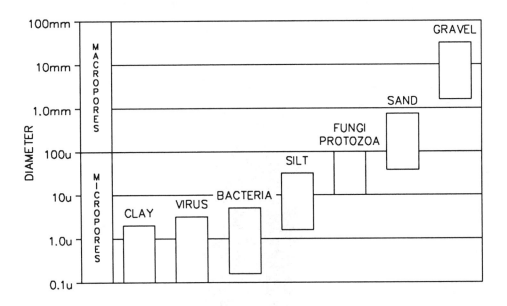

FIGURE 2.11   Ranges of diameters for soil particles and biota.

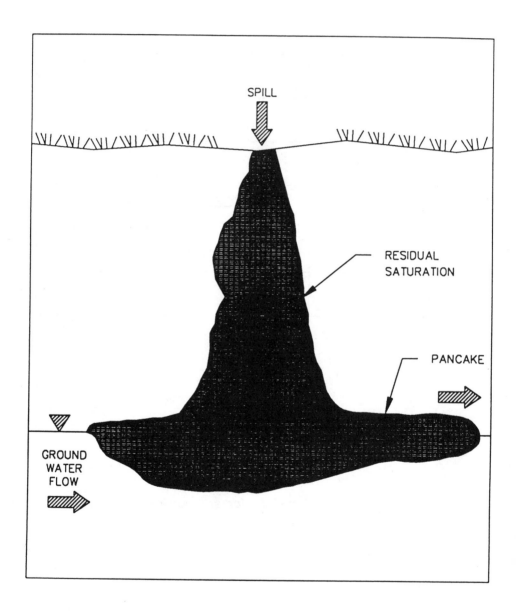

FIGURE 2.12   Light bulk hydrocarbon distribution resulting
from a major spill.

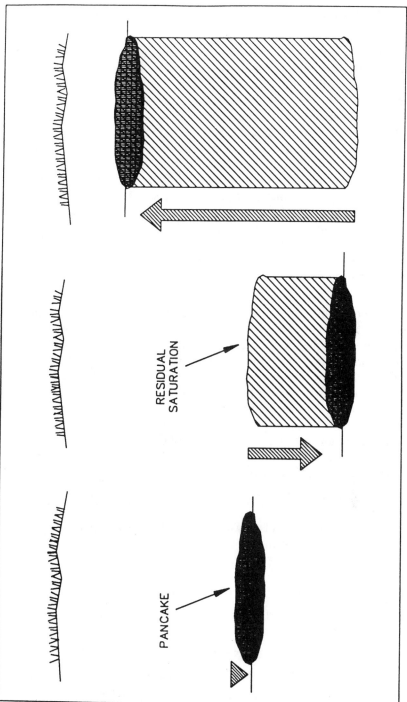

Figure 2.13 Fluctuation of a water table and pancake resulting in transformation of mobile hydrocarbon into residual saturation.

bons released during spills, leaks, and tank and pipe ruptures may enter soil and replace water as the primary solvent in localized situations. The basic principles governing the migration of water in soil are generally the same ones governing the migration of bulk hydrocarbons. However, because bulk hydrocarbons are generally not miscible with water, their distribution in soil in the presence of water should be addressed.

This section of Chapter 2 will discuss the basic principles governing the migration of bulk hydrocarbons in both the saturated and unsaturated zones. The term bulk hydrocarbon will be used to refer to unrefined petroleum (i.e. crude oil), refined petroleum (e.g. gasoline, kerosene, mineral oil), and petrochemicals (e.g. TCA, TCE, chloroform). The term light hydrocarbons will refer to bulk hydrocarbons having a density less than the density of water. These light bulk hydrocarbons are commonly known as "floaters." The term dense hydrocarbon will refer to bulk hydrocarbons having a density greater than the density of water.

Infiltration of Bulk Hydrocarbons into Soil. Similar to water, bulk hydrocarbons which are released at the soil surface will penetrate through the soil surface via the most permeable path. Bulk hydrocarbon may temporarily pond on the soil surface if the soil has a very high clay content and a low permeability. On the other hand, a sandy soil may allow rapid infiltration. Bulk hydrocarbon viscosity also affects infiltration rates: heavy oils will infiltrate slowly relative to gasoline and kerosene.

Oil spilled on a flat permeable surface will spread in all directions as a film of constant thickness[46]. The primary factors influencing oil infiltration are surface tension, gravity, and viscosity. For the case of a continuous steady oil leak over a permeable soil, the radius R of surface spread for a mass of oil can be grossly estimated[58]:

$$R = (t)^{1/4} [(R_c g \cos A) / (6.28u)]^{1/2} \qquad (2.7)$$

where

A = contact angle between bulk hydrocarbon and soil
g = acceleration due to gravity
R_c = capillary radius
t = time
u = viscosity of bulk hydrocarbon

An alternate equation[59] for grossly estimating A_s, the spill area in square meters, in a porous soil is:

$$A_s = 53.5 \, (V_{HC})^{0.89} \qquad (2.8)$$

The actual spill area may differ from the calculated one by a factor of two or three, or possibly as much as eight in extreme cases, due to effects of vegetation, terrain, climate, etc.

For the case of a continuous steady oil leak over a relatively impermeable soil under isothermal conditions, the radius R of a mass of bulk hydrocarbon can be grossly estimated[58]:

$$R = (Q/3.142h)^{1/2} t^{1/2} \qquad\qquad (2.9)$$

where

Q = rate of hydrocarbon addition
t = time after initial hydrocarbon addition to soil
h = height of the oil film

Migration and Distribution of Light Bulk Hydrocarbon. Light bulk hydrocarbon will migrate downward in unsaturated zone soil due to gravity and capillary forces. If the volume of released hydrocarbon is large, such as those related to catastrophic spills, maximum lateral spreading and downward flow occurs with all soil pores being saturated with hydrocarbon[60]. Figure 2.12 illustrates the hydrocarbon distribution most often displayed in the published literature. The distribution illustrated in Figure 2.12 is valid for major gasoline spills or tanker ruptures, but not for slow leaks; this case will be discussed later in this chapter. The downward migration of light bulk hydrocarbon will eventually cease because (a) the mobile light bulk hydrocarbon will be transformed into residual saturation, or (b) it will encounter an impermeable bed, or (c) it will reach the capillary fringe. Each situation is described in greater detail below.

As a mass of bulk hydrocarbon migrates beyond a unit mass of unsaturated zone soil, a small amount of the total hydrocarbon mass will remain attached to these soil particles via capillary forces. The bulk hydrocarbon that is retained by soil particles is known as immobile or "residual saturation." The maximum amount of bulk hydrocarbon that can be retained by a soil is known as residual saturation capacity. Residual saturation can potentially reside in soil in this state for years. If the migrating mass of bulk hydrocarbon is small relative to the soil surface area, the mass of bulk hydrocarbon will be eventually exhausted as it is converted into residual saturation. When conversion is complete, downward migration ceases.

The volume of soil required to immobilize a mass of bulk hydrocarbon depends upon the porosity of the soil and the physical properties of the bulk hydrocarbon. The number of yd^3 of soil required to immobilize a volume (barrels) of bulk hydrocarbon, V_{HC}, can be grossly estimated[61]:

$$V_s = 0.2 \ V_{HC}/P \ (RS) \qquad\qquad (2.10)$$

where

V_{HC} = volume of discharged hydrocarbon, in barrels
(44 gal = 1 barrel)
V_s = yd^3 of soil required to attain residual saturation

P = soil porosity
RS = residual saturation capacity

In general, the residual saturation capacity of soils is about 33 percent of their water-holding capacity[62]. The maximum residual saturation for light oil and gasoline is 0.1; for diesel and light fuel oil, 0.15; for lube and heavy fuel oil, 0.20. The maximum possible depth of penetration, D, in yards, can be grossly estimated using the equation[63,64]:

$$D = V_s / A \qquad (2.11)$$

where A is the area of infiltration. There is an alternative equation which grossly estimates the maximum depth of penetration of a volume of bulk hydrocarbon released on a soil[65]:

$$D = KV_{HC} / A \qquad (2.12)$$

where K is a constant dependent upon the soil's retentive capacity for oil and upon oil viscosity (see Table 2.3).

Another non-rigorous formula for grossly estimating the maximum depth of penetration of a volume of bulk hydrocarbon into the unsaturated zone is[66]:

$$D = 1000 \, V_{HC} / ARC \qquad (2.13)$$

where

R = soil retention capacity
C = approximate correction factor based on the bulk hydrocarbon viscosity (0.5 for gasoline to 2.0 for light fuel oil).

Recommended values of R in liters per cubic meter are: 5 for stone to coarse gravel, 8 for gravel to coarse sand, 15 for coarse to medium sand, 25 for medium to fine sand, and 40 for fine sand to silt. If the soil has an intermediate texture relative to those listed above, an intermediate value of R should be used. Also, these values for R are for soils with an average moisture content. For dry soils, R will be greater than the values listed above.

If a mass of bulk hydrocarbon which is migrating downward encounters an empermeable layer, it will spread laterally until (a) the bulk hydrocarbon is transformed into residual saturation, or (b) it migrates past the lateral extent of the impermeable layer. If the latter situation occurs, vertical migration will commence at the point where the lateral extent of the impermeable layer has ceased. Downward migration will continue until (a) the bulk hydrocarbon is transformed into residual saturation, (b) another impermeable barrier is encountered, or (c) the bulk hydrocarbon encounters the capillary fringe.

Percolating water, in unsaturated zone soil containing residual saturation, can initiate the downward migration of hydrocarbon[58,67]. In a laboratory study involving uniformly packed beds comprised of 90 percent sand and 10 percent soil, the movement of surface-applied kerosene by intermittently-applied water was quantified[65]:

$$Q_v = 19t^{-0.81} \tag{2.14}$$

where

Q_v = mean rate of increase of the oil contaminated soil
  (cm^3 min^{-1})

  t = time (days).

This phenomenon is expected to continue until the hydrocarbon which can migrate by this process is depleted from soil pores[60]. Then, percolating water will generally move around the hydrocarbon with minimal disturbance.

As light hydrocarbon enters the capillary fringe, it will bypass the smaller, water-filled pores and continue migrating downward through larger pores which do not contain water. Downward migration will end when the light bulk hydrocarbon encounters water-saturated large pores. Then, the light bulk hydrocarbon begins to migrate laterally over the water table in a layer roughly as thick as the capillary fringe. This layer of light hydrocarbon will assume the shape of a "pancake"; this layer of hydrocarbon is commonly known as the pancake layer (see Figure 2.12).

If a relatively large volume of bulk hydrocarbon reaches the water table, its weight will be sufficient to collapse the capillary zone and depress the water table[60]. The amount of depression will depend upon the amount of light bulk hydrocarbon present. Since the specific gravity of gasoline and light oils is approximately 0.70 to 0.80, about 75 percent of the hydrocarbon pancake will be below the depth of the original water table. Due to the force of buoyancy from below and the force of additional light hydrocarbon descending from above, the pancake will tend to spread laterally as rapidly as soil conditions will permit. Initially there may be sufficient head pressure to cause the light hydrocarbon to move a small distance up gradient, but the greatest spread will occur in the down gradient direction. The pancake will migrate until it reaches residual saturation or until it reaches a zone of ground water discharge. As the pancake migrates laterally, water in the capillary fringe will impede its movement because water occupies pore space. In the upper section of the capillary fringe where relatively small amounts of water are present, bulk light hydrocarbon comprising the pancake will migrate laterally. However, in the lower section of the capillary fringe where relatively large amounts of water are present, the pancake migrates laterally at a negligible rate. Light hydrocarbon migration over groundwater can be measured directly[68].

The maximum spread of the pancake over the groundwater table can be grossly estimated[65]:

$$S = (1000/F) (V - [Ad/K])$$ (2.15)

where

S = maximum spread of the pancake, meters^2
F = thickness of the pancake, millimeters
V = volume of infiltrating bulk hydrocarbon, meters^3
A = area of infiltration, meters^2
d = depth to groundwater, meters
K = constant dependent upon the soil's retention capacity for oil and upon oil viscosity (see Table 2.3)

TABLE 2.3   Typical Values for K for Various Soil Textures[64,65]

Soil Texture	K		
	Gasoline	Kerosene	Light Fuel Oil
Stone & Coarse Gravel	400	200	100
Gravel & Coarse Sand	250	125	62
Coarse & Medium Sand	130	66	33
Medium & Fine Sand	80	40	20
Fine Sand & Silt	50	25	12

The pancake will fluctuate vertically as the water table fluctuates vertically in response to seasonal changes and to short-term rainfall events. The total amount of mobile hydrocarbon in the pancake will decrease as fluctuating mobile hydrocarbon coats soil particles and transforms into residual saturation (see Figure 2.13).

It is important to note that the thickness of light hydrocarbon in the capillary fringe is usually less than the thickness of light hydrocarbon measured in a monitoring or recovery well. This situation occurs because the layer of mobile light hydrocarbon in the capillary fringe is some distance above the water table. When the mobile light hydrocarbon encounters a well, it pours into and accumulates on the groundwater surface. As it accumulates, its weight depresses the water surface. The mobile light hydrocarbon will continue to enter the well until the top of the light hydrocarbon in the well is level with the top of the mobile light hydrocarbon in the capillary fringe. Several approaches exist to correct measurements from monitoring wells affected by this phenomenon[69,70,71].

The hydrocarbon distribution in the unsaturated zone resulting from a small leak occurring over a longer period of time is substantially different from the distribution illustrated in Figure 2.12. Figure 2.14 and 2.15 illustrate the

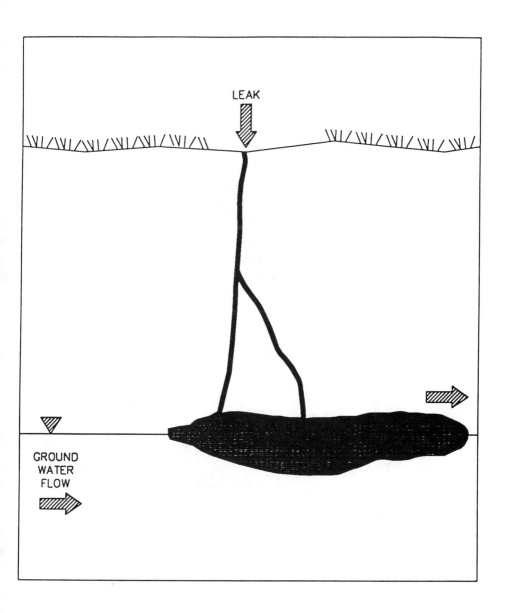

FIGURE 2.14   Light bulk hydrocarbon distribution resulting from a slow leak into soil macropores.

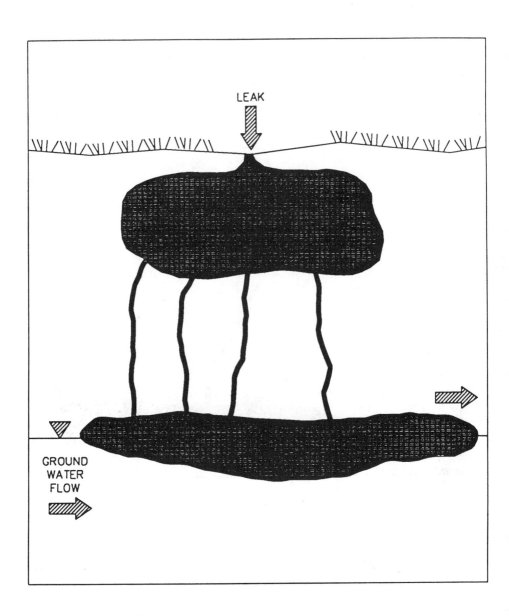

FIGURE 2.15   Light bulk hydrocarbon distribution resulting from a slow leak into soil micropores.

light bulk hydrocarbon distribution resulting from a slow leak into soil macropores and micropores, respectively, based on studies utilizing capillary networks[72]. In the former case, the bulk hydrocarbon has a soil macropore through which it can vertically migrate. As a result, lateral spreading is very small. The light bulk hydrocarbon will flow in this preferred channel as long as flow volume is not significantly increased. The flow distribution may be single pore or dendritic. An analysis of Figure 2.14 will reveal that far less bulk hydrocarbon will attain residual saturation in the unsaturated zone in this situation relative to the major spill illustrated in Figure 2.12.

Figure 2.15, based on studies utilizing capillary networks[72], illustrates the distribution of light bulk hydrocarbon when a preferred flow channel such as a macropore is not initially present. Upon initial infiltration, the bulk hydrocarbon will spread laterally and vertically to form an oval-shaped body. If the body encounters macropores or causes macropore development as it grows, then these macropores will serve as the preferred flow channels. An analysis of Figure 2.15 will reveal that far less bulk hydrocarbon will attain residual saturation in the unsaturated zone in this illustrated case relative to the major spill illustrated in Figure 2.12.

Migration and Distribution of Dense Bulk Hydrocarbons. The basic principles governing the downward migration of light bulk hydrocarbons is applicable to dense bulk hydrocarbons as well. Dense hydrocarbons will migrate downward in unsaturated zone soil due to gravity and capillary forces. If the volume of dense hydrocarbon which is released is large, such as that related to a catastrophic event, maximum lateral spreading and downward migration occurs with all soil pores being saturated with hydrocarbon (see Figure 2.16). Because the specific gravity (i.e. density) of dense hydrocarbons is greater than that of water, they will continue to migrate downward after they encounter groundwater, as illustrated in Figure 2.16. This phenomenon is sometimes referred to as density flow.

Table 2.4 lists the densities and kinematic viscosities of several light and dense bulk hydrocarbons. An analysis of the information presented in Table 2.4 will reveal that halogenated hydrocarbons and coal tar are the principle bulk hydrocarbons possessing densities greater than the density of water. Also, crude oil and distilled petroleum products are the bulk hydrocarbons possessing specific gravities less than that of water. The velocities of percolation may be considered to be approximately inversely proportional to the kinematic viscosities. In unsaturated zone soil, crude oil may migrate from 3 to 35 times slower than water, while trichloroethylene will migrate 2.5 times faster than water.

Density flow in the saturated zone can occur by "front displacement," the uniform, downward movement of the leading edge of a dense bulk hydrocarbon plume via the displacement of ground water. However, "fin-

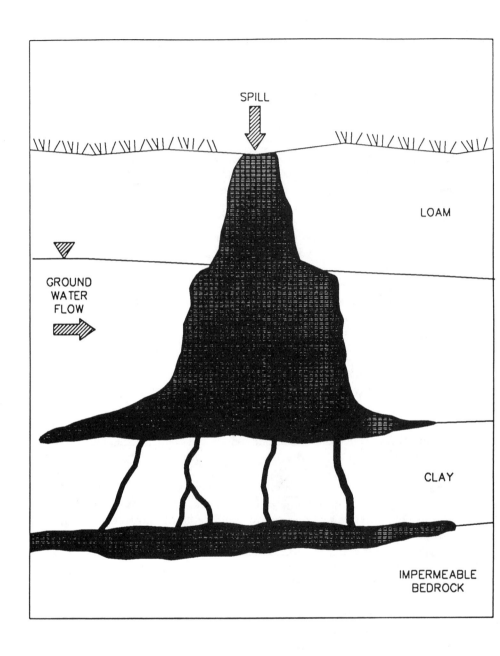

FIGURE 2.16   Dense bulk hydrocarbon distribution resulting from
a major spill.

TABLE 2.4   Densities and Kinematic Viscosities of Light and Dense Bulk Hydrocarbons

Hydrocarbon	Density (gram/ml)	Kinematic Viscosity (mm² / sec)
African, Near East, North Sea Crude	0.80 − 0.88	3 − 35
light heating oil, diesel	0.82 − 0.86	2 − 8
gasoline	0.72 − 0.78	0.5 − 0.7
jet fuel	0.77 − 0.83	2 − 4
coal tar	1.028	—
methylene chloride	1.33	0.32
chloroform	1.49	0.38
carbon tetrachloride	1.59	0.61
1,1,1-trichloroethane	1.32	0.65
trichloroethylene	1.46	0.40
tetrachloroethylene	1.62	0.54

gering'' of the dense hydrocarbon into the water may occur. Fingers will form if the mobility ratio, M, where:

where:

$$M = (K_1 V_2) / (K_2 V_1) \qquad (2.16)$$

and

K_1 & K_2 = variable transmissivities of the aquifer
V_1 & V_2 = fluid viscosities

is greater than unity[73]. A finger of dense bulk hydrocarbon will migrate downward with an extra vertical flow component due to density and viscosity differences. The flow component due to density and viscosity differences can be grossly estimated[73]:

$$V_{dz} = (K_Z / V_{dh}) (d_{dh} g - d_w g) (1/P) (V_w / V_{dh}) \qquad (2.17)$$

where

V_{dz} = vertical flow component due to density and viscosity differences
d_{dh} = density of dense hydrocarbon
d_w = density of groundwater
g = gravity
K_Z = intrinsic vertical permeability
P = porosity
V_{dh} = viscosity of dense hydrocarbon
V_w = viscosity of groundwater

The downward migration of a major spill of dense bulk hydrocarbon will eventually cease for two reasons. First, as the dense bulk hydrocarbon flows, it will be transformed into residual saturation. When conversion is complete, downward migration ceases. The volume of soil required to immobilize a volume of dense bulk hydrocarbon can be grossly estimated by utilizing the equations discussed earlier.

Second, downward migration will eventually cease if the dense bulk hydrocarbon encounters an impermeable bed. It is most important to recognize that clays and tills, which often possess very low permeabilities to water, can possess higher permeabilities to hydrocarbons, as illustrated in Figure 2.16. The reasons why the permeability of clays to some bulk hydrocarbons changes will be discussed in greater detail in a future section of this chapter.

Figure 2.17 illustrates the dense hydrocarbon distribution resulting from a small continuous leak into a soil macropore. In the unsaturated zone, lateral spreading is small, and the flow distribution may be single pore or dendritic. Upon reaching the groundwater table, the dense hydrocarbon may spread laterally to form a small pancake since it is not able to immediately penetrate the water surface. However, the dense hydrocarbon will displace water and continue to migrate downward as soon as the dense hydrocarbon attains sufficient mass[65,74].

In Figure 2.17, the dense bulk hydrocarbon continues to migrate downward through the saturated zone via several macropores, and eventually encounters a clay bed. The hydrocarbon accumulates on the surface of the clay bed since dense hydrocarbon migration through the dense clay is slightly slower than dense hydrocarbon migration through the loam. Also, dense bulk hydrocarbon accumulates in a bedrock depression.

In summary, dense bulk hydrocarbon distribution is dependent not only on the rate of hydrocarbon addition, but also on the size and distribution of soil pores and on the permeability of the soil to the dense hydrocarbon. Because all of these factors may vary considerably from one site to the next, bulk hydrocarbon distribution patterns can become very complex.

DISSOLUTION OF CHEMICALS COMPRISING BULK HYDROCARBONS INTO WATER

Bulk hydrocarbons are generally not miscible with water. However, bulk hydrocarbons are comprised of numerous individual chemicals that are soluble to some extent in water. When percolating soil water and groundwater contact bulk hydrocarbons, some of the chemicals are released from the bulk hydrocarbon to the soil or groundwater.

Very few studies have been published addressing the source strength of a bulk hydrocarbon. Source strength has been defined as the intensity with which dissolved chemicals are released from a bulk hydrocarbon to water

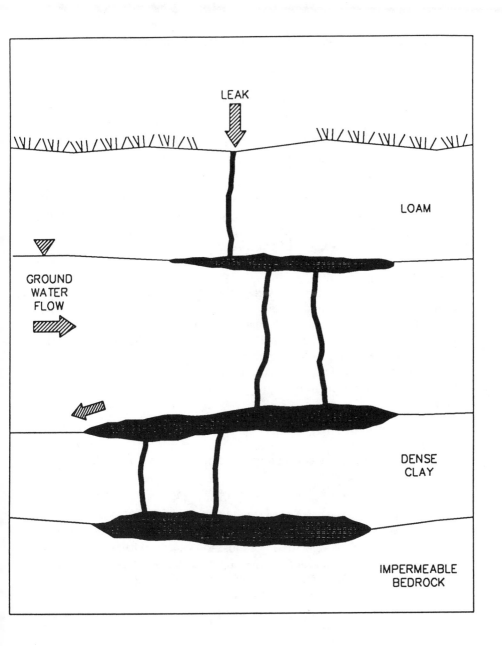

FIGURE 2.17 Dense hydrocarbon distribution resulting from a small leak into a soil macropore.

over time, expressed as mass/time per unit contact area[75]:

$$S = K_m A \qquad (2.18)$$

where   S = source strength, mass/t/meter^2
         k_m = mass exchange coefficient, mg/meter^2/sec
         A = contact area, meter^2

The contact area, A, is the interface across which mass exchange occurs. It governs the actual mass that is exchanged in a given period of time. Due to the complexity of hydrocarbon distribution in soil pores, the quantification of A is, at best, exceedingly difficult. As a result, studies have focused on quantifying K_m using several approaches. Unfortunately, these values exhibit substantial variation.

Mass exchange coefficients were grossly estimated to be 1.0 mg meter^2/sec for gasoline and tar oil, 0.01 for fuel oil, diesel, and kerosene, and 0.001 for lube oils and heavy fuel oil[76]. However, these should be considered as maximum mass exchange coefficients since field values could be as much as one to two orders of magnitude lower[76].

Another approach to characterizing mass exchange of chemicals from bulk hydrocarbon to water utilizes the Sherwood number and the Peclet number[76]:

$$Sh = 0.55 + 0.025 \ Pe^{3/2} \qquad (2.19)$$

where

         Sh = the Sherwood number, a dimensionless transfer
              coefficient
         Pe = the Peclet number, a dimensionless flow velocity

For Pe < 1, the chemical exchange should be independent of flow velocities and essentially diffusion-controlled. For Pe > 10, the chemical exchange should depend on flow velocities. For most soil systems, chemical exchange should be diffusion controlled. If the rate of diffusion of one chemical is known, the rate of diffusion of a structurally similar chemical can be estimated from Graham's law of diffusion:

$$D_1 / D_2 = (d_2 / d_1)^{0.5} = (M_2 / M_1)^{0.5} \qquad (2.20)$$

where

         d_1, d_2 = densities of two chemicals
         M_1, M_2 = molecular weights of two chemicals

The larger the molecule, the slower it diffuses and dissolves. Table 2.5 lists the range of the number of carbon atoms found in hydrocarbons which comprise various petroleum products. In general, the time required for a hydrocarbon sphere with a 5u radius and a 10^{-6} cm^2/sec diffusion coefficient, which occupies 25 percent of the volume of a hydrocarbon-water system, to dissolve in an infinite stagnant medium is[78]:

$$t = 0.6/C_o \qquad\qquad (2.21)$$

where

t = time, sec

C_o = water solubility, gram/cm^3

For example, t would equal 45 min. if a hydrocarbon had a 230 ppm water solubility and would equal 30 hr. if a hydrocarbon had a 5 ppm water solubility. An analysis of these data reveal that residual saturation would quickly dissolve in water if it were not for the fact that the maximum concentration of a chemical which diffuses from the bulk hydrocarbon to water does not exceed the water solubility of the chemical. Under some conditions, a contact time between hydrocarbon and water of only 15 min. is needed in order to reach maximum concentration[77]. Under other conditions, longer time periods may be needed. For example, a maximum concentration of diesel oil of 6 mg/liter was reached in 5 days; for lubricating oil 4.85 mg/liter in 7 days; for fuel oil, 2.5 mg/liter in 18 days; for crude oil, 9.61 mg/liter in 18 days[77].

Once a chemical has been released from the bulk hydrocarbon into soil or groundwater, density is no longer a factor governing the chemical's migration. It is most important to recognize that density is a physical property affecting the potential migration of a bulk hydrocarbon and not of dilute chemicals. Therefore, the terms "floater" and "sinker" should be applied to bulk hydrocarbons and not to the dissolved chemicals that have been released from the bulk hydrocarbon. The factors governing the migration and degradation of dilute chemicals in soil and groundwater will be discussed in great detail in the remaining chapters of this book.

HOW BULK HYDROCARBONS INFLUENCE SOIL HYDRAULIC CONDUCTIVITY

The ranges of soil hydraulic conductivity illustrated in Figure 2.1 are applicable for soils having water as the primary solvent. However, it is most important to recognize that these values may not be applicable for silty and clayey soils in which bulk hydrocarbons are the primary solvent. Many bulk hydrocarbons can change the hydraulic conductivity of these soils.

Hydraulic conductivity changes are caused by changes in the size of the interparticle spacing which separates adjacent clay particles. In order to adequately describe how bulk hydrocarbons affect this spacing, this section must first describe factors that control the size of the spacing, factors such as coulombic forces, fluid-phase pressures, and geostatic loads. Then this section will discuss how the interparticle spacing changes when bulk hydrocarbons replace water as the primary solvent.

The basic principles discussed in this section will apply not only to the hydraulic conductivity of field soils but also to landfill and surface impoundment liners which utilize clay liners, to clay caps, and to confining beds of deepwell injection systems.

There are two theories describing the increases and decreases in the magnitude of the interparticle spacing. The first attributes spacing changes to changes in the ionic composition within the spacing. Figure 2.18 illustrates the typical chemical structures of clay minerals commonly found in U.S. soils. Soil clay minerals are comprised primarily of layers of tetrahedral and octahedral sheets. Each tetrahedral sheet is comprised of SiO_4 tetrahedra, while each octahedral sheet is comprised of aluminum oxyhydroxide octahedra. The typical clay minerals found in U.S. soils are comprised of octahedral-tetrahedral sheets (i.e. 1:1 clays) and of tetrahedral-octahedral-tetrahedral sheets (i.e. 2:1 clays).

If the octahedral sheets contained only Al^{3+} and if the tetrahedral sheets contained only Si^{4+}, the clays would be electrically neutral since the total number of positive and negative charges would be equal. However, when clays precipitate, the presence of insufficient amounts of Al^{3+} result in the substitution of Fe^{2+} and Mg^{2+} for Al^{3+} in the octahedral layer of some clays. Also, insufficient amounts of Si^{4+} result in the substitution of Al^{3+} for Si^{4+} in the tetrahedral layer of some clays. This substitution process of replacement is called isomorphous substitution and is responsible for the negative charges on clay mineral surfaces.

The negative charges on clay surfaces attract ions to the surface. The magnitude or size of the interparticle spacing has been related to the type and amount of cations and anions in the water occupying the interparticle spacing as well as the type and amount of cations and anions adjacent to the surface. In general, the following relationships have been identified[79,80]:

(a) As the electrolyte concentration in the interparticle spacing increases, the size of the spacing decreases.
(b) As ionic valence of ions in the interparticle spacing increases, the size of the spacing decreases.
(c) As the dielectric constant increases, the size of the spacing decreases.
(d) As the pH of water in the interparticle spacing decreases (i.e. as H^+ concentration increases), the size of the spacing decreases.
(e) As anion adsorption to interparticle surfaces decreases, the size of the spacing decreases.
(f) As the amount of Na adsorbed onto interparticle surfaces decreases, the size of the spacing decreases.
(g) Spacing decreases and hydraulic conductivity increases caused by factors listed above are less pronounced in soils with low amounts of swelling

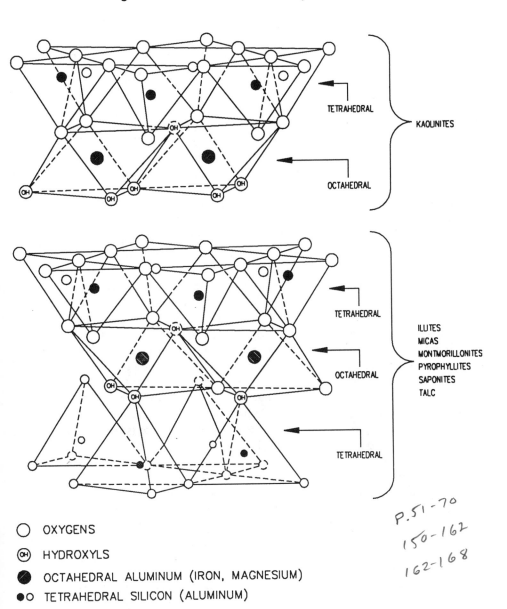

TETRAHEDRAL

OCTAHEDRAL

} KAOLINITES

TETRAHEDRAL

OCTAHEDRAL

TETRAHEDRAL

} ILLITES
MICAS
MONTMORILLONITES
PYROPHYLLITES
SAPONITES
TALC

P. 51-70
150-162
162-168

○ OXYGENS

⊕ HYDROXYLS

● OCTAHEDRAL ALUMINUM (IRON, MAGNESIUM)

●○ TETRAHEDRAL SILICON (ALUMINUM)

Figure 2.18 Structure of clay minerals commonly found in the U.S.

clay minerals than in soils with large amounts of swelling clay minerals.
(h) The changes in spacings caused by all the factors listed above may be negligible in soils containing high amounts of binding agents such as iron or aluminum hydrous oxides or organic matter.

With regard to (f) above, it is interesting to note that due to the dipolar nature of water, water will hydrate ions such as Na^+ and cause their effective diameter to be large relative to their non-hydrated ionic diameter. The effectiveness of Na-bentonite as a liner is due to the unit ionic valence and large diameter of the hydrated Na^+; this combination allows a maximum interparticle spacing due to repulsion between adjacent bentonite particles. If the hydrated Na^+ were replaced by Ca^{2+}, which has a smaller hydrated diameter and two positive charges, the interparticle spacing would significantly decrease, and cracks and fissures would occur in the liner due to clay shrinkage.

The second theory accounting for the expansion and contraction of the interparticle spacings deals with the structure of water in the interparticle spacing. Figure 2.19a illustrates the chemical structural formula for water. The physical shape of the water molecule is illustrated in Figure 2.19b. An analysis of these figures will reveal that the water molecule is not linear but angular. This causes a charge imbalance within the molecule because the centers for positive and negative charge are displaced from the center of the molecule. As a result, the water molecule is dipolar. Dipolar water molecules can arrange themselves with the hydrogen atoms of one molecule being adjacent to the oxygen atoms of another water molecule (see Figure 2.20). This hydrogen-oxygen arrangement is known as the hydrogen bond. Through this arrangement, water is capable of existing as a bulk liquid or as very thin water films.

Through hydrogen bonding between water molecules and the surface oxygens of clay particles, an ordered water structure develops at the particle surface and extends out into the interparticle spacing (Figure 2.21). The structure of the water is strained to match that of the clay particle. Several researchers have found that the b dimension, a measure of clay lattice structure, was related to the size of the water structure: as the b dimension of montmorillonite increased, the interparticle spacing also increased [81,82,83,84,85,86].

Other factors probably affect the size of the interparticle spacing. For example, geostatic and hydrostatic forces have been identified as forces affecting the spacing[87]. Based on a theoretical analysis of a number of forces affecting interparticle spacing, reductions in spacing by geostatic loads were estimated to occur at depths as shallow as 1.0 to 0.0001 meters or less[87].

Also, changes in environmental conditions may cause changes in hydraulic

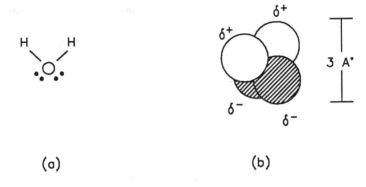

FIGURE 2.19   The chemical structural formula (a) and the
physical shape of the water molecule (b).
Hydrogen atoms occupy the light-colored
spheres while the two unshared pair of
electrons associated with the oxygen atom
occupy the dark-colored spheres.

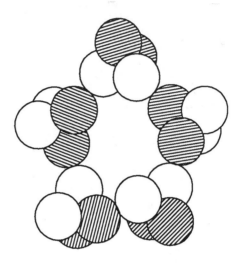

FIGURE 2.20   The formation of liquid water through hydrogen bonding of randomly oriented water molecules.

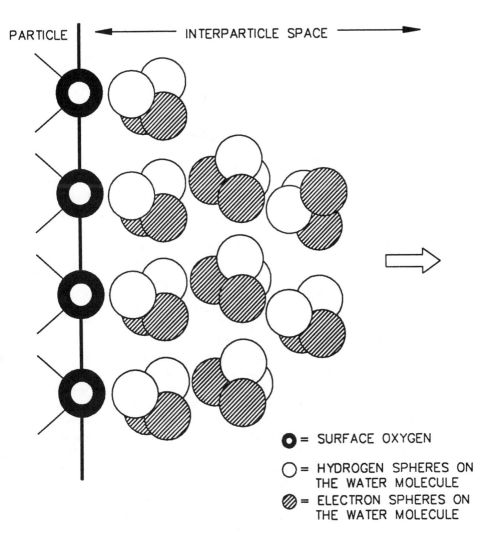

FIGURE 2.21   Orderly water structure emanating from a particle surface.

TABLE 2.5 Range of the Number of Carbon Atoms for Hydrocarbons Comprising Various Petroleum Products[77]

Product	Carbon Number
gas, liquified gas	$C_2 - C_4$
petroleum ether	$C_5 - C_6$
gasoline	$C_6 - C_{12}$
kerosene	$C_6 - C_{16}$
diesel oil	$C_{15} - C_{25}$
fuel oil nos. 1-6	$C_{10} - C_{60}$
lubricating oil	$C_{17} - C_{70}$
bitumen, coke	$>> C_{70}$

conductivity and interparticle spacing. For example, the oxidation of structural Fe^{2+} to Fe^{3+} in the octahedral layer transformed a low-swelling smectite, from a paddy soil derived from marine alluvium, into a high-swelling smectite[88].

A number of studies have been performed on the changes of soil and clay hydraulic conductivity due to the introduction of bulk hydrocarbons as the primary solvent (see Table 2.6). There are two mechanisms by which bulk hydrocarbons cause hydraulic conductivity increases. One mechanism involves the dielectric constant of the bulk hydrocarbon. The dielectric constant represents the ability of a liquid to transmit charge. Since most bulk hydrocarbons possess dielectric constants which are less than the dielectric constant of water (Table 2.7), the entry of a bulk hydrocarbon into the interparticle spacing forces water and ions out and may act as an insulator between electrostatic repulsion forces emanating from two adjacent particles. As a result of the lack of ions and the presence of an insulator, the interparticle spacing would significantly decrease and cause major cracks and fissures to occupy space previously occupied by soil particles. These cracks and fissures would act as major flow channels for bulk hydrocarbons and result in substantial soil hydraulic conductivity increases.

An alternate mechanism involves the effect of bulk hydrocarbon on the structure of water in the interparticle spacing. When bulk hydrocarbon enters the interparticle spacing, it pushes dipolar water, via mass action, out of the interparticle space and destroys the water structure which extends out from the surface of the particle. If the bulk hydrocarbon has a very low dielectric constant, it exhibits no propensity to align with surface oxygens of clay minerals. As a result, no solvent structure extends out from the particle surface, and the interparticle spacing is very small. As the dielectic constant increases, the propensity of the bulk hydrocarbon to align with the surface oxygens of clay particles should increase. Then, solvent structure emanating from the particle surface will extend out; the larger the dielectric constant,

TABLE 2.6 Effect of Bulk Hydrocarbons on the Hydraulic Conductivity and Intrinsic Permeability of Soil and Clay.

Bulk Hydrocarbon	Media	Effect[a]	Reference
acetone	10% bentonite-sand	1400 × hc incr	89
	16% bentonite-sand	55 × hc incr	89
	20% bentonite-sand	43 × hc incr	90
	Houston black clay soil[b]	100 × hc incr	91
	kaolinite	2 × hc incr	92
	24.5% kaolinite-sand	336 × hc incr	90
	kaolinite-sandy clay loam soil	100 × hc incr	93
	Lufkin clay soil[b]	> 1000 × hc incr	91
	micaceous sandy clay loam soil	10 × hc incr	93
	22% mica/kaolinite-sand	3 × hc incr	90
	26% mica-sand	16 × hc incr	89
acetone waste	20% bentonite-sand	10 × hc incr	90
	24.5% kaolinite-sand	2 × hc incr lab	90
		7 × hc incr fld	90
	22% mica/kaolinite-sand	2 × hc incr lab	90
		7 × hc incr fld	90
aniline	7% bentonite-sand	1000 × hc incr	79
	calcareous smectitic clay soil	17 × hc incr	94
	Houston black clay soil	10 × hc incr	91
	illitic clay soil	breakthrough	94
	kaolinitic clay soil	breakthrough	94
	Lufkin clay soil	100 × hc incr	91
	noncalcareous smectitic clay soil	1000 × hc incr	94
benzene	kaolinite	2700 × hc decr	92
carbon tetrachloride	7% bentonite-sand	100 × hc incr	79
diesel fuel	10% bentonite-sand	1400 × hc incr	89
	16% bentonite-sand	1800 × hc incr	89
	26% mica-sand	40 × hc incr	89
ethanol	smectite	100 × hc incr	95
ethanol & methanol	illite	10 × hc incr	95
	kaolinite	no change	95
ethylene glycol	Anthony silt loam soil[c]	1.2 × ip incr	96
	Canelo clay loam soil[d]	6 × ip incr	96
	Chalmers sandy clay loam soil[h]	2 × ip incr	96
	Fanno clay soil[c]	39 × hc incr	80
		41 × ip incr	96
	Houston black clay soil	3 × hc incr	91
	lake bottom clay[e]	47 × hc incr	80
		5 × ip incr	96
	Lufkin clay soil	100 × hc incr	91

TABLE 2.6 Effect of Bulk Hydrocarbons on the Hydraulic Conductivity and Intrinsic Permeability of Soil and Clay. (cont.)

Bulk Hydrocarbon	Media	Effect[a]	Reference
ethylene glycol	Nicholson clay soil[f]	3 × hc incr	80
(contd)		3 × ip incr	96
	river bottom sand[g]	1.1 × ip decr	96
gasoline	10% bentonite-sand	5400 × hc incr	89
	16% bentonite-sand	3500 × hc incr	89
	26% mica-sand	3900 × hc incr	89
heptane	Houston black clay soil	100 × hc incr	91
	Lufkin clay soil	100 × hc incr	91
isopropyl alcohol	Anthony silt loam soil	2.8 × ip incr	96
	Canelo clay loam soil	21 × ip incr	96
	Chalmers sandy clay		
	loam soil	3.7 × ip incr	96
	Fanno clay soil	58 × hc incr	80
		60 × ip incr	96
	lake bottom clay	13 × hc incr	80
		13 × ip incr	96
	Nicholson clay soil	16 × hc incr	80
		16 × ip incr	96
	river bottom sand	1.2 × ip incr	96
kerosene	Anthony silt loam soil	6 × ip incr	96
	10% bentonite-sand	4100 × hc incr	89
	16% bentonite-sand	3500 × hc incr	89
	Canelo clay loam soil	33 × ip incr	96
	Chalmers sandy clay		
	loam soil	6.2 × ip incr	96
	Fanno clay soil	100 × hc incr	80
		67 × ip incr	96
	lake bottom clay	2 × hc incr	80
		20 × ip incr	96
	26% mica-sand	3600 × hc incr	89
	Nicholson clay soil	67 × hc incr	80
		67 × hc incr	96
	river bottom sand	1.3 × ip incr	96
methanol	bentonite-calcareous		
	smectitic clay soil	200 × hc incr	97
	calcareous smectitic clay		
	soil	100 × hc incr	94
	illitic clay soil	breakthrough	94
	kaolinite	5 × hc incr	98
	kaolinite clay soil	40 × hc incr	94
	noncalcareous smectitic		
	clay soil	1000 × hc incr	94
motor oil	10% bentonite-sand	170 × hc incr	89
	16% bentonite-sand	83 × hc incr	89
	26% mica-sand	60 × hc incr	89

TABLE 2.6   Effect of Bulk Hydrocarbons on the Hydraulic Conductivity and Intrinsic Permeability of Soil and Clay. (cont)

Bulk Hydrocarbon	Media	Effect[a]	Reference
naphtha	Ca-montmorillonite	40,000 × hc incr	99
	Na-montmorillonite	730,000 × hc incr	99
nitrobenzene	kaolinite	3700 × hc decr	92
oil (unspecified)	clay loam soil	5-100 × hc incr	100
paraffin oil	mica-sand	10-100 × hc incr	101
phenol	kaolinite	2 × hc incr	92
trichloroethylene	10% bentonite-sand	4900 × hc incr	89
	16% bentonite-sand	28,200 × hc incr	89
	bentonite-calcareous		
	smectitic clay soil	200 × hc incr	97
xylene	Anthony silt loam soil	3.5 × ip incr	96
	20% bentonite-sand	42,900 × hc incr	90
	calcareous smectitic		
	clay soil	1000 × hc incr	94
	Canelo clay loam soil	23 × ip incr	96
	Chalmers sandy clay		
	loam soil	8 × ip incr	96
	Fanno clay soil	94 × hc incr	80
		94 × ip incr	96
	Houston black clay soil	>1000 × hc incr	91
	illitic clay soil	1000 × hc incr	94
	24.5% kaolinite-sand	10,000 × hc incr	96
	kaolinitic clay soil	100 × hc incr	94
	kaolinitic sandy clay		
	loam soil	1000 × hc incr	93
	lake bottom clay	27 × hc incr	80
		27 × ip incr	96
	Lufkin clay soil	>1000 × hc incr	91
	micaceous sandy clay		
	loam soil	10 × hc incr	93
	22% mica/kaolinite-sand	1500 × hc incr	90
	26% mica-sand	6400 × hc incr	89
	montmorillonitic sandy		
	clay loam soil	10 × hc incr	93
	Nicholson clay soil	67 × hc incr	80
		67 × ip incr	96
	noncalcareous smectitic		
	clay soil	400 × hc incr	94
	river bottom sand	no change	96
xylene waste	20% bentonite-sand	243 × hc incr	90
	24.5% kaolinite-sand	555 × hc incr	90
	22% mica/kaolinite-sand	426 × hc incr	90
	micaceous sandy clay		
	loam soil	100 × hc incr	93
	mica-sand	100 × hc incr	101

TABLE 2.6 Effect of Bulk Hydrocarbons on the Hydraulic Conductivity and Intrinsic Permeability of Soil and Clay. (cont)

a abbreviations: breakthrough = particle spacing decrease resulted in major cracks and fissures with no measurable equilibrium hydraulic conductivity, decr = decrease, fld = field measurement, hc = hydraulic conductivity, incr = increase, ip = intrinsic permeability, lab = laboratory measurement.

b dominant clay is montmorillonite.

c dominant clays are montmorillonite and mica.

d dominant clay is kaolinite.

e dominant clays are illite and kaolinite.

f dominant clay is vermiculite.

g dominant clays are mica and kaolinite.

h dominant clays are montmorillonite and vermiculite.

TABLE 2.7 Dielectric Constants of Bulk Hydrocarbons

acetone	21.5	n-hexane	1.9
acetophenone	17.4	hydrogen cyanide	115
aniline	6.9	methanol	32.6
benzene	2.3	methylene chloride	9.1
d-camphene	2.3	naphthalene	2.5
carbon tetrachloride	2.2	nitrobenzene	34.8
chloroform	4.8	n-nonane	2.0
chloromethane	12.6	n-octane	1.9
o-cresol	11.5	n-pentane	1.8
cyclohexane	2.0	phenol	9.8
n-decane	2.0	1,2-propanediol	32
o-dichlorobenzene	9.9	1-propanol	20.1
1,1-dichloroethylene	4.6	styrene	2.4
n-dodecane	2.0	toluene	2.4
ethanol	24.3	o-toluidine	6.3
ethylbenzene	2.4	trichloroethylene	3.4
ethylene glycol	37.0	n-undecane	2.0
formamide	110.0	water	80.1
glycerol	42.5	o-xylene	2.6

the greater the ability for the bulk hydrocarbon to align and the farther the solvent structure extends into the interparticle spacing. As a result, a significant interparticle spacing exists and major cracks and fissures do not form.

There are several other points to remember concerning the effect of bulk hydrocarbons on soil hydraulic conductivity. First, the ability of bulk hydrocarbons to increase soil hydraulic conductivity is a physical process which for most cases does not permanently alter the particle surface. As a result, it is a reversible process. If water replaces bulk hydrocarbon in the interparticle spacing, the soil hydraulic conductivity to water should significantly decrease.

Second, some published information indicates that not all solvents cause hydraulic conductivity increases. An analysis of information presented in

Table 2.6 will reveal that two solvents—benzene and nitrobenzene—caused substantial hydraulic conductivity decreases. At this time, however, strong conclusions regarding which bulk hydrocarbons cause hydraulic conductivity decreases cannot be made; published information in this area is not adequate to support any strong conclusions or to distinguish actual hydraulic conductivity changes from experimental artifacts.

Third, if the concentration of the bulk hydrocarbon is too dilute, hydraulic conductivity or interparticle spacing changes will not occur. An analysis of the information presented in Table 2.8 will reveal that concentrations of water-miscible bulk hydrocarbon greater than 60 percent in H_2O should affect either hydraulic conductivity or interparticle spacing. Also, concentration of water-miscible bulk hydrocarbon ranging from 1 percent to 60 percent in H_2O may or may not affect hydraulic conductivity or interparticle spacing. However, concentrations less than 1 percent bulk hydrocarbon in water (i.e. less than 10,000 ppm) apparently have no effect on soil hydraulic conductivity. It is important to note that the majority of landfill leachates and liquids in surface impoundments contain less than 10,000 ppm bulk hydrocarbons in water; as a result, these leachates and liquids should not be expected to affect the hydraulic conductivity of the soil and soil liners at these facilities.

TABLE 2.8   Effect of Bulk Hydrocarbon Concentration on Soil/Clay Hydraulic Conductivity or Interparticle Spacing.

Bulk Hydrocarbon	Clay/Soil	Concentration[a]	Effect[b]	Ref.
acetone	kaolinite	0.1%	slight hc decr	102
		100%	2× hc incr	
	mica-sand	2%	no hc change	101
		12.5%	no hc change	
		25%	no hc change	
		50%	no hc change	
		75%	10× hc incr	
		100%	67× hc incr	
	Ca-mont-morillonite	1-10%	260% is incr	103
		20-50%	39% is incr	
		60-90%	16% is incr	
		100%	9% is incr	
aniline	bentonite-sand	1.5%	no hc change	101
		3.0%	no hc change	
		100%	1000× hc incr	
benzene	kaolinite	0.1%	slight hc decr	102
		100%	100× hc decr	
carbon tetrachloride	7% bentonite-sand	720 ppm	no hc change	79
		100%	100× hc incr	

TABLE 2.8   Effect of Bulk Hydrocarbon Concentration on Soil/Clay Hydraulic Conductivity or Interparticle Spacing. (cont)

Bulk Hydrocarbon	Clay/Soil	Concentration[a]	Effect[b]	Ref.
contaminated groundwater	clay soil[c]	35 ppm[d]	no hc change	104
diethylene glycol	kaolinite	20%	no hc change	105
		40%	slight hc decr	
dioxane	Ca-mont-morillonite	1-10%	21% is decr	103
		20-50%	21% is decr	
		60-90%	21% is decr	
		100%	21% is decr	
ethanol	Ca-mont-morillonite	1-10%	55% is incr	103
		20-50%	11% is incr	
		60-90%	11% is decr	
		100%	11% is decr	
ethylene glycol	Ca-mont-morillonite	1-10%	2% is incr	103
		20-50%	11% is decr	
		60-90%	11% is decr	
		100%	12% is decr	
landfill leachate	clay soil[e]	—	no hc change	106
	clay soil[c]	0.1%[d]	no hc change	99
	clay soil[c]	0.14%[d]	no hc change	104
	silt soil[e]	—	no hc change	106
methanol	Ca-mont-morillonite	1-10%	no is change	103
		20-50%	no is change	
		60-90%	13% is decr	
		100%	14% is decr	
morpholine	Ca-mont-morillonite	1-10%	21% is decr	103
		20-50%	21% is decr	
		60-90%	21% is decr	
		100%	21% is decr	
nitrobenzene	kaolinite	0.1%	slight hc decr	102
		100%	100× hc decr	
1,5-pentanediol	Ca-mont-morillonite	1-10%	16% is incr	103
		20-50%	5% is decr	
		60-90%	5% is decr	
		100%	26% is decr	
phenol	kaolinite	0.1%	slight hc decr	102
		100%	2× hc incr	
n-propanol	Ca-mont-morillonite	1-10%	21% is incr	103
		20-50%	5% is decr	
		60-90%	5% is decr	
		100%	6% is decr	

[a] in H_2O
[b] abbreviations: decr = decrease, hc = hydraulic conductivity, incr = increase, is = interparticle spacing.
[c] unspecified soil type and mineralogy.
[d] ppm total organic carbon.
[e] dominant clays are kaolinite and Ca-nontronite.

REFERENCES

1. UOP Inc. Ground Water and Wells. St. Paul, MN: Johnson Division, UOP Inc. (1982).
2 Van Schilfgaarde, J. (ed). Drainage for Agriculture. Madison, WI: American Society of Agronomy, Inc. (1974).
3. Todd, D. K. Groundwater Hydrology. Second Edition. New York: John Wiley & Sons, Inc. (1980).
4. Freeze, R. A., and Cherry, J. A. Groundwater. Englewood Cliffs, NJ: Prentice Hall, Inc. (1979).
5. Chorley, R. J. (ed). Introduction to Physical Hydrology. London: Methuen & Co., Ltd. (1969).
6. Buckman, H. O., and Brady, N. C. The Nature and Properties of Soils. New York: The Macmillan Co. (1972).
7. Martin, J. P., and Koerner, R. M. The Influence of Vadose Zone Conditions on Groundwater Pollution. Part I: Basic Principles and Static Conditions. Journal of Hazardous Materials 8:349-366 (1984).
8. Martin, J. P., and Koerner, R. M. The Influence of Vadose Zone Conditions in Groundwater Pollution. Part II: Fluid Movement. Journal of Hazardous Materials 9:181-207 (1984).
9. Corey, A. T., and Klute A. Application of the Potential Concept to Soil Water Equilibrium and Transport. Soil Science Society of America Journal 49:3-11 (1985).
10. Baver, L. D., Gardner, W. H. and Gardner, W. R. Soil Physics. Fourth Edition. New York: John Wiley & Sons, Inc. (1972).
11. Fairbridge, R. W., and Finkl, C. W. Jr. (eds). The Encyclopedia of Soil Science Part I. Physics, Chemistry, Biology, Fertility, and Technology. Stroudsburg, PA: Dowden, Hutchinson & Ross, Inc. (1979).
12. Pettyjohn, W. A., Kent, D. C., Prickett, H. E., LeGrand, H. E., and Witz, F. E. Methods for the Prediction of Leachate Plume Migration and Mixing. U.S. EPA Municipal Environmental Research Laboratory, Cincinnati, OH (1982).
13. Lutton, R. J. Predicting Percolation through Waste Cover by Water Balance. *In* Disposal of Hazardous Waste. Proceedings of the Sixth Annual Research Symposium, Chicago, IL, March 17-20, 1980. EPA-600/9-80-010. Cincinnati, OH: U.S. Environmental Protection Agency (1980).
14. Goring, C. A. I., and Hamaker, J. W. (eds). Organic Chemicals in the Soil Environment. New York: Marcel Dekker, Inc. (1972).
15. Fuller, W. H., and Warrick, A. W. Soils in Waste Treatment and Utilization. Vols. I & II. Boca Raton, FL: CRC Press, Inc. (1985).
16. Parlange, J. Y. A Note on the Use of Infiltration Equations. Soil Science Society of America Journal 41:654-655 (1977).
17. Haverkamp, R., Vauclin, M., Touma, J., Wierenga, P. J., and Vachaud, G. A Comparison of Numerical Simulation Models for One-Dimensional Infiltration. Soil Science Society of America Journal 41:285-293 (1977).
18. Ghosh, R. K. Modeling Infiltration. Soil Science 130:297-302 (1980).
19. Thomas, G. W. and Phillips, R. E. Consequences of Water Movement in Macropores. Journal of Environmental Quality 8:149-152 (1979).

20. Coon, J. J., and Pehrson, J. E. Avocado Irrigation. California Agricultural Extension Leaflet No. 50 (1955).
21. Shaffer, K., Fritton, D. D., and Baker, D. E. Drainage Water Sampling in a Wet, Dual-Pore Soil System. Journal of Environmental Quality **8**:241-246 (1979).
22. Beven, K. , and Germann, P. Macropores and Water Flow in Soils. Water Resources Research **18**:1311-1325 (1982).
23. Quisenberry, V. L., and Phillips, R. E. Percolation of Surface-Applied Water in the Field. Soil Science Society of America Journal **40**:484-489 (1976).
24. Shuford, J. W., Fritton, D. D., and Baker, D. E. Nitrate-Nitrogen and Chloride Movement Through Undisturbed Field Soil. Journal of Environmental Quality **6**:255-259 (1977).
25. LaFleur, K. S. Sorption of Pesticides by Model Soils and Agronomic Soils: Rates and Equilibria. Soil Science **127**:94-101 (1979).
26. Greenland, D. J., and Hayes, M. H. B. (eds). The Chemistry of Soil Processes. New York: John Wiley & Sons, Inc. (1981).
27. Parry, G. D. R., Bell, R. M., and Jones, A. K. Degraded and Contaminated Land Reuse—Covering Systems. *In* Proceedings of the National Conference on Management of Uncontrolled Hazardous Waste Sites, November 29 - December 1, 1982, Washington, D. C. Silver Spring, MD: Hazardous Materials Control Research Institute (1982).
28. Mutch, R. D. Jr., Daigler, J., and Clarke, J. H. Clean-up of Shope's Landfill, Girard, PA. *In* Proceedings of the National Conference on Management of Uncontrolled Hazardous Waste Sites, October 31 - November 2, 1983, Washington, D.C. Silver Spring, MD: Hazardous Materials Control Research Institute (1983).
29. McAneny, C. C., and Hatheway, A. W. Design and Construction of Covers for Uncontrolled Landfill Sites. *In* Proceedings of the 6th National Conference on Management of Uncontrolled Hazardous Waste Sites, November 4-6, 1985, Washington, D.C. Silver Spring, MD: Hazardous Materials Control Research Institute (1985).
30. Furman, C., and Cochran, S. R. Case Study Investigations of Remedial Response Programs at Uncontrolled Hazardous Waste Sites. *In* Proceedings of the Eleventh Annual Research Symposium on the Land Disposal of Hazardous Waste. EPA-600/9-85/013. Cincinnati, OH: U.S. Environmental Protection Agency (1985).
31. Emrich, G.H., and Beck, W. W. Jr. Top Sealing to Minimize Leachate Generation-Status Report. *In* Proceedings of the Seventh Annual Research Symposium on Land Disposal: Hazardous Waste. EPA-600/9-81-0026. Cincinnati, OH: U.S. Environmental Protection Agency (1981).
32. Beck, W. W. Jr., Dunn, A. L., and Emrich, G. H. Leachate Quality Improvements After Top Sealing. *In* Proceedings of the Eighth Annual Research Symposium on the Land Disposal of Hazardous Waste. EPA-600/9-82/002. Cincinnati, OH: U.S. Environmental Protection Agency (1982).
33. Jones, A. K., Bell, R. M., Bakker, L. J., and Bradshaw, A. D. Coverings for Metal Contaminated Land. *In* Proceedings of the National Conference on Management of Uncontrolled Hazardous Waste Sites, November 29 - Decem-

ber 1, 1982, Washington, D.C. Silver Spring, MD: Hazardous Materials Control Research Institute (1982).

34. Matthews, M. L. UMTRA Project: Implementation of Design. *In* Proceedings of the 8th Annual Symposium on Geotechnical and Geohydrological Aspects of Waste Management, Fort Collins, CO, 5-7 February, 1986. Boston: Balkema (1986).

35. Thiers, G. R., Watken, T. R., and Farnes, L. L. Tailings Stabilization Experience at the Canonsburg UMTRA Site. *In* Proceedings of the 8th Annual Symposium on Geotechnical and Geohydrological Aspects of Waste Management, Fort Collins, CO, 5-7 February 1986. Boston: Balkema (1986).

36. Neely, N. S., Walsh, J. J., Gillespie, D. P., and Schauf, F. J. Remedial Actions at Uncontrolled Hazardous Waste Sites. *In* Proceedings of the Seventh Annual Research Symposium on Land Disposal: Hazardous Waste. EPA-600/9-81-002b. Cincinnati, OH: U.S. Environmental Protection Agency (1981).

37. Gushue, J. J., and Cummings, R. S. On-site Containment of PCB-Contaminated Soils at Aerovox, Inc., New Bedford, MA. *In* Proceedings of the Fourth National Symposium on Aquifer Restoration and Groundwater Monitoring, May 23-25, 1984, Columbus, OH. Worthington, OH: National Water Well Association (1984).

38. Bracken, B. D., and Theisen, H. M. Cleanup and Containment of PCBs: A Success Story. *In* Proceedings of the National Conference on Management of Uncontrolled Hazardous Waste Sites, November 29 - December 1, 1982, Washington, D.C. Silver Spring, MD: Hazardous Materials Control Research Institute (1982).

39. Dowiak, M. J., Lucas, R. A., Nazar, A., and Theelfall, D. Selection, Installation, and Post-Closure Monitoring of a Low Permeability Cover over a Hazardous Waste Disposal Facility. *In* Proceedings of the National Conference on Management of Uncontrolled Hazardous Waste Sites, November 29-December 1, 1982, Washington, D.C. Silver Spring, MD: Hazardous Materials Control Research Institute (1982).

40. Solym, P. The Case Story of the BT-Kemi Dumpsite. *In* Proceedings of the National Conference on Management of Uncontrolled Hazardous Waste Sites, October 31 - November 2, 1983, Washington, D.C. Silver Spring, MD: Hazardous Materials Control Research Institute (1983).

41. Hatch, N. N. Jr., and Hayes, E. State-of-the-Art Remedial Action Technologies Used for the Sydney Mine Waste Disposal Site Cleanup. *In* Proceedings of the 6th National Conference on Management of Uncontrolled Hazardous Waste Sites, November 4-6, 1985, Washington, D.C. Silver Spring, MD: Hazardous Materials Control Research Institute (1985).

42. Kunze, R. J., Parlange, J. Y., and Rose, C. W. A Comparison of Numerical and Analytical Techniques for Describing Capillary Rise. Soil Science **139**:491-496 (1985).

43. Doering, E. J. A Direct Method for Measuring the Upward Flow of Water From the Water Table. Soil Science **96**:191-195 (1963).

44. Doering, E. J., Reeve, R. C., and Stockinger, K. R. Salt Accumulation and Salt Distribution as an Indicator of Evaporation From Fallow Soils. Soil Science

97:312-319 (1964).

45. Kovda, V. A. U.S.S.R. Nomenclature and Classifications of Saline Soils. In Kovda, V. A. (ed). Irrigation, Drainage, Solidity, and International Sources Book. London: Hutchinson/F.A.O./U.N.E.S.C.O. (1973).

46. Richards, L. A. Gardner, W. R., and Ogata, G. Physical Processes Determining Water Loss From Soil. Soil Science Society of America Proceedings **20**:310-314 (1956).

47. Saleh, H. H., and Troech, F. R. Salt Distribution and Water Consumption From a Water Table With and Without a Crop. Agronomy Journal **74**:321-324 (1982).

48. Matthess, G., and Pekdeger, A. Concepts of a Survival and Transport Model of Pathogenic Bacteria and Viruses in Groundwater. The Science of the Total Environment **21**:149-159 (1981).

49. U.S. EPA. Municipal Environmental Research Laboratory. Compatibility of Grouts with Hazardous Wastes. EPA-600/2-84-015. Cincinnati, OH: U.S. Environmental Protection Agency (1984).

50. Hagedorn, C., McCoy, E. L., and Rahe, T. M. The Potential for Groundwater Contamination from Septic Effluents. Journal of Environmental Quality **10**:1-7 (1981).

51. Wilkinson, H. T., Miller, R. D., and Millar, R. L. Infiltration of Fungal and Bacterial Propagules into Soil. Soil Science Society of America Journal **45**:1034-1039 (1981).

52. Wang, D., Gerba, C. P., and Lance, J. C. Effect of Soil Permeability on Virus Removal Through Soil Columns. Applied and Environmental Microbiology **42**:83-88 (1981).

53. Tan, K. H. Principles of Soil Chemistry. New York: Marcel Dekker, Inc. (1982).

54. Tessens, E. Clay Migration in Upland Soils of Malaysia. Journal of Soil Science **35**:615-624 (1984).

55. Moore, R. S., Taylor, D. H., Sturman, L. S., Reddy, M. M. and Fuchs, G. W. Poliovirus Adsorption by 34 Minerals and Soils. Applied and Environmental Microbiology **42**:963-975 (1981).

56. Hendricks, D. W. Post, F. J., and Khairnar, D. R. Adsorption of Bacteria on Soils: Experiments, Thermodynamic Rationale, and Application. Water, Air, and Soil Pollution **12**:219-232 (1979).

57. Lance, J. C., and Gerba, C. P. Virus Movement in Soil During Saturated and Unsaturated Flow. Applied and Environmental Microbiology **47**:335-337 (1984).

58. Raisbeck, J. M., and Mohtadi, M. F. The Environmental Impacts of Oil Spills on Land in the Arctic Regions. Water, Air, and Soil Pollution **3**:195-208 (1974).

59. MacKay, D., and Mohtadi, M. 1975. The Area Affected by Oil Spills on Land. Canadian Journal of Chemical Engineering **53**:140-143 (1975).

60. Farmer, V. E. Jr. Behavior of Petroleum Contaminants in an Underground Environment. *In* Proceedings of a Seminar on Ground Water and Petroleum Hydrocarbons, June 26-28, 1983. Toronto, Ontario, Canada: Petroleum Association for Conservation of the Canadian Environment (1983).

61. American Petroleum Institute. The Migration of Petroleum Products in Soil and Ground Water. Principles and Counter Measures. Washington, D.C.: American Petroleum Institute (1972).

62. Bossert, I., and Bartha, R. The Fate of Petroleum in Soil Ecosystems. *In* Atlas, R. M. (ed.). Petroleum Microbiology. New York: Macmillan Co. (1984).
63. Van Dam, J. The Migration of Hydrocarbons in a Water Bearing Stratum. *In* Hepple, P. (ed.). The Joint Problems of the Oil and Water Industries. London: Institute of Petroleum (1967).
64. Dietz, D. N. Pollution of Permeable Strata by Oil Components. *In* Hepple, P. (ed.). Water Pollution by Oil. New York: Elsevier Publishing Co. and the Institute of Petroleum (1970).
65. CONCAWE Secretariat. Inland Oil Spill Clean-up Manual. Report No. 4/74. The Hague, Netherlands: CONCAWE (1974).
66. CONCAWE. Protection of Groundwater from Oil Pollution. The Hague, Netherlands: CONCAWE (1979). NTIS PB82-174608.
67. Vanloocke, R., DeBorger, R., Voets, J. P., and Verstraete, W. Soil and Groundwater Contamination by Oil Spills; Problems and Remedies. International Journal of Environmental Studies **8**:99-111 (1975).
68. Kerfoot, W. B., and Massard, V. Direct Measurement of Gasoline Flow. *In* Proceedings of the Third National Symposium on Aquifer Restoration and Groundwater Monitoring, The Fawcett Center, Columbus, OH, May 25-27, 1983. Worthington, OH: National Water Well Association (1983).
69. Yaniga, P. M., and Warburton, J. G. Discrimination between Real and Apparent Accumulation of Immiscible Hydrocarbons on the Water Table: A Theoretical and Empirical Analysis. *In* Proceedings of the Fourth National Symposium on Aquifer Restoration and Groundwater Monitoring, The Fawcett Center, Columbus, OH, May 23-25, 1984. Worthington, OH: National Water Well Association (1984).
70. Blake, S. B., and Hall, R. A. Monitoring Petroleum Spills with Wells: Some Problems and Solutions. *In* Proceedings of the Fourth National Symposium on Aquifer Restoration and Groundwater Monitoring, The Fawcett Center, Columbus, OH, May 23-25, 1984. Worthington, OH: National Water Well Association (1984).
71. Hall, R. A., Blake, S. B., and Champlin, S. C. Jr. Determination of Hydrocarbon Thicknesses in Sediments Using Borehole Data. *In* Proceedings of the Fourth National Symposium on Aquifer Restoration and Groundwater Monitoring, The Fawcett Center, Columbus, OH, May 23-25, 1984. Worthington, OH: National Water Well Association (1984).
72. Schwille, F. Migration of Organic Fluids Immiscible with Water in the Unsaturated Zone. *In* Yaron, B., Dagan, G., and Goldshmid, J. (eds.). Pollutants in Porous Media. Ecological Studies Vol. 47. pp. 27-48. New York: Springer-Verlag (1984).
73. Van Duijvenbooden, W., and Kooper, W. F. Effects on Groundwater Flow and Groundwater Quality of a Waste Disposal Site in Noordwijk, The Netherlands. The Science of the Total Environment **21**:85-92 (1981).
74. Villaume, J. F. Investigations at Sites Contaminated with Dense, Non-Aqueous Phase Liquids (NAPLS). Groundwater Monitoring Review **5**:60-74 (1985).
75. Pfannkuch, H. O. Determination of the Contaminant Source Strength From Mass Exchange Processes at the Petroleum-Groundwater Interface in Shallow

Aquifer Systems. *In* Proceedings of the NWWA/API Conference on Petroleum Hydrocarbons and Organic Chemicals in Groundwater—Prevention, Detection, and Restoration. November 5-7, 1984, Houston, TX. Worthington, OH: National Water Well Association (1984).

76. U.S. Geological Survey. Groundwater Contamination by Crude Oil at the Bemidji, Minnesota, Research Site: U.S. Geological Survey Toxic Waste—Groundwater Contamination Study. Water-Resources Investigations Report 84-4188 (1984).

77. Yang, W-P. Volatilization, Leaching, and Degradation of Petroleum Oils In Sand and Soil Systems. Phd Thesis, Dept. of Civil Engineering, North Carolina State University (1981).

78. Hartley, G. S. Physico-Chemical Aspects of the Availability of Herbicides in Soils. *In* Woodford, E. K. (ed.). Herbicides in the Soil. Oxford: Blackwell Scientific Publications (1961).

79. Evans, J. C., Fang, H-Y, and Kugelman, I. J. Containment of Hazardous Materials with Soil-Bentonite Slurry Walls. *In* Proceedings of the 6th National Conference on the Management of Uncontrolled Hazardous Waste Sites, November 4-6, 1985, Washington, D.C. Silver Spring, MD: Hazardous Materials Control Research Institute (1985).

80. Anderson, D. C., and Jones, S. G. Clay Barrier-Leachate Interaction. *In* Proceedings of the National Conference on Management of Uncontrolled Hazardous Waste Sites, October 31-November 2, 1983, Washington, D.C. Silver Spring, MD: Hazardous Materials Control Research Institute (1983).

81. Ravina, I., and Low, P. F. Relation Between Swelling, Water Properties, and b-Dimension in Montmorillonite - Water Systems. Clays and Clay Minerals **20**:109-123 (1972).

82. Odom, J. W., and Low, P. F. Relation Between Swelling, Surface Area, and b Dimension of Na-Montmorillonites. Clays and Clay Minerals **26**:345-351 (1978).

83. Low, P. F. Nature and Properties of Water in Montmorillonite-Water Systems. Soil Science Society of America Journal **43**:651-658 (1979).

84. Low, P. F., and Margheim, J. F. The Swelling of Clay: I. Basic Concepts and Empirical Equations. Soil Science Society of America Journal **43**:473-481 (1979).

85. Low, P. F. The Swelling of Clay: II. Montmorillonites. Soil Science Society of America Journal **44**:667-676 (1980).

86. Low, P. F. The Swelling of Clay: III. Dissociation of Exchangeable Cations. Soil Science Society of America Journal **45**:1074-1078 (1981).

87. Parker, J. C., and Zelazny, L. W. The Influence of Geostatic and Hydrostatic Forces on Double-Layer Interactions in Soils. Soil Science Society of America Journal **47**:191-195 (1983).

88. Egashira, K., and Ohtsubo, M. Swelling and Mineralogy of Smectites in Paddy Soils Derived From Marine Alluvium, Japan. Geoderma **29**:119-127 (1983).

89. Brown, K. W. and Thomas, J. C. Conductivity of Three Commercially Available Clays to Petroleum Products and Organic Solvents. Hazardous Waste **1**:545-553 (1984).

90. Brown, K. W., Thomas, J. C. and Green, J. W. Field Cell Verification of the

Effects of Concentrated Organic Solvents on the Conductivity of Compacted Soils. Hazardous Waste and Hazardous Materials 3:1-19 (1986).

91. Anderson, D. A., and Brown, K. W. Organic Leachate Effects on the Permeability of Clay Liners. *In* Proceedings of the Seventh Annual Research Symposium on Land Disposal: Hazardous Waste. EPA - 600/9-81-002b Cincinnati, OH: U.S. Environmental Protection Agency (1981).

92. Acar, Y. B., Olivieri, I., and Field, S. D. Organic Fluid Effects on the Structural Stability of Compacted Kaolinite. *In* Proceedings of the Tenth Annual Research Symposium on the Land Disposal of Hazardous Waste. EPA -600/9-84-007. Cincinnati, OH: U.S. Environmental Protection Agency (1984).

93. Brown, K. W., Green, J. W., and Thomas, J. C. The Influence of Selected Organic Liquids on the Permeability of Clay Liners. *In* Proceedings of the Ninth Annual Research Symposium on the Land Disposal of Hazardous Waste. EPA-600/9-83-018. Cincinnati, OH: U.S. Environmental Protection Agency (1983).

94. Anderson, D. C., Brown, K. W., and Green, J. Organic Leachate Effects on the Permeability of Clay Liners. *In* Proceedings of the National Conference on Management of Uncontrolled Hazardous Waste Sites, October 28-30, 1981, Washington, D.C. Silver Spring, MD: Hazardous Materials Control Research Institute (1981).

95. Mesri, G. and Olson, R. E. Mechanisms Controlling the Permeability of Clays. Clays and Clay Minerals 19:151-158 (1971).

96. Schramm, M. Warrick, A. W., and Fuller, W. H. Permeability of Soils to Four Organic Liquids and Water. Hazardous Waste and Hazardous Materials 3:21-27 (1986).

97. Anderson, D. C., Crawley, W., and Zabcik, J. D. Effects of Various Liquids on Clay Soil: Bentonite Slurry Mixtures. *In* Hydraulic Barriers in Soil and Rock. ASTM STP 874. Philadelphia, PA: American Society for Testing and Materials (1985).

98. Foreman, D. E., and Daniel, D. E. 1984. Effects of Hydraulic Gradient and Method of Testing on the Hydraulic Conductivity of Compacted Clay to Water, Methanol, and Heptane. *In* Proceedings of the Tenth Annual Research Symposium on the Land Disposal of Hazardous Waste. EPA - 600/9-84-007. Cincinnati, OH: U.S. Environmental Protection Agency (1984).

99. Buchanan, P. N. Effect of Temperature and Adsorbed Water on Permeability and Consolidation Characteristics of Sodium and Calcium Montmorillonite. Ph.D. Dissertation, Texas A & M University, College Station, TX (1964).

100. Van Schaik, J. C. Oil:Water Permeability Ratios as a Measure of the Stability of Soil Structure. Canadian Journal of Soil Science 54:331-332 (1974).

101. Brown, K. W., Thomas, J. C., and Green, J. W. Permeability of Compacted Soils to Solvent Mixtures and Petroleum Products. *In* Proceedings of the Tenth Annual Research Symposium on the Land Disposal of Hazardous Waste. EPA-600/9-84-007. Cincinnati, OH: U.S. Environmental Protection Agency (1984).

102. Acar, Y. B., Hamidon, A., Field, S. D., and Scott, L. The Effect of Organic Fluids on Hydraulic Conductivity of Compacted Kaolinite. *In* Hydraulic Bar-

riers in Soil and Rock. ASTM STP 874. Philadelphia, PA: American Society for Testing and Materials (1985).

103. Griffin, R. A., Hughes, R. E., Follmer, L. R., Stohr, C. J., Morse, W. J., Johnson, T. M., Bartz, J. K., Steele, J. D., Cartwright, K., Killey, M. M., and DuMontelle, P. B. Migration of Industrial Chemicals and Soil-Waste Interactions at Wilsonville, Illinois. *In* Proceedings of the Tenth Annual Research Symposium on the Land Disposal of Hazardous Waste. EPA-600/9-84-007. Cincinnati, OH: U.S. Environmental Protection Agency (1984).

104. Wuellner, W. W., Wierman, D. A., and Koch, H. A. Effect of Landfill Leachate on the Permeability of Clay Soils. *In* Proceedings of the Eighth Annual Madison Waste Conference, September 18-19, 1985, Madison, WI. Madison, WI: University of Wisconsin (1985).

105. Whittle, G. P., Carlton, T. A., and Henry, H. R. Permeability Changes in Clay Liners of Hazardous Waste Storage Pits. *In* Seventh Annual Madison Waste Conference Proceedings, September 11-12, 1984, Madison, WI. Madison, WI: University of Wisconsin (1984).

106. Eklund, A. G. A Laboratory Comparison of the Effects of Water and Waste Leachate on the Performance of Soil Liners. *In* Hydraulic Barriers in Soil and Rock. ASTM STP 874. Philadelphia, PA: American Society for Testing and Materials (1985).

3

Element Fixation in Soil

INTRODUCTION

The inorganic fraction of soil is comprised of numerous sparingly soluble inorganic chemicals known as minerals. Soil minerals are products of an extremely complex chain of events involving the action of weathering, topography, and biota on a parent geologic material over some period of time. In general, minerals form in soil via three processes. First, soil minerals are derived from the physical disintegration of minerals originally present in the soil or by the physical disintegration of minerals deposited at the site by geological processes. Second, soil minerals are derived from the chemical transformation of minerals that are susceptible to weathering into those that are resistant to weathering. Third, soil minerals are formed via the subsequent precipitation of water-soluble elements which are released during the first and second processes. The first and second processes have a negligible impact on the ultimate fate of an element added to soil; however, the third has a major impact on an element's ultimate fate in soil.

Fixation refers to the soil chemical reactions which immobilize an element within the structure of a mineral or at the mineral surface. There are three types of fixation reactions. First, chemisorption is the formation of a covalent bond between an adsorbed element and a mineral surface which results in element immobilization. Second, solid state diffusion refers to the irreversible penetration of an element into the pore spaces of a mineral's structure. Third, precipitation refers to the formation of an insoluble solid comprised of elements which were previously dissolved in water.

The fixation of elements by soil minerals is very similar to the fixation, stabilization, or solidification methods utilized for some types of hazardous wastes containing metals. On the other hand, fixation of elements by soil minerals is unlike encapsulation, which encapsulates the metal within an insoluble, impermeable shell. In the waste fixation process, binders or fixatives such as fly-ash, soluble silicates, calcium and sulfur compositions, and cements and concretes are mixed with a hazardous waste. The fixative dissolves and releases anions which react with the metals to form precipitation products. Stabilized wastes are produced if the metal is "grafted" into an insoluble crystalline structure with the fixative via strong chemical bonds. A stable, fixed waste will bind and hold metals under natural environmental conditions. Likewise, fixation in soils should bind and hold elements under natural environmental conditions.

Relatively little is known about the first two types of soil mineral fixation reactions discussed above. However, these are not considered to be extensively occurring reactions. On the other hand, the fixation of elements via incorporation into the structure of soil minerals during mineral precipitation is an extremely important reaction. This chapter will focus on the types and amounts of elements found in soil, how these elements are fixed into mineral structures, and how some remedial actions have utilized element fixation.

ELEMENT CONCENTRATIONS IN SOIL

Eleven of the elements listed in Table 3.1, along with carbon, hydrogen, and oxygen, constitute over 99 percent of the total elemental content of soil: Al, Ca, Fe, K, Mg, Mn, Na, P, S, Si, and Ti. The remaining one percent is comprised of elements known commonly as the "trace elements." The word "trace" identifies the fact that they occur in soil in minute amounts; it has no bearing or relationship to any concentration limit protecting human health or biota.

Table 3.1 lists the mean concentrations, typical ranges, and observed limits of several elements in natural soil (i.e. background concentrations). The total concentration of any element, C_{Total}, in a soil is equal to:

$$C_{Total} = C_{Fixed} + C_{Adsorbed} + C_{Water} \qquad (3.1)$$

where:

C_{Fixed} = concentration of fixed element comprising part of the structure of clay and soil minerals, in mg element/kg soil.

$C_{Adsorbed}$ = concentration of element adsorbed onto the surface of soil minerals and onto organic matter exchange sites, in mg element/kg soil.

C_{Water} = concentration of element in soil water or groundwater in equilibrium with $C_{Adsorbed}$, in mg soluble element/kg soil. (See Table 3.2 for natural background levels found in groundwater).

C_{Fixed} represents the "immobile" fraction of C_{Total}. The sum of $C_{Adsorbed}$ and C_{Water} represents the potentially mobile portion of C_{Total}; these will be discussed in detail in the next chapter.

There are four important facts that should be understood concerning the data listed in Table 3.1, the parameters listed in Equation 3.1, and the interrelationships of these parameters. First, C_{Total} should not be expected to be uniform with depth. Natural processes involved in the distribution of elements in the soil profile include:

TABLE 3.1 Native Soil Concentrations of Various Elements

Element	Concentration (ppm)	
	Typical Range	Extreme Limits
Ag	0.1 - 5.0	0.1 - 50
Al	10,000 - 300,000	—
As	1.0 - 40	0.1 - 500
B	2.0 - 130	0.1 - 3000
Ba	100 - 3500	10 - 10,000
Be	0.1 - 40	0.1 - 100
Br	1.0 - 10	—
Ca	100 - 400,000	—
Cd	0.01 - 7.0	0.01 - 45
Ce	30 - 50	—
Cl	10 - 100	—
Co	1.0 - 40	0.01 - 500
Cr	5.0 - 3000	0.5 - 10,000
Cs	0.3 - 25	—
Cu	2.0 - 100	0.1 - 14,000
F	30 - 300	—
Fe	7,000 - 550,000	—
Ga	0.4 - 300	—
Ge	1.0 - 50	—
Hg	0.01 - 0.08	—
I	0.1 - 40	—
K	400 - 30,000	—
La	1.0 - 5000	—
Li	7.0 - 200	1.0 - 3000
Mg	600 - 6000	—
Mn	100 - 4000	1.0 - 70,000
Mo	0.2 - 5.0	0.1 - 400
Na	750 - 7500	400 - 30,000
Ni	5.0 - 1000	0.8 - 6200
P	50 - 5000	—
Pb	2.0 - 200	0.1 - 3000
Ra	$10^{-6.5} - 10^{-5.7}$	—
Rb	20 - 600	3.0 - 3000
S	30 - 10,000	—
Sb	0.6 - 10	—
Sc	10 - 25	—
Se	0.1 - 2.0	0.01 - 400
Si	230,000 - 350,000	—
Sn	2.0 - 200	0.1 - 700
Sr	50 - 1000	10 - 5000
Th	0.1 - 12	—
Ti	1000 - 10,000	400 - >10,000
U	0.9 - 9.0	< 250
V	20 - 500	1.0 - 1000
Y	10 - 500	—
Zn	10 - 300	3.0 - 10,000
Zr	60 - 2000	10 - 8000

a Based on an Analysis of Data Presented in References 1,2,3,4,5, and 6.

- Leaching of mobilized elements such as calcium, boron, lithium, iron, magnesium, manganese, selenium, or sodium (a) out of the soil profile, or (b) into zones of accumulation.
- Translocation, in the course of soil-forming processes such as podzolization, of trace elements together with iron and aluminum.
- Mobilization of trace elements through breakdown of soil minerals as a result of alternate wetting and drying.
- Mechanical translocation of clay, which increases trace element concentrations in those soil horizons having higher amounts of clay particles.
- Surface accumulation of relatively soluble elements such as boron, calcium, and sodium in arid regions.
- Mobilization or fixation arising from chemical and/or microbiological activity.
- Surface enrichment due to trace element uptake by plants.

Second, analytical data derived from the chemical analysis of the total element content of a soil (i.e. C_{Total}) relays no information regarding C_{Fixed}, $C_{Adsorbed}$, and C_{Water} other than the magnitude of their combined concentrations. In other words, if a laboratory report states that a soil contains 125 ppm total Cu, this datum cannot reveal if 0.1 percent is potentially mobile (i.e. $C_{Adsorbed} + C_{Water}$) or if 99 percent is potentially mobile. At background concentrations, the relative magnitudes of the parameters listed in Equation 3.1 for cations generally are:

$$C_{Fixed} >> C_{Adsorbed} > C_{Water}$$

The greater part of C_{Total} exists as C_{Fixed} and is immobile. However, this relative ranking may or may not change as C_{Total} increases above the background concentration.

Third, the background concentrations listed in Table 3.1 represent the total concentration of an element present after the soil was formed and weathered. This concentration gives no information on the element-loading capacity of a soil. The element-loading capacity can be defined as the maximum amount of an element that can be added to soil which does not cause water migrating through this soil to contain a harmful concentration of that element. In other words, knowing that a soil contains 125 ppm total background Cu will not reveal if soil will or will not completely convert an additional loading of 500 ppm Cu into C_{Fixed}.

Soil cleanup standards that specify the excavation or treatment of soil containing concentrations of an element over a background concentration are usually based on an incorrect premise that the background concentration of an element in soil represents a maximum concentration of an element which the soil can immobilize. The background concentration represents the total concentration present after the soil was formed and undergone some degree

TABLE 3.2 Natural Concentrations of Various Elements in Groundwater.[a]

	Concentration	
Element	Typical Value	Extreme Value
———— Major Elements (ppm) ————		
Ca	1.0 - 150[b]	95,000[c]
	< 500[d]	
Cl	1.0 - 70[b]	200,000[c]
	< 1000[d]	
F	0.1 - 5.0	70
		1600[c]
Fe	0.01 - 10	> 1000[c,e]
K	1.0 - 10	25,000[c]
Mg	1.0 - 50[b]	52,000[c]
	< 400[d]	
Na	0.5 - 120[b]	120,000[c]
	< 1000[d]	
NO_3	0.2 - 20	70
SiO_2	5.0 - 100	4,000[c]
SO_4	3.0 - 150[b]	200,000[c]
	< 2000[d]	
Sr	0.1 - 4.0	50
———— Trace Elements (ppb) ————		
Ag	< 5.0	
Al	< 5.0 - 1000	
As	< 1.0 - 30	4,000
B	20 - 1000	5,000
Ba	10 - 500	
Br	< 100 - 2000	
Be	< 10	
Bi	< 20	
Cd	< 1.0	
Co	< 10	
Cr	< 1.0 - 5.0	
Cu	< 1.0 - 30	
Ga	< 2.0	
Ge	< 20 - 50	
Hg	< 1.0	
I	< 1.0 - 1000	48,000[c]
Li	1.0 - 150	
Mn	< 1.0 - 1000	10,000[e]
Mo	< 1.0 - 30	10,000
Ni	< 10 - 50	
PO_4	< 100 - 1000	
Pb	< 15	
Ra	< 0.1 - 4.0[f]	720[c, f]
Rb	< 1.0	
Se	< 1.0 - 10	
Sn	< 200	
Ti	< 1.0 - 150	
U	0.1 - 40	
V	< 1.0 - 10	70
Zn	< 10 - 2000	
Zr	< 25	

[a] based on an analysis of data presented in references 7,8, and 9.
[b] in relatively humid regions.
[c] in brine.
[d] in relatively dry regions.
[e] in thermal springs and mine areas.
[f] picocuries/liter (i.e. 0.037 disintegrations/sec).

of weathering; it gives no indication of the maximum concentration of an element which a soil can immobilize, i.e., the element loading capacity of the soil.

Fourth, a number of established, accepted laboratory methods exist for determining the magnitude of C_{Total}, C_{Fixed}, $C_{Adsorbed}$, and C_{Water} in soil. C_{Total} is usually measured by dry ashing at 500 to 550°C for 3 to 4 hours or by wet ashing with a mixture of perchloric and nitric or sulfuric acids. $C_{Adsorbed}$ and C_{Water} are usually determined by using mineral acids (e.g. 0.1 N HCl), organic acids, and chelating agents (e.g. EDTA, DTPA); hot water extractions as usually utilized for elements that exist as anions (e.g. B, Mo, Se). It is most important to note that the test method employed is dependent upon the individual element to be tested, the parameter to be tested (e.g. C_{Fixed} versus $C_{Adsorbed}$), and the soil type. There is no "universal" analytical method which is applicable for all forms of an element in all soils.

It is most important to note that there are test methods, which are similar to the ones mentioned above, that are utilized to determine the amount of extractable chemical from wastes; these are used to determine if the waste should be classified as a hazardous waste and must be disposed in a Class I landfill. These test methods include the U.S. EPA's EP Toxicity, the TCLP test method, and the State of California's CAM-WET procedure (California Assessment Manual - Wet Extraction Procedure). These methods should not be utilized for soil cleanup criteria from spills of hazardous materials for two reasons. First, these methods, when applied to soils, provide a value, $C_{Extract}$, where:

$$C_{Extract} = a\, C_{Fixed} + bC_{Adsorbed} + C_{Water} \qquad (3.2)$$

where:

$C_{Extract}$ = concentration of an element extracted from a soil;
$$C_{Total} > C_{Extract}$$
a, b = fractions

Since a and b are not determined, it is not possible to relate the parameters of Equation 3.2 to those of Equation 3.1. In other words, $C_{Extract}$ provides no information regarding the magnitude of C_{Fixed}, $C_{Adsorbed}$, and C_{Water}, information which is needed to determine the potential migration and transformation of an element in soil.

Second, when soils are exposed to the extractants utilized by these test methods (acid and citrate or acid and acetate), gross alterations can occur in soil mineralogy, in naturally occurring soil chemical reactions, and in soil physical and chemical properties. These gross alterations result from the fact that the extractants can (a) selectively dissolve soil minerals, (b) impede crystallization and formation of aluminum hydroxides and other soil minerals while causing structural distortions in newly formed minerals, (c) perturb hydrolytic reactions of aluminum, and (d) desorb, via mass action, elements and organics from soil adsorption sites which may not normally be desorbed. Because the extractants cause gross alterations in the chemical and mineralogical properties of soil systems, the data derived from these test methods, when soil is utilized as the solid phase, cannot be extrapolated to actual field conditions. Therefore, they should not be used for soil cleanup criteria from spills of hazardous materials.

The cleanup of contaminated soil should be engineered on a case-by-case basis using published and appropriate soil testing methods; the scientific literature contains at least seven test methods addressing the leaching potential of chemicals in soils[10], and numerous methods addressing the biological, chemical, and physical properties of soils[11,12,13,14].

ELEMENT FIXATION IN CLAY AND SOIL MINERALS

The quantity C_{Fixed} in Equation 3.1 was defined as the concentration of fixed element which has been incorporated into the structure of clay and soil minerals. This section of Chapter 3 will discuss how an element that enters a soil system can be immobilized by fixation into the structure of clay and soil minerals.

Some chemicals such as HCl and water are miscible in all proportions. In other words, continual additions of HCl into a beaker filled with water will not result in the precipitation of a solid in a beaker nor the formation of a separate phase of HCl. For most chemicals, however, there is a limit to the amount that can be added before a solid will precipitate. For example, when a small amount of $MgCO_3$ is added to water, it dissolves. As more $MgCO_3$ is added, it dissolves. However, a point will be reached where additions of $MgCO_3$ will not dissolve but will settle at the bottom of the beaker as a crystalline solid. An equilibriuim is established in which the rate of precipitation of $MgCO_3$ (solid, s) equals the rate of dissolution of $MgCO_3$ (s) into dissolved Mg^{2+} and CO_3^{2-}:

$$Mg^{2+} + CO_3^{2-} \; \underset{\text{Dissolution}}{\overset{\text{Precipitation}}{\rightleftharpoons}} \; MgCO_3 \text{ (s)} \tag{3.3}$$

Now suppose that in addition to Mg^{2+} and CO_3^{2-}, dissolved Ni^{2+} was also present in water at a concentration equal to that of dissolved Mg^{2+}. Ions such as Ni^{2+} which have the same valence and similar size as Mg^{2+} can replace Mg^{2+} in the crystal structure of the precipated structure. In other words, Ni^{2+} will substitute for part of the Mg^{2+} in the precipitated solid as a new equilibrium is established:

$$Mg^{2+} + Ni^{2+} + CO_3^{2-} \rightleftharpoons (Mg, Ni)CO_3 \ (s) \qquad (3.4)$$

A second form of substitution or proxying involves ions of different valence with similar size and identical coordination number. The "coordination number" is the number of ions or molecules that can surround an ion in a crystal structure. Table 3.3 lists the coordination numbers of various cations. For example, Mg^{2+} and Fe^{2+} can substitute for Al^{3+} in minerals where Al^{3+} is in sixfold coordination with oxygen. The transition from one group of coordination numbers to the adjacent one is not abrupt, and ions listed at the boundaries of each group in Table 3.3 are often capable of exhibiting two coordination numbers. For example, Al^{3+} can exist in both fourfold and sixfold coordination; Al^{3+} can substitute for Si^{4+} at the center of an oxygen tetrahedron.

The proxying of elements into the structure of clay minerals found in soil is very prevalent, as evidenced from an analysis of the clay mineral formulae listed in Table 3.4. When an element is added to a soil system, it may be fixed via proxying not only into the structure of clay minerals, but also into the structure of other minerals such as carbonates, phosphates, sulfates, sulfides, and silicates.

ELEMENT FIXATION AS SOIL HYDROXIDES, OXIDES, AND OXYHYDROXIDES

An analysis of the mineral formulae presented in Table 3.4 as well as the mineral structures presented in Figure 2.18 will reveal that oxygen (O) and hydroxyl (OH) play prominent roles in soil mineral structures. Metal ions in soil water and groundwater may react with water to form reaction products which are incorporated into the structures of soil minerals such as those listed in Table 3.4. In addition, it is most important to recognize that these reaction products may also precipitate in soil as oxide and oxyhydroxide minerals or to form oxide and oxyhydroxide coatings on soil minerals.

The process in which elements are converted into oxide and oxyhydroxide soil minerals will be illustrated via a series of simplified, stepwise reactions involving Al^{3+} and Fe^{3+}. When Al^3 or Fe^{3+} enter soil water or ground water, they attract polar water molecules and form aquo complexes:

$$Al^{3+} + 6H_2O \rightleftharpoons Al(H_2O)_6^{3+} \qquad (3.5)$$

$$Fe^{3+} + 6H_2O \rightleftharpoons Fe(H_2O)_6^{3+} \qquad (3.6)$$

TABLE 3.3 Coordination Numbers and Geometrical Configurations For Various Cations

Coordination Number	Geometrical Configuration	r_c/r_o* Range	Cation ($r_c r_o$)	
4	Tetrahedron	0.225 - 0.414	Si^{4+} (0.30)	Mn^{4+} (0.39)
			As^{5+} (0.35)	
6	Octahedron	0.414 - 0.732	Al^{3+} (0.43)	Li^{+} (0.59)
			As^{3+} (0.44)	Mg^{2+} (0.59)
			Sb^{5+} (0.47)	Ni^{2+} (0.59)
			Ti^{4+} (0.48)	Zr^{4+} (0.60)
			Pd^{4+} (0.49)	In^{3+} (0.61)
			Fe^{3+} (0.51)	Pd^{2+} (0.61)
			Ru^{4+} (0.51)	Co^{2+} (0.62)
			Nb^{5+} (0.52)	Fe^{2+} (0.63)
			Rh^{3+} (0.52)	Zn^{2+} (0.63)
			Mn^{3+} (0.53)	Pb^{2+} (0.64)
			Cu^{3+} (0.53)	Pu^{4+} (0.67)
			Te^{4+} (0.53)	Mn^{2+} (0.69)
			Sn^{4+} (0.54)	Ce^{4+} (0.70)
			Re^{4+} (0.55)	Sn^{2+} (0.70)
			Sb^{3+} (0.58)	
8	Cubic	> 0.732	Sm^{3+} (0.73)	Pu^{3+} (0.83)
			Na^{+} (0.74)	Th^{4+} (0.85)
			Pm^{3+} (0.74)	Po^{2+} (0.88)
			Nd^{3+} (0.75)	Ag^{+} (0.95)
			La^{3+} (0.77)	Eu^{2+} (0.95)
			Ce^{3+} (0.78)	Hg^{+} (0.96)
			Eu^{3+} (0.78)	Sr^{+} (0.96)
			Am^{4+} (0.78)	K^{+} (1.01)
			Ca^{2+} (0.80)	Ba^{2+} (1.08)
			U^{4+} (0.82)	Ra^{2+} (1.08)
			Am^{3+} (0.82)	Cs^{+} (1.27)
			Bi^{3+} (0.83)	
			Hg^{2+} (0.83)	
			Np^{3+} (0.83)	

* Ratio of Cation Radius to Oxygen Radius. The Geometry of Soil Minerals is dominated by Silicates which are Dominated by Oxygen.

In general, many metal ions will form complexes with 4 or 6 water molecules. Due to the polarizing ability of Al and Fe, the aquo complexes then form mononuclear (i.e. monohydroxyl) species[15]:

$$Al\,(H_2O)_6^{3+} \rightleftharpoons [Al(OH)(H_2O)_5]^{2+} + H^+ \qquad (3.7)$$

$$Fe\,(H_2O)_6^{3+} \rightleftharpoons [Fe(OH)(H_2O)_5]^{2+} + H^+ \qquad (3.8)$$

The mononuclear species are stable in waters up to about pH 5. At higher pHs, however, the mononuclear species follow two reaction pathways. First,

TABLE 3.4   Approximate Formulae for Selected Soil Clay Minerals.

Type	Mineral	Approximate Formula*
1:1	Kaolinite	$Al_4Si_4O_{10}(OH)_8$
2:1	Beidellite	$0.67M^+Al_4(Si_{7.33}Al_{0.67})O_{20}(OH)_4$
	Biotite	$K_2[Mg,Fe,Mn]_6(Si_6Al_2)O_{20}(OH)_4$
	Glauconite	$(K,H_3O)_2[Fe,Al,Mg]_4(Si,Al)_8O_{20}(OH)_4$
	Illite	$(K,H_3O)_2[Al,Fe,Mg]_4(Si,Al)_8O_{20}(OH)_4$
	Montmorillonite	$0.67M^+(Al_{3.33}Mg_{0.67})Si_8O_{20}(OH)_4$
	Pyrophyllite	$Al_4Si_8O_{20}(OH)_4$
	Sauconite	$0.67M^+[Mg,Zn]_6(Si_{7.33}Al_{0.67})O_{20}(OH)_4$
	Talc	$Mg_6Si_8O_{20}(OH)_4$
	Volkonskoite	$0.67M^+[Fe,Cr,Al]_4(Si,Al)_8O_{20}(OH)_4$
2:2	Dioctahedral Chlorite	$Al_8[Si,Al]_8O_{20}(OH)_{16}$
	Trioctahedral Chlorite	$[Mg,Fe,etc.)_{12}(Si,Al)_8O_{20}(OH)_{16}$

* Cationic Proxying Identified by Brackets.

the mononuclear species form polynuclear species that precipitate as soil hydroxides, oxides, and oxyhydroxides[15]:

$$[Al(OH)(H_2O)_5]^{2+} \rightleftharpoons [Al_n^{3+}(OH)_{2(n-1)}(H_2O)_{2n+2}]^{n+2} \rightleftharpoons \text{Al hydroxides,} \quad (3.9)$$
$$\text{oxides, &}$$
$$\text{oxyhydroxides}$$

$$[Fe(OH)(H_2O)_5]^{2+} \rightleftharpoons [Fe_n^{3+}(OH)_{2(n-1)}(H_2O)_{2n+2}]^{n+2} \rightleftharpoons \text{Fe hydroxides,} \quad (3.10)$$
$$\text{oxides, &}$$
$$\text{oxyhydroxides}$$

The initial structures of the hydroxide, oxide, and oxyhydroxide minerals are amorphous (i.e. lack crystallinity). However, the structures change over time periods spanning months to years into more crystalline structures. Table 3.5 lists common hydroxide, oxide, and oxyhydroxide minerals found in soil. It is important to note that the majority of elements form hydroxide, oxide and oxyhydroxide minerals. Also, the hydrous oxides of Fe and Mn may hold significant proportions of some elements such as Co, Mo, and Pb[15].

Second, the mononuclear species form polynuclear species that adsorb onto the surfaces of soil minerals and form coatings on the mineral surface[15]. The mineral then possesses two types of surfaces: one comprised of hydroxides, oxides, and oxyhydroxides, and a second comprised of atoms from the original structure of the mineral. The hydroxide/oxide/oxyhydroxide surface may grow and spread over the surfaces of adjacent sand and silt particles; by associating with these particles, aluminum oxides, oxyhydroxides, iron oxides, silica, calcium carbonate and other amorphous minerals act as inorganic bonding agents for adjacent particles.

TABLE 3.5 Common Soil Hydroxide, Oxide, and Oxyhydroxide Minerals.

Mineral	Formula
Amorphous silica	SiO_2
Coesite	SiO_2
Cristobalite	SiO_2
Quartz	SiO_2
Stishovite	SiO_2
Tridymite	SiO_2
Bayerite	alpha — $Al(OH)_3$
Boehmite	gamma — $Al(OH)_3$
Corundum	alpha — Al_2O_3
Diaspore	alpha — $AlOOH$
Gibbsite	gamma — $Al(OH)_3$
Nordstrandite	$Al(OH)_3$
Hercynite	$FeAl_2O_4$
Hogbomite	$Mg(Al,Fe,Ti)_4 O_7$
Spinel	$MgAl_2O_4$
Chrysoberyl	$BeAl_2O_4$
Brucite	$Mg(OH)_2$
Akaganeite	beta — $FeOOH$
Goethite	alpha — $FeOOH$
Hematite	alpha — Fe_2O_3
Lepidocrocite	gamma — $FeOOH$
Maghemite	gamma — Fe_2O_3
Bixbyite	$(Mn,Fe)_2 O_3$
Feitknechtite	beta — $MnOOH$
Groutite	alpha — $MnOOH$
Manganite	gamma — $MnOOH$
Chromite	$FeCr_2 O_4$
Jacobsite	$MnFe_2 O_4$
Magnesioferrite	$MgFe_2O_4$
Magnetite	$FeFe_2O_4$
Ulvospinel	$TiFe_2O_4$
Birnessite	delta — MnO_2
Cryptomelane	KMn_8O_{16}
Nsutite	gamma — MnO_2
Psilomelane	$(Ba,Mn,R)_3 Mn_8O_{16} (O,OH)_6$
Pyrolusite	beta — MnO_2
Ramsdellite	MnO_2
Braunite	$(Mn,Si) Mn_2O_4$
Hausmannite	$MnMn_2O_4$
Anatase	TiO_2
Baddeleyite	ZrO_2
Brookite	TiO_2
Cassiterite	SnO_2
Rutile	TiO_2
Ilmenite	$FeTiO_3$
Pseudobrookite	Fe_2TiO_5
Perovskite	$CaTiO_3$

Eh, pH, AND ELEMENT FIXATION

An analysis of the elements listed in Table 3.3 will reveal that several elements can exist as part of the structure of soil minerals in more than one oxidation state. For example, manganese (Mn) can exist in soil as Mn^{2+}, Mn^{3+}, or Mn^{4+}. Arsenic (As) can exist as As^{5+} or As^{3+}. The particular species which will be present will depend upon the oxidation-reduction (i.e. redox) status of the soil.

The redox status of a soil refers to the presence or absence of electrons in soil. The electron (e^-) is a species present in soil that participates in chemical reactions and can be viewed in the same manner as H^+. pH is expressed as the negative log of H^+ in moles/liter; the concentration of e^- is sometimes expressed as pe, the negative log of electron concentration. pH is the measure of proton presssure: the lower the pH, the greater the pressure. Likewise, pe is a measure of electron pressure: the lower the pe, the greater the electron pressure. Redox reactions respond to electron pressure just as acid-base reactions respond to proton pressure:

$$Fe^{3+} + e^- \rightleftharpoons Fe^{2+} \tag{3.11}$$

$$OH^- + H^+ \rightleftharpoons H_2O \tag{3.12}$$

Although pe probably represents the most simplified expression for electron activity in soil, it is not commonly used. The most commonly used expression for redox relationships in the soils literature is Eh, the redox potential, where:

$$Eh = pe\,(2.3RT/F) \tag{3.13}$$

and

$$Eh = \text{redox potential, V}$$

$$R = \text{gas constant, 0.001987 kcal/}^\circ K$$

$$T = \text{temperature, }^\circ K$$

$$F = \text{faraday constant, 23.06 kcal/Vg equivalent}$$

or

$$pe = 16.9\,(Eh) \tag{3.14}$$

The principles governing redox reactions and relations are relatively simple when a reaction is comprised of only a few reactants and products such as the ones illustrated in Equation 3.11. However, Eh in soil systems is the net result of a complex combination of oxidation and reduction species and reactions:

$$Eh = E^\circ + (RT/nF)\ln([ox]/[red]) \tag{3.15}$$

where

E° = standard reference potential

n = number of electrons

[ox] = activity of species in the oxidized state

[red] = activity of species in the reduced state

The soil system is complex since soil Fe, Mn, S, organic matter, and numerous other ions at equilibrium in the soil each have some finite ratio of oxidation states and/or molecular species that combine to form [ox] and [red]. Also, the observed Eh may vary with factors such as water content, oxygen activity in the soil, and pH. Therefore, the contribution of any particular soil phase to the total potential cannot be deduced.

It is important to note that Eh measures electron activity in water. It is not a measure of redox capacity. Redox capacity is analogous to pH buffer capacity, the ability of a system to maintain a given pH upon addition of acid or base. The redox capacity is the maximum quantity of electrons that can be added or removed from the entire soil-water system while maintaining a given Eh or pe. It can be measured and expressed as the quantity of oxidant or reductant (in moles) that must be added to a soil-water system in order to cause a unit change in Eh. In soil, the redox capacity is usually very large since electrons can quickly be supplied or removed via migration through water and through solids via solid state diffusion.

Since electrons neutralize protons (i.e. H^+), many naturally-occurring reactions are both Eh and pH dependent. The fixation of elements in soil minerals is also dependent upon the Eh-pH status of the soil system. Figure 3.1 illustrates the approximate Eh-pH limits for naturally-occurring soil. The upper limit of water stability identifies those Eh-pHs at which water will decompose according to the reaction:

$$H_2O = 1/2\ O_2(g) + 2H^+ + 2e^- \tag{3.16}$$

The lower limit of water stability identifies those Eh-pHs at which water will decompose according to the reaction:

$$2H^+ + 2e = H_2\ (g) \tag{3.17}$$

The predominant dissolved species or minerals of the elements in aqueous systems are commonly deduced from simplified geochemical models of equilibrium. For illustrative purposes, the results of these calculations are commonly expressed as pe-pH or Eh-pH diagrams.

Eh-pH diagrams for several elements are illustrated in Figures 3.2 through

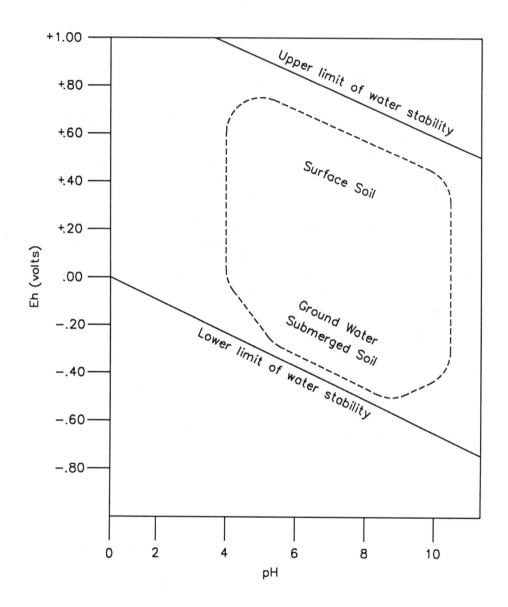

FIGURE 3.1   Approximate Eh—pH limits observed for
naturally—occurring soil.

3.45. Eh-pH diagrams illustrate three different relationships between elemental species. The first relationship is between predominant dissolved species; for example, Figure 3.7 illustrates Eh-pH regions where $BiOH^{2+}$ is predominant, which is different from the region in which BiO^+ is predominant. The second relationship is between a predominant dissolved species and a predominant mineral species; for example, Figure 3.7 illustrates Eh-pH regions where Bi_2O_3 is predominant which is different from the region in which BiO^+ is predominant. The third relationship is between predominant mineral species; for example, Figure 3.7 illustrates Eh-pH regions where Bi_2O_3 is predominant, which is different from the region in which Bi is predominant.

Eh-pH diagrams give important information regarding the potential fixation of an element in soil. To illustrate their usefulness, the Eh-pH diagram for Am (Figure 3.2) will be discussed. Below pH 5, Am will exist primarily as Am^{3+}; it does not exhibit a propensity to exist as an immobile hydroxide, oxide, or oxyhydroxide, mineral; it should prefer to exist as a mobile, dissolved species. Above pH 5, Am will readily form AmO_2 or $Am(OH)_3$. If the Eh is relatively high and the pH is above 5, Am will exist predominantly as Am^{4+} in the mineral AmO_2. If the Eh is relatively low (i.e. more electrons are present) and the pH is above 5, Am will exist predominantly as the reduced ion Am^{3+} in the mineral $Am(OH)_3$. By utilizing Eh-pH diagrams in this manner, one can qualitatively estimate if soil conditions are conducive to the fixation of an element added to the soil via the discharge of a hazardous material.

In general, the published literature has shown that the minerals influencing the behavior of elements depicted in Eh-pH diagrams also influence the behavior of elements in soil systems. For example, in well-aerated soils, hydrous ferric oxide minerals and coatings are considered to be the soil minerals controlling aqueous Fe^{3+} concentrations[21]. The chemistry of Cr in well-aerated soils resembles that of Fe; the aqueous Cr^{3+} concentration is controlled by chromic oxides and hydroxides[22]. The aqueous Mn^{2+} concentrations in neutral pH soil is primarily controlled by Mn oxides and carbonates[2]. The aqueous Cd^{2+} concentration in calcareous soil is apparently controlled by Cd carbonates[23,24]. The retention of Ag, Am, Bi, Cd, Mo, Np, Pd, Po, Pu, Rh, Ru, Tc, Te, Th, U, Y, and Zn in strata at the Oklo fossil nuclear reactor was described in terms of Eh-pH diagrams[25,26].

Several important facts should be remembered during an analysis and interpretation of the Eh-pH diagrams presented in this chapter. First, the boundary lines between the fields of mineral stability do not represent abrupt boundaries. The boundary lines indicate the general location of a domain of transition that exists between two fields of mineral stability.

Second, the boundary lines may shift somewhat as the concentrations of

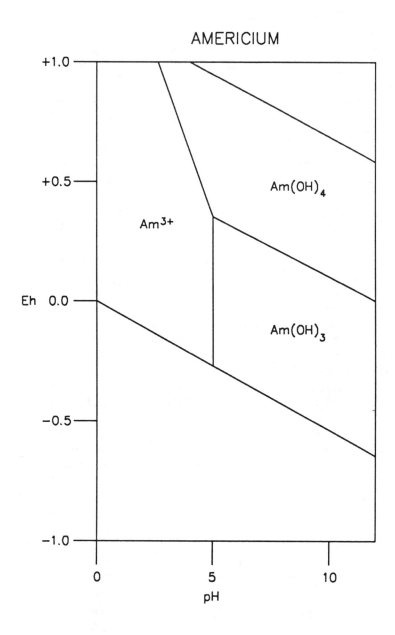

FIGURE 3.2   Eh–pH Diagram for an Am–H_2O system, with dissolved Am activity assumed to be 10^{-6}. Constructed from data presented in Ref. 16, 17 and 18.

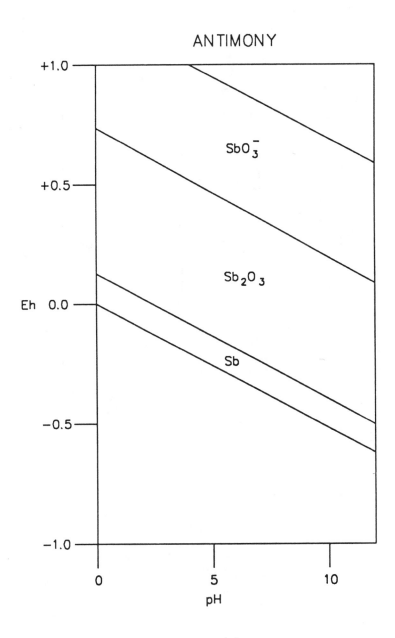

FIGURE 3.3 Eh–pH Diagram for a Sb–H_2O system, where Sb activity is 10^{-6}. Constructed from data from Ref. 16, 17 and 18.

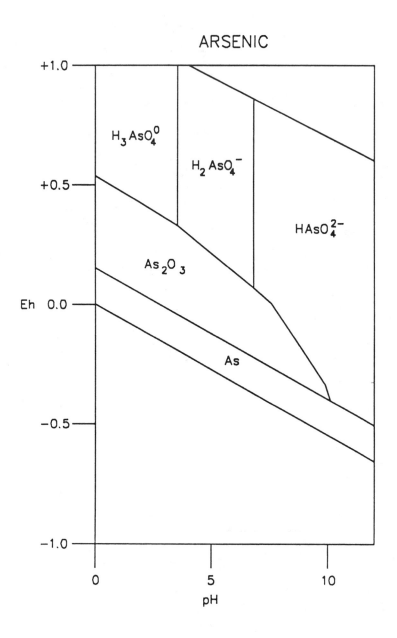

FIGURE 3.4   Eh–pH Diagram for an As–H_2O system, where
As activity is 10^{-6}. Constructed from data
from Ref. 16, 17 and 18.

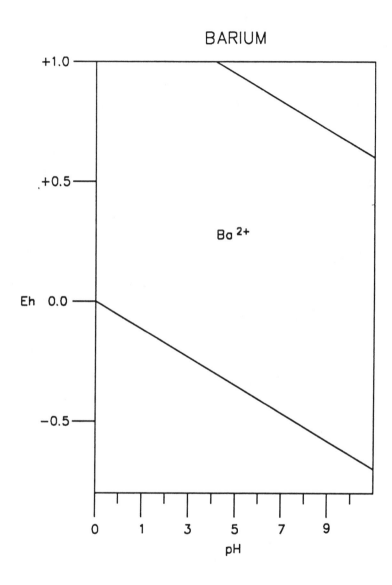

BARIUM

FIGURE 3.5 Eh−pH Diagram for a Ba−H$_2$O system.
Calculated from data presented in
Ref. 16, 17 and 18.

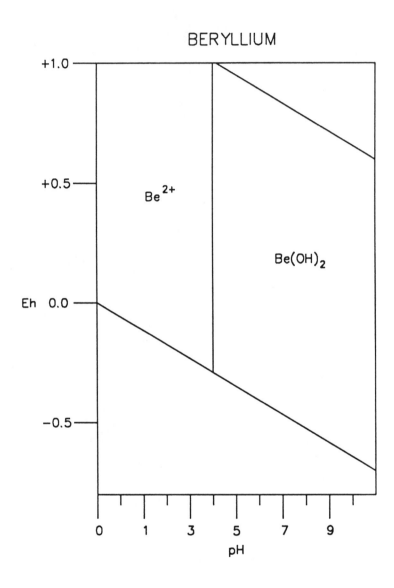

FIGURE 3.6   Eh–pH Diagram for a Be–H$_2$O system, where
Be activity is 10^{-6}. Calculated from data
presented in Ref. 16, 17 and 18

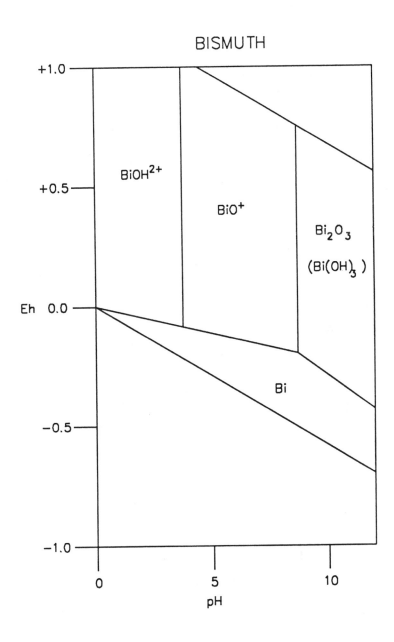

FIGURE 3.7 Eh–pH Diagram for a Bi–S–H$_2$O system, with dissolved Bi activity assumed to be 10^{-6}. Constructed from data presented in Ref. 16, 17 and 18.

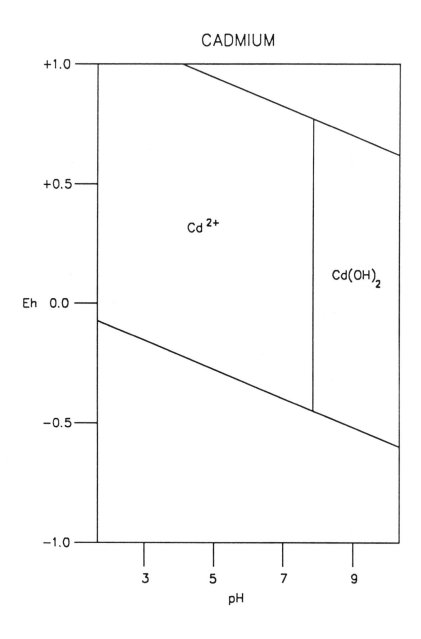

FIGURE 3.8   Eh–pH Diagram for a Cd–H$_2$O system, where
Cd activity is 10^{-4}. Constructed from data
from Ref. 16, 17 and 18.

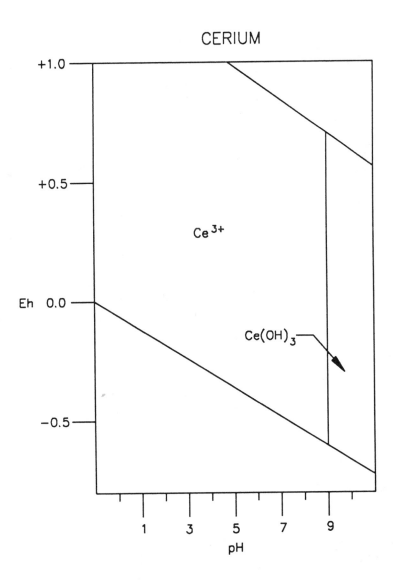

FIGURE 3.9 Eh–pH Diagram for a Ce–H$_2$O system, where activity of Ce is 10^{-6}. Constructed from data from Ref. 16, 17 and 18.

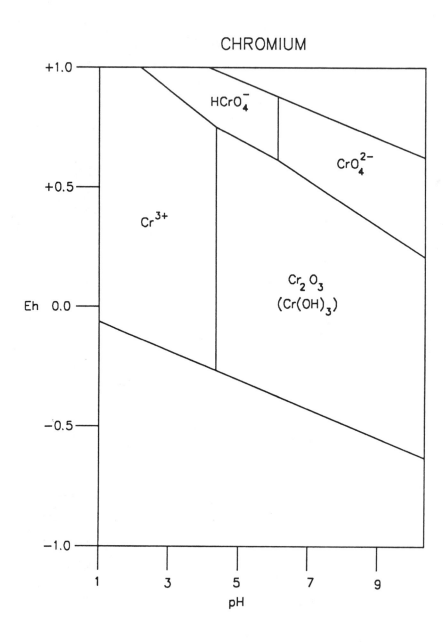

FIGURE 3.10   Eh–pH Diagram for a Cr–H$_2$O system, where Cr activity is 10^{-6}. Constructed from data from Ref. 16, 17 and 18.

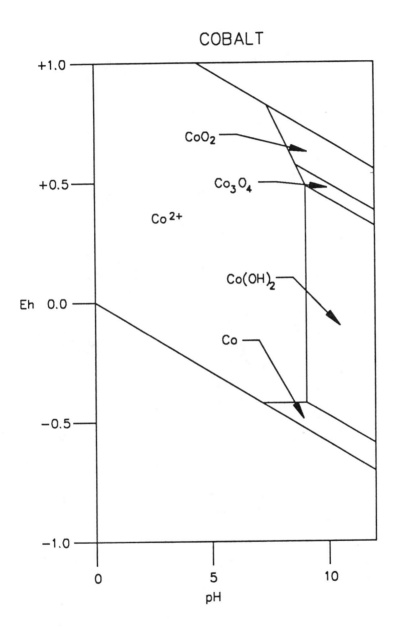

FIGURE 3.11   Eh–pH Diagram for a $Co-H_2O$ system, where
Co activity is 10^{-6}. Constructed from data
from Ref. 16, 17 and 18.

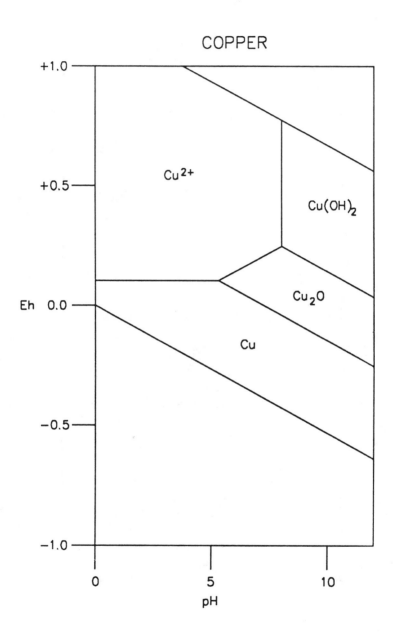

FIGURE 3.12   Eh–pH Diagram for a Cu–H$_2$O system, where Cu activity is 10^{-6}. Constructed from data from Ref. 16, 17 and 18.

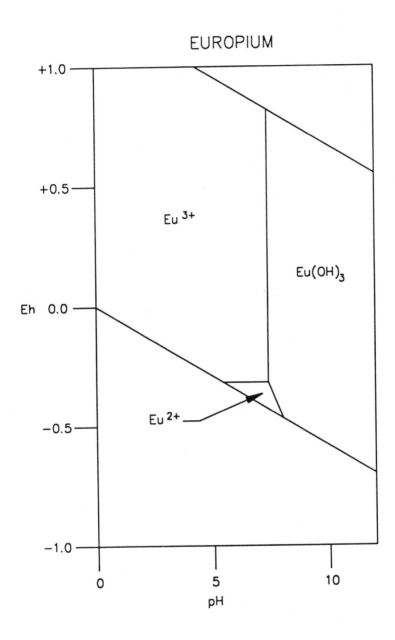

FIGURE 3.13 Eh−pH Diagram for an Eu−H_2O system , where Eu activity is 10^{-6}. Constructed from data from Ref. 16 and 18.

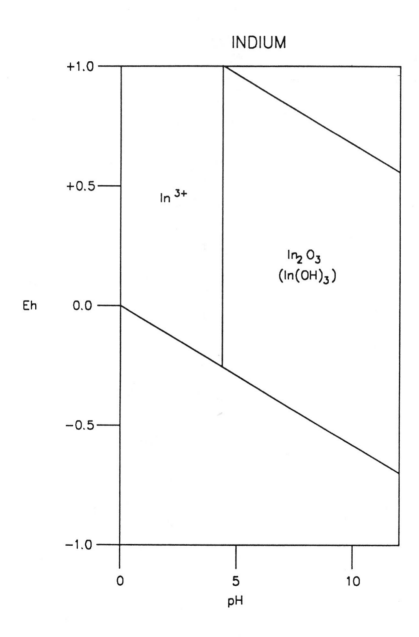

FIGURE 3.14   Eh–pH Diagram for an In–H$_2$O system, where
In activity is 10^{-6}. Constructed from data
from Ref. 16, 17 and 18.

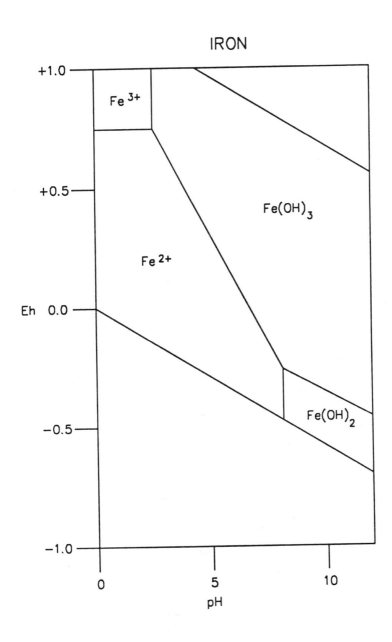

FIGURE 3.15   Eh–pH Diagram for an Fe–H$_2$O system, where Fe activity is 10^{-4}. Constructed from data from Ref. 16, 17 and 18.

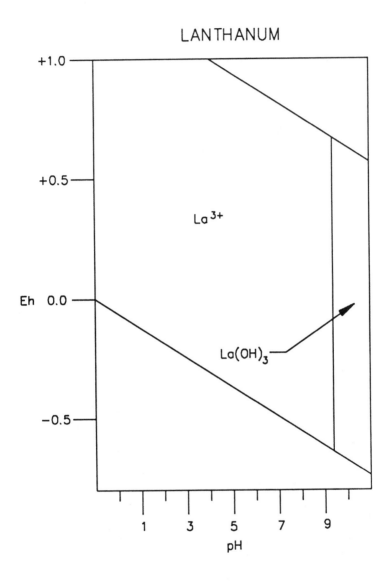

FIGURE 3.16   Eh–pH Diagram for a La–H_2O system, where
La activity is 10^{-6}. Constructed from data
from Ref. 16, 17 and 18.

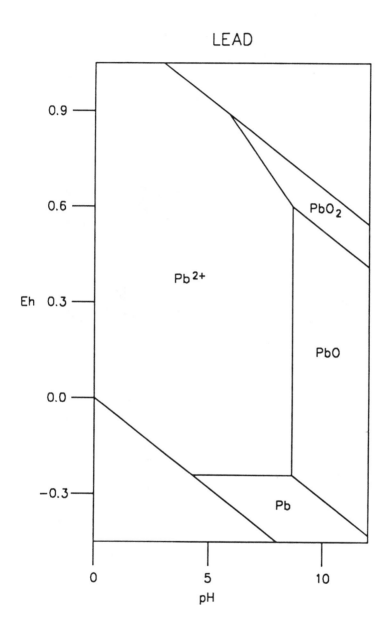

FIGURE 3.17   Eh–pH Diagram for a Pb–H$_2$O system, where
total Pb = 10^{-4}M. Constructed from data
presented in Ref. 16, 17 and 18.

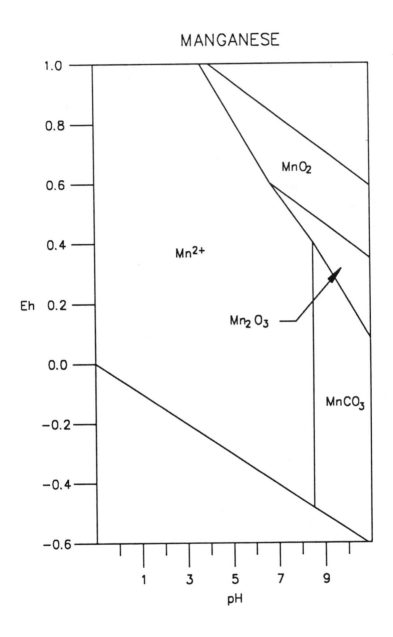

FIGURE 3.18   Eh–pH Diagram for a $Mn–CO_2–H_2O$ system[7], where CO_2 pressure is 10^{-4}atm, and Mn activity is 0.01.

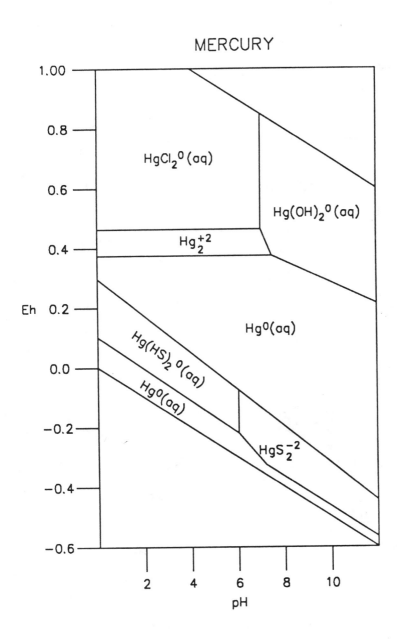

FIGURE 3.19 Eh–pH Diagram for Aqueous species of Hg[19], with 36ppm Cl⁻ and 96ppm S–SO_4^{-2}.

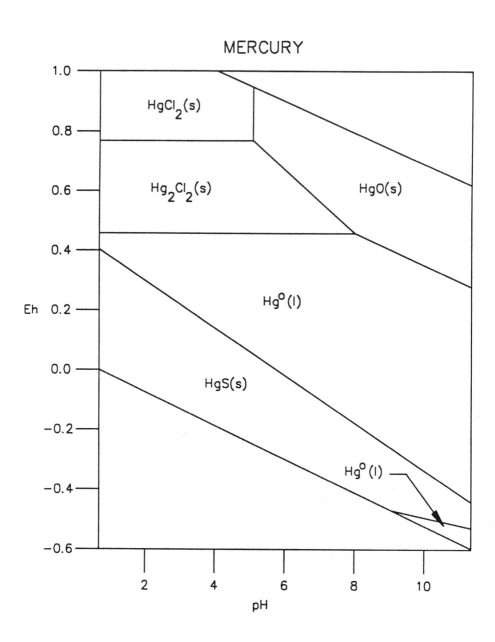

FIGURE 3.20   Eh–pH Diagram for liquid (l) and solid (s)
Hg[19], with 36 ppm Cl and 96 ppm S-SO$_4^{-2}$.

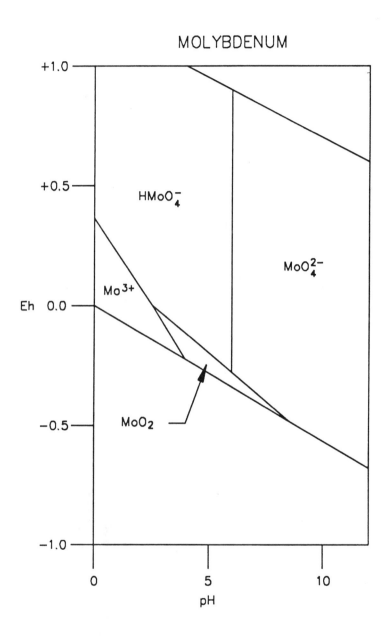

FIGURE 3.21   Eh–pH Diagram for a Mo–H$_2$O system, where
Mo activity is 10^{-6}. Constructed from data
from Ref. 16, 17 and 18.

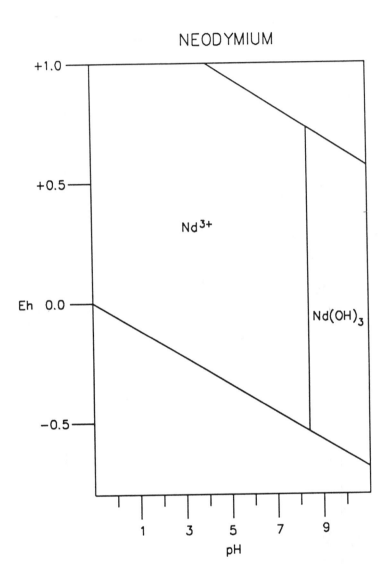

FIGURE 3.22   Eh–pH Diagram for a Nd–H₂O system, where
Nd activity is 10⁻⁶. Constructed from data
from Ref. 16 and 18.

NEPTUNIUM

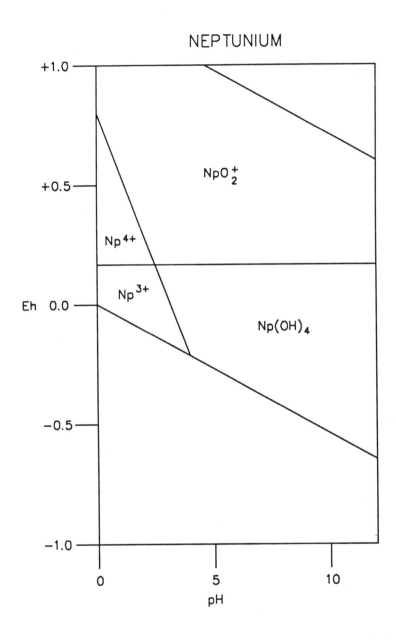

FIGURE 3.23 Eh—pH Diagram for a Np—H_2O system, where Np activity is 10^{-6}. Constructed from data from Ref. 16, 17 and 18.

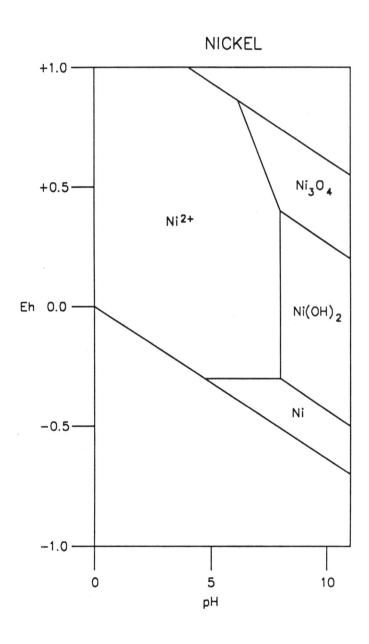

FIGURE 3.24   Eh—pH Diagram for a Ni—H$_2$O system, where Ni activity is 10^{-4}. Constructed from data from Ref. 16, 17 and 18.

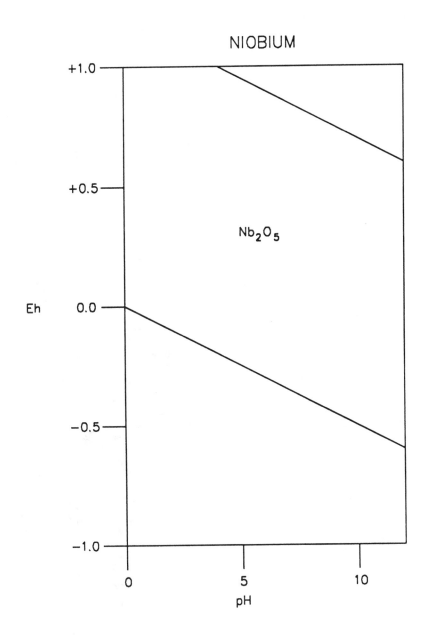

FIGURE 3.25 Eh–pH Diagram for a Nb–H$_2$O system.
Constructed from data from Ref. 16 and 17.

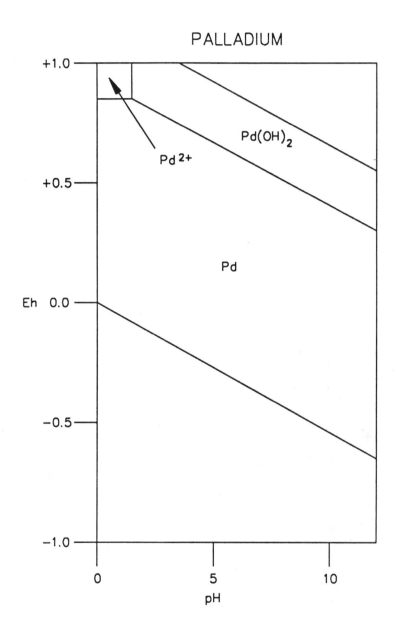

FIGURE 3.26   Eh–pH Diagram for a Pd–H$_2$O system, where
Pd activity is 10^{-6}. Constructed from data
from Ref. 16, 17 and 18.

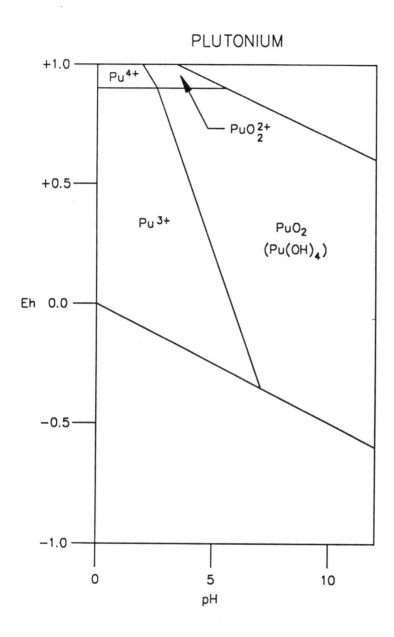

FIGURE 3.27 Eh–pH Diagram for a Pu–H$_2$O system, where Pu activity is 10^{-6}. Constructed from data from Ref. 16, 17 and 18.

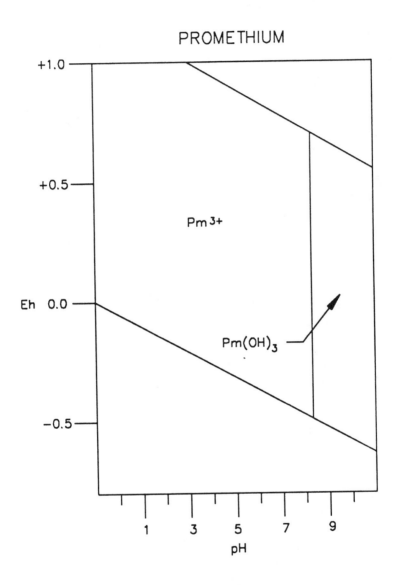

FIGURE 3.28   Eh–pH Diagram for a Pm–H$_2$O system, where Pm activity is 10^{-6}. Constructed from data from Ref. 16.

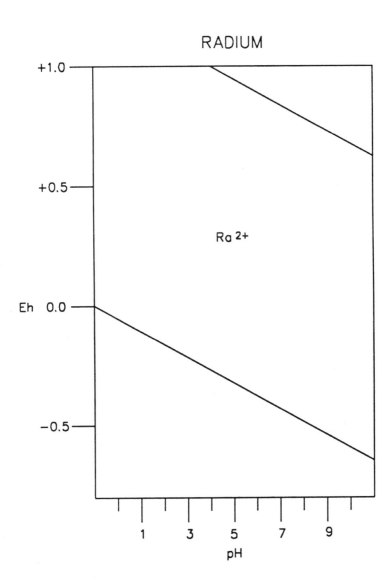

FIGURE 3.29   Eh–pH Diagram for a Ra–H$_2$O system.
Constructed from data from Ref. 16 and 17.

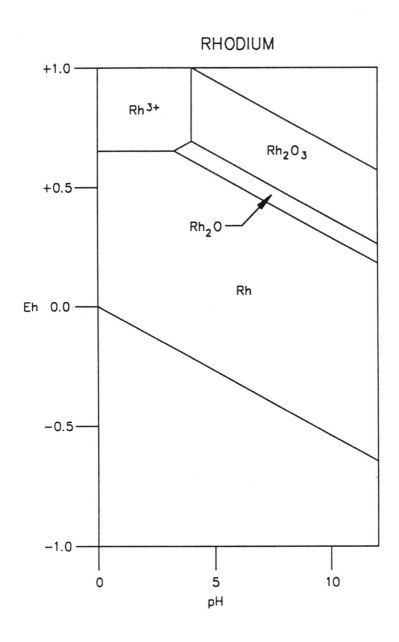

FIGURE 3.30   Eh—pH Diagram for a Rh—H_2O system, where Rh activity is 10^{-6}. Constructed from data from Ref. 16, 17 and 18.

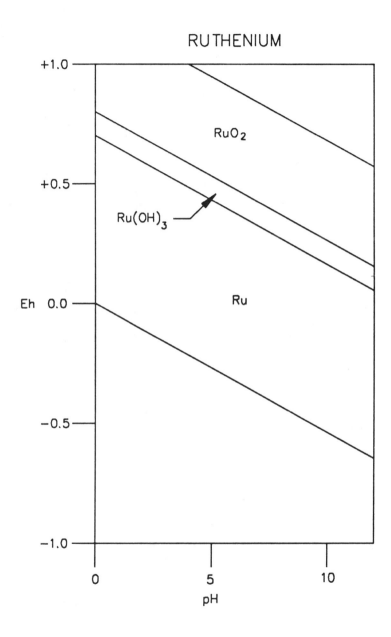

RUTHENIUM

FIGURE 3.31 Eh−pH Diagram for a Ru−H₂O system.
Constructed from data from Ref. 16 and 17.

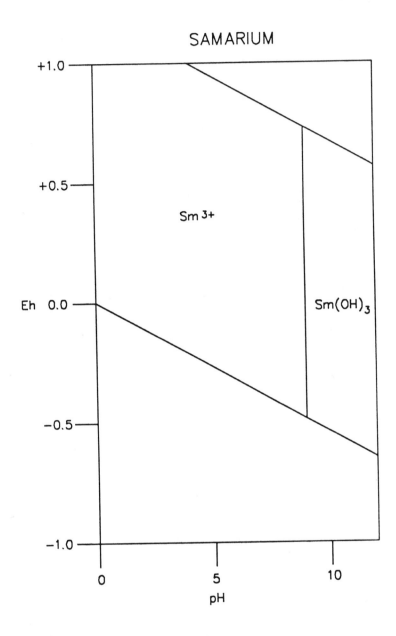

SAMARIUM

Sm^{3+}

Sm(OH)$_3$

FIGURE 3.32   Eh—pH Diagram for a Sm—H$_2$O system, where Sm activity is 10^{-6}. Constructed from data from Ref. 16 and 18.

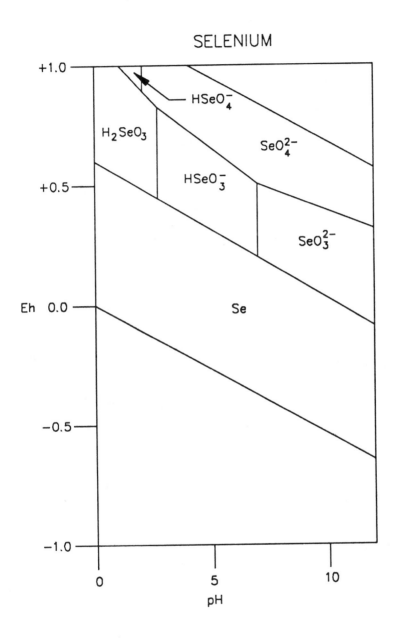

SELENIUM

FIGURE 3.33 Eh–pH Diagram for a Se–H$_2$O system, where Se activity is 10^{-6}. Constructed from data from Ref. 16, 17 and 18.

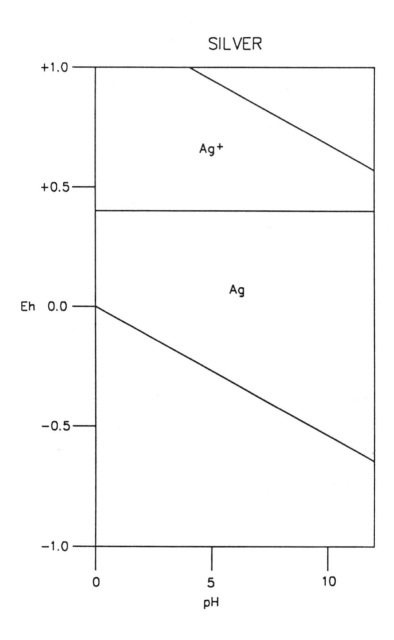

FIGURE 3.34   Eh–pH Diagram for a Ag–H$_2$O system, where Ag activity is 10^{-6}. Constructed from data from Ref. 16, 17 and 18.

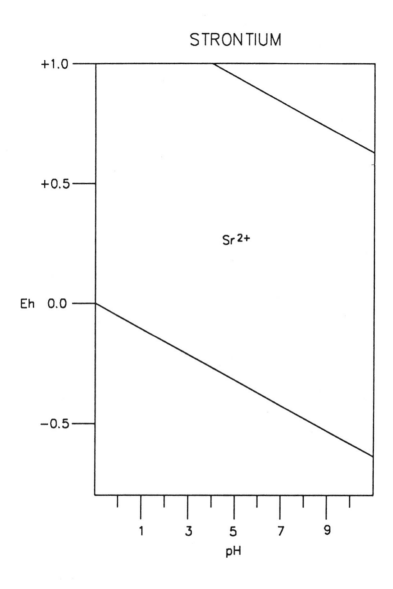

FIGURE 3.35   Eh—pH Diagram for a Sr—H$_2$O system.
Constructed from data from Ref. 16,
17 and 18.

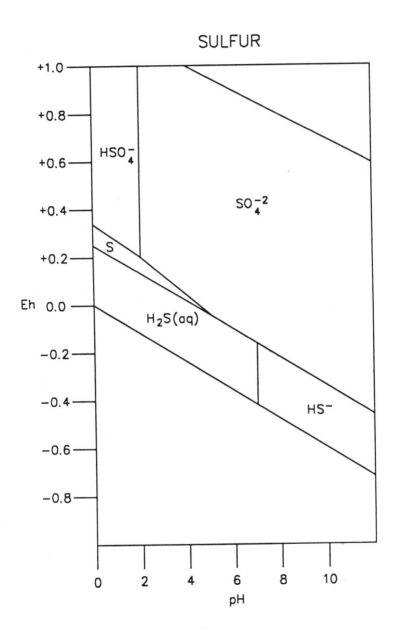

FIGURE 3.36 Eh–pH Diagram for a S–H$_2$O system[7], where total Sulfur activity is 96 mg/l.

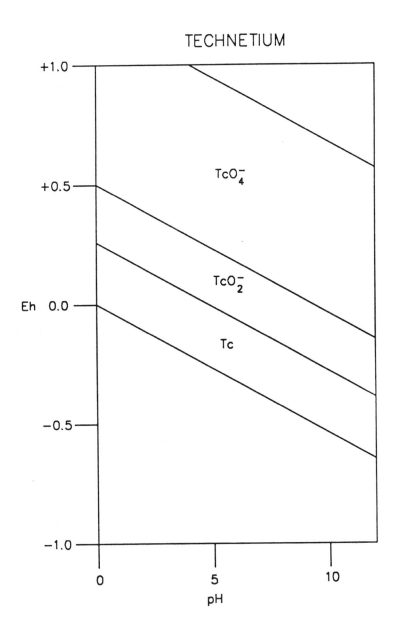

FIGURE 3.37 Eh–pH Diagram for a Tc–H_2O system, where Tc activity is 10^{-6}. Constructed from data from Ref. 16 and 20.

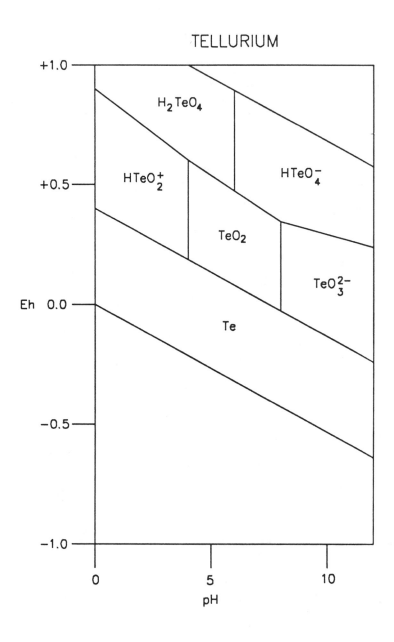

FIGURE 3.38   Eh–pH Diagram for a Te–H_2O system, where
Te activity is 10^{-6}. Constructed from data
from Ref. 16, and 17.

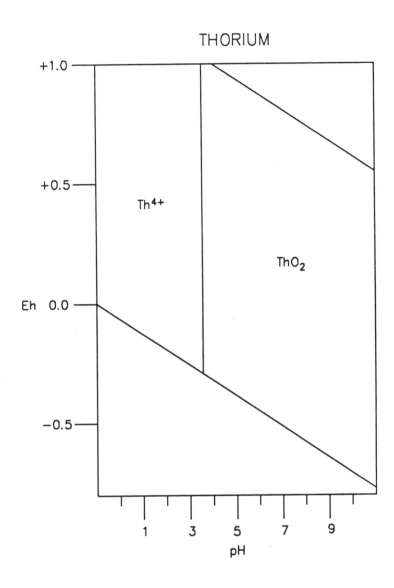

FIGURE 3.39 Eh–pH Diagram for a Th–H$_2$O system, where
Th activity is 10^{-6}. Constructed from data
from Ref. 16, 17 and 18.

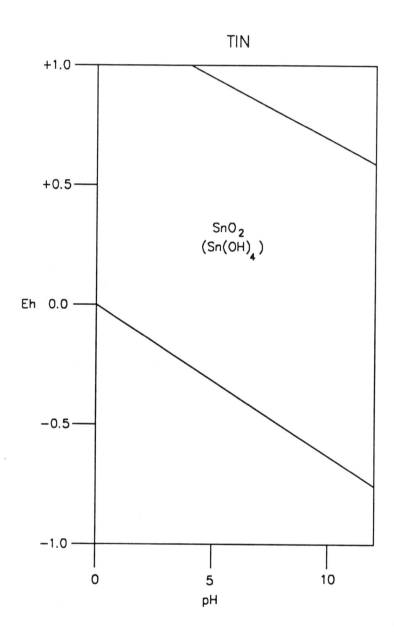

FIGURE 3.40   Eh−pH Diagram for a Sn−H$_2$O system, where
Sn activity is 10^{-6}. Constructed from data
from Ref. 16, 17 and 18.

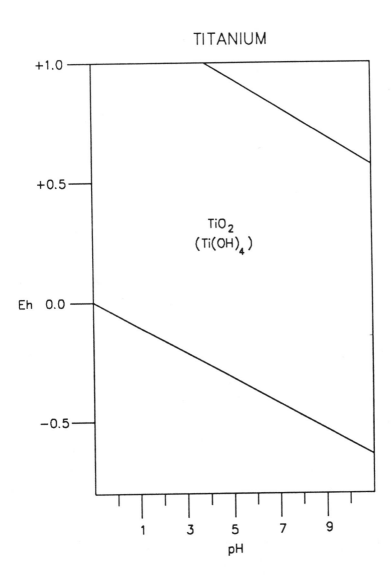

FIGURE 3.41   Eh–pH Diagram for a Ti–H$_2$O system, where Ti activity is 10^{-6}. Constructed from data from Ref. 16, 17 and 18.

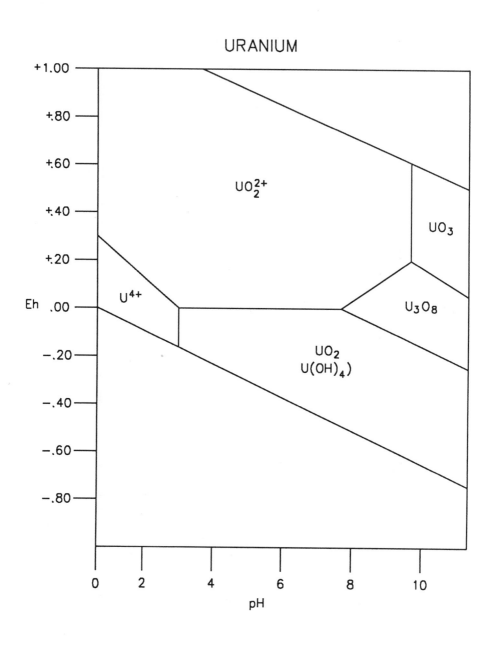

FIGURE 3.42   Eh–pH Diagram for a U–H_2O system , where U activity is 10^{-6}. Constructed from data from Ref. 16, 17 and 18.

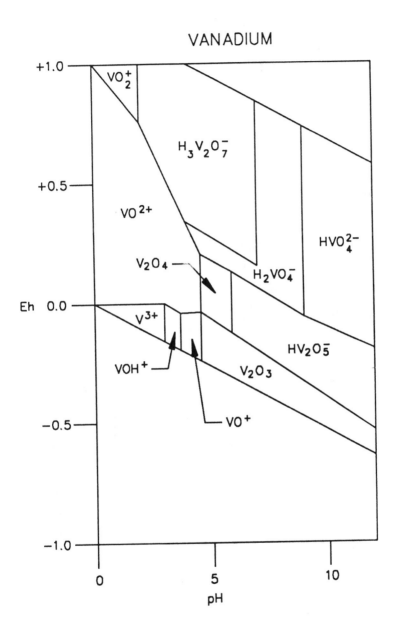

FIGURE 3.43 Eh–pH Diagram for a V–H_2O system, where V activity is 10^{-4}. Constructed from data from Ref. 16, 17 and 18.

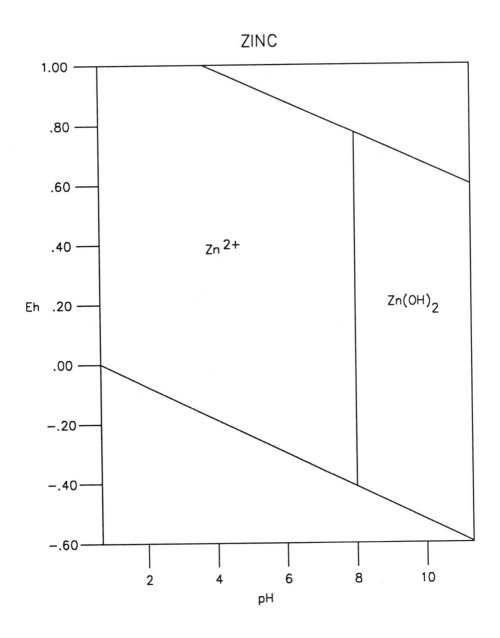

FIGURE 3.44   Eh–pH Diagram for a Zn–H$_2$O system, where
Zn activity is 10^{-6}. Constructed from data
from Ref. 16, 17 and 18.

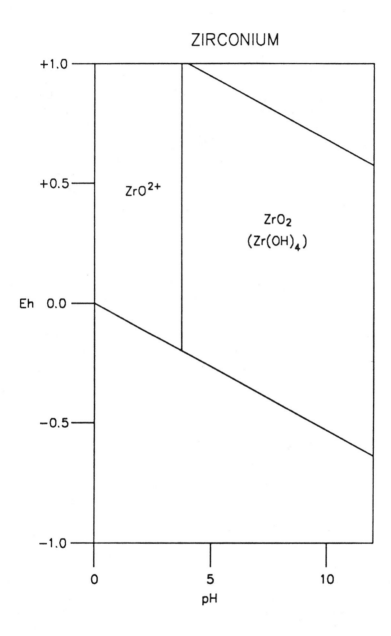

FIGURE 3.45 Eh–pH Diagram for a Zr–H$_2$O system, where Zr activity is 10^{-6}. Constructed from data from Ref. 16, 17 and 18.

the elements, which were utilized to calculate the position of the boundary, changes. For example, in Figure 3.2, the boundary line between Am^{3+} and AmO_2 as well as the boundary line between Am^{3+} and $Am(OH)_3$ shifts to the right as the dissolved Am decreases from 10^{-6} to 10^{-8}.

Third, the magnitude and position of the field of stability for a mineral depend upon the presence or absence of other ions. When applying the Eh-pH diagrams presented in this chapter to a site specific problem, one must verify that the conditions assumed during the calculations of stability fields of an Eh-pH diagram are applicable to the conditions present at the site of concern. If the conditions are not applicable, one can either utilize another published diagram or construct one that is applicable to a specific site. The standard methods available for the calculation of Eh-pH diagrams are discussed by several authors[27,28,29].

Fourth, the validity of an Eh-pH diagram is dependent upon the quality of the thermodynamic data utilized to calculate the mineral stability fields. For many elements, good thermodynamic data exists. However, good data does not exist for some elements.

Fifth, Eh-pH diagrams are based on calculations in which an element's concentration or activity remains constant at some predetermined level. This is not the minimum aqueous concentration or activity of that element that must be present in order for fixation to occur. A mineral will precipitate when the concentration of the ions dissolved in water exceeds the solubility product.

THE SOLUBILITY PRODUCT CONSTANT AND ELEMENT FIXATION

Consider the dissolution-precipitation reactions of the mineral AB comprised of elements A and B in water:

$$A + B \underset{\text{Dissolution}}{\overset{\text{Precipitation}}{\rightleftharpoons}} AB(s) \qquad (3.18)$$

Equilibrium is established when the rate of precipitation equals the rate of dissolution. After equilibrium is established, the aqueous concentrations of A and B do not change but remain constant. The water phase comprised of A and B in equilibrum with AB(s) is known as a saturated solution. The solubility of AB is defined as the concentration of AB in a saturated solution at a defined temperature. The solubility product constant, K_{SP}, is defined as:

$$K_{sp} = [A^{c+}]^d [B^{d-}]^c \qquad (3.19)$$

where

$$[A] \quad = \text{equilibrium molar concentration of A}$$

$$[B] \quad = \text{equilibrium molar concentration of B}$$

$$c \quad = \text{cationic charge of A}$$

$$d \quad = \text{anionic charge of B}$$

The K_{sp}s for a number of minerals are listed in Table 3.6. These values can be utilized to grossly estimate the element concentration that should be present in order for soil mineral fixation to occur. For illustrative purposes, consider a hypothetical situation where Zn is leaching from acid dredgings from a steel pickling facility. The soil pH and soil buffer capacity are sufficient to buffer the leachate-soil system at pH 6. The Eh under the dredging pile is approximately -0.2. The soil under this pile contains 10^{-3}M sulfide. Using the K_{sp} for ZnS from Table 3.6 and Equation 3.19, one can grossly estimate that if the Zn concentration in the leachate-soil system were to exceed $10^{-9.6}$ M, then ZnS should precipitate.

It is important to note that the K_{sp} values for two different minerals may not be directly comparable because the K_{sp} values may have different dimensions. For example, consider data for $Fe(OH)_2$ and $Fe(OH)_3$ from Figure 3.15 and Table 3.6:

$$K_{sp} [Fe(OH)_2] \quad = 10^{-15.10} \text{ M}^3/\text{liter}^3 \qquad (3.20)$$

$$K_{sp} [Fe(OH)_3] \quad = 10^{-37.50} \text{ M}^4/\text{liter}^4 \qquad (3.21)$$

It would be technically incorrect to infer that $Fe(OH)_3$ is less soluble than $Fe(OH)_2$.

The extent to which element fixation will occur in a soil cannot be accurately predicted. Data presented in Table 3.7 illustrates this point. All soils listed in Table 3.7 contained 125 ppm total Cu. Yet, an analysis of the data presented in Table 3.7 will reveal that the range of fixed Cu, expressed as percent of total Cu, was 12.0 percent to 98.4 percent. When addressing site specific situations, the amount of fixed element in soil should be determined via use of the appropriate soil chemical analytical method.

In general, K_{sp} should be limited to use in generating gross estimations because it fails to account for several factors affecting the precipitation and dissolution of minerals. First, the age of the mineral may determine the crystallinity of the mineral. Some freshly precipitated minerals have disordered crystalline structures with relatively higher solubility products compared to aged minerals with more ordered structures and relatively lower solubility products. Soils contain a variety of minerals possessing varying degrees of crystallinity.

TABLE 3.6 Solubility Product Constants for Several Minerals.[a]

Mineral	pK_{sp}[b, c]
Ag_2CO_3	11.21
$Ag(OH)_2$	7.82 (20 °C)
Ag_3PO_4	16.48
Ag_2S	48.80 (18 °C)
Ag_2SO_4	4.77
$Am(OH)_3$	19.57
$AlAsO_4$	15.80
$1/2 As_2O_3$	15.02
$Ca_3(AsO_4)_2$	18.17
$FeAsO_4$	20.24
$Mg_3(AsO_4)_2$	19.68
$Mn_3(AsO_4)_2$	28.72
$BaCO_3$	8.09
$Ba(OH)_2$	2.30
$Ba_3(PO_4)_2$	38.22
$BaSO_4$	9.97
$BeCO_3$	3.00
$Be(OH)_2$	17.70
$Bi(OH)_3$	39.50
$BiO(OH)$	12.00
$BiPO_4$	22.89
Bi_2S_3	96.00
$CdCO_3$	11.60
$Cd(OH)_2$	14.30
$Cd_3(PO_4)_2$	38.10
CdS	28.44
CeO_2	36.10
$Ce(OH)_3$	19.82
$Ce(OH)_4$	47.70
Ce_2S_3	10.22
$Ce_2(SO_4)_3$	1.70
$CoCO_3$	12.85
$Co(OH)_2$	14.80
$Co(OH)_3$	44.50 (19 °C)
CoS	25.52 (18 °C)
$Cr(OH)_3$	30.20
$CrPO_4$ (green)	22.62 (18-20 °C)
$CrPO_4$ (violet)	17.00 (18-20 °C)
$CrSO_4$	0.50
$CuCO_3$	9.85
$Cu(OH)_2$	18.30
$Cu_3(PO_4)_2$	36.89
CuS	44.07 (18 °C)
Cu_2S	46.70 (16-18 °C)
$Eu(OH)_3$	21.47
$FeCO_3$	10.50

TABLE 3.6   Solubility Product Constants for Several Minerals.[a] (cont)

Mineral	pK_{sp}[b, c]
$Fe(OH)_2$	15.10
$Fe(OH)_3$	37.50 (18 °C)
$FePO_4$	21.89 (18-20 °C)
FeS	18.70
FeS_2	30.20
$HgCl_2$	14.60
Hg_2Cl_2	17.70
Hg_2CO_3	16.05
HgO	23.75
Hg_2O	22.80
$Hg(OH)_2$	25.52
HgS	48.7 − 52.4 (18 °C)
Hg_2S	51.52
$HgSO_4$	1.43
Hg_2SO_4	6.17
$In(OH)_3$	32.16
$La(OH)_3$	19.00
La_2S_3	13.5
$La_2(SO_4)_3$	4.50
$MnCO_3$	10.74
$Mn(OH)_2$	13.40 (18 °C)
MnS	14.85 (18 °C)
$CaMoO_4$	4.91
$Mo(OH)_4$	50.00
$Nd(OH)_3$	23.30
$NiCO_3$	8.18
$Ni(OH)_2$	13.80
NiS	23.85 (18 °C)
$NpO_2(OH)_2$	21.60
$PbCl_2$	4.80
$PbCO_3$	13.48 (18 °C)
PbO	15.32
PbO_2	65.50
$Pb(OH)_2$	19.90
$Pb_4O(PO_4)_2$	65.17
$Pb_3(PO_4)_2$	44.60
$Pb_5(PO_4)_3Cl$	84.40
$Pb_5(PO_4)_3OH$	76.80
PbS	27.47 (18 °C)
$PbSO_4$	7.97 (18 °C)
$Pd(OH)_4$	70.20
$Pm(OH)_3$	21.00
$Pu(OH)_4$	55.15
$PuO_2(OH)_2$	20.50
$RaSO_4$	14.38
$1/2\ Rh_2O_3$	47.70
Ru_2O_3	45.00

TABLE 3.6 Solubility Product Constants for Several Minerals.[a] (cont)

Mineral	pK_{sp}[b, c]
RuS_2	65.50
$1/2\ Sb_2O_3$	41.40
Sb_2S_3	92.80
Al_2Se_3	24.40
$CaSeO_3$	5.53
$CaSeO_4$	0.30
$FeSe$	26.00
$Fe_2(SeO_3)_3$	30.70
$MgSeO_3$	4.89
$MnSeO_3$	6.90
$Sm(OH)_3$	21.17
$Sn(OH)_2$	27.85
$Sn(OH)_4$	57.00
SnS	25.00
$SrCO_3$	8.80
$Sr(OH)_2$	3.50
$Sr_3(PO_4)_2$	27.80
SrS	−2.70
$SrSO_4$	6.49
$Te(OH)_4$	53.52
$Th(OH)_4$	44.50
$Th_2(PO_4)_3$	57.61
$Th_3(PO_4)_4$	78.59
$Th(SO_4)_2$	2.40
$Ti(OH)_4$	53.10
$K(UO_2)PO_4$	23.11
$U(OH)_4$	45.00
$UO_2(OH)_2 \cdot H_2O$	22.00
$(VO)_3(PO_4)_2$	24.10
$ZnCO_3$	10.85
$Zn(OH)_2$ amorphous	15.40
$Zn_3(PO_4)_2$	32.04 (18-20 °C)
ZnS	22.92 (18 °C)
$Zr(OH)_4$	55.80
$Zr_3(PO_4)_4$	132.00

[a] compiled from information presented in references 30,32,33,49, and 50.
[b] the negative logarithm of the solubility product constant K_{sp}.
[c] at 25 °C, unless specified otherwise.

Second, complexes may form among dissolved inorganic ions and between dissolved inorganic ions and dissolved organic chemicals or organic matter. These are usually not accounted for during the derivation of the Ksp.

Third, ions may undergo acid-base reactions in water which may not be accounted for in the derivation of the K_{sp}.

Fourth, surface area of the mineral influences its solubility. For example, although K_{sp} for CuO is lower than that of $Cu(OH)_2$, $Cu(OH)_2$ is the more stable structure at very small particle sizes[35].

Fifth, the K_{sp} is measured at a specified pH. It can be expected to change somewhat as pH shifts.

Sixth, solubility is decreased somewhat by the common-ion effect. The solubility of a mineral is decreased by the addition to its saturated solution of a second chemical possessing an ion in common with the first compound. For example, silver acetate is less soluble in an aqueous solution containing sodium acetate than in pure water.

Eh AND SOIL COLOR

The redox status of a soil significantly affects the natural color of soil primarily due to its influence on Fe and Mn in soil minerals. Colors such as red, yellow, and reddish brown predominate under oxidizing conditions[36]. If a soil remains wet for long periods of time, it can undergo reduction. This process is commonly referred to as gleying. The color of reduced or gleyed soil depends upon the type of minerals present and on the redox potential. The majority of soils subject to strong reduction are uniformly gray in color[2]. Other soils can exhibit subdued colors or black[2]. In rarer situations, deep soils, especially under peat, may be greenish gray or bluish gray due to the presence of minerals containing high amounts of iron.

In the zone where the water table fluctuates, soil may exhibit a gray background with spots and streaks of yellow-brown, yellow-red, or red. This condition wherein alternate streaks or spots of oxidized and reduced soil colors occurs is known as mottling.

It is most important to note that in the majority of site investigations the presence of various colors and color patterns in soil is not caused by a hazardous material. Conclusions regarding the presence of a hazardous material in soil should never be based solely on color because the color of uncontaminated soil varies broadly, both areally across the soil surface and vertically through the soil profile.

Also, it is important to recognize that the typical descriptions of gross soil color found in boring logs and drilling logs are too subjective and are usually of little value. In other words, a soil that is "dark brown" to one individual may be "brownish black" or "black" to another. Soil colors should be measured by comparison with a color chart. The one generally used with soil is the Munsell color chart which consists of approximately 175 different colored chips. These chips are systematically arranged according to hue, value, and chroma, the three parameters that combine to form all colors. Hue is the dominant spectral (rainbow) color: red (R), yellowish red (YR), yellow

(Y). Value is the relative lightness of color and is a function of the total amount of light. Value is identified on a relative scale from O (absolute black) to 10 (absolute white) with intermediate identifications of very dark gray, dark gray, gray, and light gray. Chroma (or saturation) is the relative purity or strength of the spectral color, and it increases with decreasing grayness. The notation for chroma consists of numbers beginning at O for neutral grays and increasing to approximately 8 for soils. For absolute achromatic colors which have zero chroma and no hue—pure grays, white, and black—the letter N (neutral) replaces the hue designation.

In the soil color chart, all colors on a given card are of a constant hue, designated by the symbol in the upper right-hand corner of the card. Vertically, the colors become successively lighter by visually equal steps; their value increases. Horizontally, they increase in chroma to the right and become grayer to the left. The value and chroma of each color in the chart is printed immediately beneath the color. The first number is the value, and the second is the chroma. As arranged in the chart, the colors form three scales: (a) radial, or from one card to the next, in hue; (2) vertical in value; and (3) horizontal in chroma.

In writing the Munsell notation, the order is hue, value, and chroma, with a space between the hue letter and the succeeding value number, and a virgule between the two numbers for value and chroma. If expression beyond the whole number is desired, decimals are always used, never fractions. Thus the notation for a color of hue 5YR, value 5, chroma 6, is 5YR 5/6, a yellowish-red. The notation for a color midway between the 5YR 5/6 and 5YR 6/6 chips is 5YR 5.5/6; for one midway between 2.5YR 5/6 and 5YR 6/8, it is 3.75YR 5.5/7. The notation is decimal and capable of expressing any degree of refinement desired. Since color determinations cannot be made precisely in the field—generally no closer than half the interval between colors in the chart—the designation of soil color should ordinarily be that of the nearest color chip.

In using the color chart, accurate comparison is obtained by holding the soil sample above the color chips being compared. Rarely will the color of the sample be perfectly matched by any color in the chart. The probability of having a perfect matching of the sample color is less than one in one hundred. However, the closest match should be evident. The principal difficulties encountered in using the soil color chart are (1) in selecting the appropriate hue card, (2) in determining colors that are intermediate between the hues in the chart, and (3) in distinguishing between value and chroma where chromas are strong. In addition, the chart does not include some extreme dark, strong (low value, high chroma) colors occasionally encountered in moist soils. With experience, these extreme colors lying outside the range of the chart can be estimated. Furthermore, the ability to sense color differences varies among people, even among those not regarded as color blind.

Soil color changes with the moisture content, very markedly in some soils and comparatively little in others. Between dry and moist, soil colors commonly are darker by 1/2 to 3 steps in value and may change from $-1/2$ to $+2$ steps in chroma. Seldom are they different in hue. Some of the largest differences in value between the dry and moist colors occur in gray and grayish-brown horizons having moderate to moderately low contents of organic matter. Reproducible quantitative measurements of color are obtained at two moisture contents: (1) air dry, and (2) field capacity. The latter may be obtained with sufficient accuracy for color measurements by moistening a sample and reading the color as soon as visible moisture films have disappeared. In most soil descriptions, unless stated otherwise, colors are given for moist soils.

TABLE 3.7   Amount of Fixed Cu in Northeastern U.S. Soils Containing 125 ppm Total Cu[34].

Soil Name	Texture	pH	Fixed Cu (% Total)
Fauquier	Silt Loam	6.0	98.4
Paxton	Sandy Loam	5.7	93.6
Caribou	Silt Loam	5.2	92.0
Christiana	Sandy Loam	5.2	64.8
Minespoil	Loam	4.2	58.4
Gilpin	Silt Loam	6.2	47.2
Mardin	Silt Loam	5.4	42.4
Lima	Loam	6.5	40.8
Pocomoke	Loamy Sand	4.4	40.8
Sassafras	Sandy Loam	5.1	35.2
Marlton	Sandy Loam	4.2	32.0
Vergennes	Silt Loam	6.0	27.2
Groveton	Silt Loam	6.4	20.8
Evesboro	Sand	5.3	19.2
Hagerstown	Silt Loam	5.5	12.0
Dekalb	Loam	4.5	12.0

MINERAL DISSOLUTION AND ELEMENT RELEASE

In order to describe how soil minerals respond to a decrease in the elemental composition of soil water or ground water, Equation 3.3 will be utilized once again:

$$Mg^{2+} + CO_3^{2-} \underset{\text{Dissolution}}{\overset{\text{Precipitation}}{\rightleftharpoons}} MgCO_3(s) \qquad (3.3)$$

Whenever the concentration of Mg^{2+} in water exceeds the equilibrium solubility of $MgCO_3(s)$, $MgCO_3(s)$ begins to precipitate. Conversely, whenever the concentration of Mg^{2+} in water drops below the equilibrium solubility of $MgCO_3(s)$, this mineral begins to dissolve as it replenishes the concentration of Mg^{2+}. In this way, $MgCO_3(s)$ as well as other soil minerals tend to buffer the concentration of Mg^{2+} and other ions in soil.

The example of $MgCO_3(s)$ dissolution in Equation 3.3 illustrates an important chemistry principle commonly known as Le Chatelier's Principle. If the conditions of a system originally at equilibrium are changed, the point of equilibrium is shifted in the direction which tends to nullify the effects of the change. For example, if the solution concentration of Mg^{2+} in Equation 3.3 was increased, the point of equilibrium changes in the direction which would cause the concentration of Mg^{2+} to decrease (i.e. more mineral precipitation would occur).

The dissolution or decomposition of minerals may occur via incongruent dissolution or congruent dissolution. Incongruent dissolution is the dissolution or breakdown of a soil mineral followed by the formation of other crystalline or amorphous minerals. Congruent dissolution is the breakdown of a soil mineral without the formation of a residual mineral. The pathway of dissolution and the type of mineral formed is dependent upon the soil type and the factor(s) that caused the depression of the ion(s) concentration in water below the equilibrium solubility.

Under natural environmental conditions, the polar nature of the water molecule is responsible for dissolving soil minerals. Referring to the simple reactions of $MgCO_3(s)$, the surface and edges of $MgCO_3(s)$ are comprised of alternating positive and negative ions. Polar water molecules are attracted to these ions comprising the solid; the negative pole of the water molecule aligns adjacent to cations while the positive pole of the water molecule aligns adjacent to anions. When a sufficient number of water molecules congregates around an ion, they can pull the ion into solution and overcome the attraction of the other ions in the crystal for the ion at the edge or surface. The ions on the edges and the corners are particularly susceptible to attack by water molecules since more water molecules can congregate around these ions than around ions on flat surfaces. Therefore, they are removed from the solid into solution more rapidly than ions on the flat surfaces of the solid.

Only those substances possessing ions whose attraction for water molecules is about as strong as the attraction of water molecules for one another will be able to force the water molecules apart and dissolve. These substances include NaCl, KNO_3, $CaCl_2$, alcohols, sugars, etc. On the other hand, if the attraction between ions in a solid is great, these ions will prefer to associate with each other rather than with water molecules. As a result, these solids will exhibit very low water solubilities, such as those exhibited by many oxides and aluminosilicate minerals discussed in Chapters One and Two. Soils are comprised primarily of these types of solids.

If a bulk hydrocarbon which possesses weak or negligible polarity replaces water as the primary solvent in a soil system, the hydrocarbon will not be able to dissolve the solid phase. The ions in the solid will have far greater attraction for each other than they will have for the nonpolar hydrocarbon molecules. For this reason, chlorinated solvents, gasoline, and other non-polar liquids will not dissolve soil minerals.

The addition of several types of hazardous materials to soil can cause congruent or incongruent dissolution of soil minerals to commence. First, the addition of an inorganic acid or organic acid may lower soil pH and cause dissolution. Second, the addition of an inorganic or organic base may raise soil pH and cause dissolution. The effects of bulk acids and bases in soil will be discussed in detail in Chapter 5.

Third, the addition of *bulk* amounts of some inorganic chemicals can cause mineral dissolution. Consider the addition of a brine solution of 40 percent NaCl to the saturated solution of $MgCO_3(s)$ illustrated by Equation 3.3. Cl^- is an inorganic liquid that will complex Mg^{2+} in solution to form $MgCl^+$ and $MgCl_2^\circ$. This complexation decreases the Mg^{2+} concentration in solution; in response, $MgCO_3(s)$ begins to dissolve in an attempt to replenish the Mg^{2+} concentration in water.

A similar response occurs when oil field brine, which is commonly used for dust control and deicing of roads, enters a soil system. Brines typically contain as much as 40 percent by weight NaCl. The Cl^- will form soluble complexes with many of the dissolved and adsorbed elements commonly found in soil; in response, soil minerals dissolve in an attempt to replenish the uncomplexed element concentrations in water. In general, the addition of bulk quantitites of liquids containing relatively large quantities of dissolved nitrates, chlorides, bromides, iodides, and sulfates of Ag, Al, Ba, Bi, Ca, Cd, Co, Cr, Cu, Fe, Hg, K, Mg, Mn, Na, NH_4, Ni, Pb, Sb, Sn, Sr, and Zn should dissolve soil minerals in a manner similar to brines.

It is most important to recognize that the areal and vertical extent of mineral dissolution depends upon the amount of chemical added to the soil system and the extent of chemical dilution that occurs within the soil system. For example, the addition of brine to soil causes the dissolution of many soil minerals; however, when the plume's aqueous Cl concentration has been diluted by soil water and groundwater, the system no longer favors mineral dissolution. Mineral precipitation predominates.

Also, it is important to recognize that the magnitude of the soil contamination problem may be increased due to mineral dissolution. If, for example, an acid brine is spilled in soil containing relatively high background concentration of As in soil minerals, mineral dissolution can result in elevated dissolved As concentrations in ground water.

Fourth, minerals dissolve if soil conditions are conducive to the formation of a new mineral having a solubility that is lower than the solubility of the original mineral. For example, if sufficient sulfide entered a soil system containing $CoCO_3$, the $CoCO_3$ dissolves to liberate Co^{2+} into solution that would subsequently react with S^- to form the relatively less soluble CoS. In general, the addition of bulk quantities of sulfides, phosphates, silicates, oxides, and hydroxides to soil results in the precipitation of relatively insoluble minerals. Minerals differ widely in the rates at which they dissolve while attempting to re-establish equilibrium. Some minerals, primarily the very soluble ones, dissolve rapidly and re-establish equilibrium within a few hours. Some require several days, others require years, and the extremely insoluble minerals dissolve so slowly that they may never reach equilibrium. For example, $Cr(OH)_3$ is a relatively low solubility mineral. If a soil's Eh changes from 0.0 to +0.7 at pH7, $Cr(OH)_3$ dissolution is extremely slow and conversion of Cr(III) to Cr(VI) proceeds at a negligible rate. K_{sp} is an equilibrium measurement; it is not a measurement of reaction kinetics.

For the primary soil clay minerals, dissolution in unsaturated zone soil is generally dependent on the length of time the mineral is in contact with water. If the rate of soil water migration is fast, contact time is slow. Likewise if the rate of soil water migration is slow, contact time is high. Dissolution rates for silica from clay minerals in water were estimated to be on the order of 10^{-3} percent per day[37,38]. Dissolution rates for clay minerals in organic acids were estimated to be on the order of 10^{-3} to 10^{-2} percent per day[37,38].

INORGANIC CHEMICALS THAT RAPIDLY REACT WITH WATER

The previous section discussed the fact that many relatively water-soluble inorganic chemicals readily dissolve in water; after dissolution in water, these inorganic chemicals may then commence dissolution of some soil minerals. It is most important to know that certain inorganic chemicals can hydrdolyze, oxidize, or reduce very quickly upon contact with water. If these chemicals are present in bulk quantities, their reaction with water may be violent because energy is released in quantities or at rates too high to be absorbed by the surrounding environment.

TABLE 3.8   Inorganic Chemicals That May Rapidly React With Soil Water and Groundwater[a]

aluminum tetrahydroborate
ammonium peroxodisulfate

barium oxide
beryllium hydride
beryllium tetrahydroborate
bis (S,S-difluoro-N-sulfimido) sulfur tetrafluoride
bismuth nitride
bismuth pentafluoride
boron azide diiodide
boron bromide diiodide
boron triazide
boron tribromide
boron triiodide
bromine pentafluoride
bromyl fluoride

caesium amide
caesium oxide
caesium trioxide
calcium tetrahydroborate
cerium hydride
cerium trihydride
chlorine fluorosulfate
chlorine pentafluoride
chlorine trifluoride
chlorogermane
chlorotrimethylsilane
chromyl chloride
chromyl fluorosulfate

diamidophosphorous acid
diboron oxide
diphosphoryl chloride
dipotassium phosphinate
disulfur dichloride
disulfur heptaoxide

germanium tetrachloride

hexaaminetitanium chloride
hydrazine
hydroxylamine

iodine pentafluoride

lanthanum oxide
lithium amide
lithium tetradeuteroaluminate
lithium tetrahydroaluminate
lithium tetrahydroborate
lithium tetrahydrogallate

TABLE 3.8   Inorganic Chemicals That May Rapidly React With Soil Water and Groundwater[a]
(cont)

magnesium hydride
magnesium tetrahydroaluminate
manganese fluoride trioxide

neptunium hexafluoride

oxygen difluoride

palladium tetrafluoride
perfluorosilanes
phosphorus pentachloride
phosphorus pentaoxide
phosphorus tribromide
phosphorus trichloride
phorphorus tricyanide
phosphorus triiodide
phosphorus trioxide
phosphoryl chloride
platinum tetrafluoride
plutonium hexafluoride
plutonium (III) hydride
potassium amide
potassium azo disulfonate
potassium hydride
potassium peroxide
potassium sulfurdiimidate
potassium triamido thallate ammoniate

rhodium tetrafluoride
rubidium hydride

selenium tetrafluoride
silicon dibromide sulfide
silicon monohydride
silicon monosulfide
silver difluoride
sodium amide
sodium dithionate
sodium germanate
sodium hydrazide
sodium hydride
sodium oxide
sodium peroxide
sodium tetrahydroaluminate
sodium tetrahydrogallate
sulfinyl chloride
sulfur dichloride
sulfur oxide-(N-fluorosulfonyl)imide
sulfur trioxide

TABLE 3.8   Inorganic Chemicals That May Rapidly React With Soil Water and Groundwater[a]
(cont)

tetrachlorosilane
tetrafluoroammonium hexafluoromanganate
tetrafluoroammonium hexafluoronickelate
tetraphosphorus decaoxide
tetraphosphorus hexaoxide-bis(borane)
tetraphosphorus tetraoxide trisulfide
titanium(II) chloride
titanium diiodide
titanium trichloride
triboron pentafluoride
B-1,3,5-trichloroborazine
1,3,5-trichloro-2,4,6-trifluoroborazine
trisilylamine

uranium hexachloride
uranium hexafluoride
uranium(III) hydride

vanadium dichloride
vanadium tribromide oxide
vanadium trichloride oxide

xenon tetrafluoride
xenon trioxide

zinc hydride
zirconium dibromide
zirconium tetrachloride
zirconium trichloride

[a] compiled following an analysis of information presented in reference 39.

Table 3.8 lists several classes of inorganic chemicals that react rapidly, and sometimes violently, with water. Because all soils contain water, as discussed in Chapter 2, these chemicals may react rapidly, and possibly violently, in soil systems.

These inorganic chemicals were grouped together so that general relationships between chemical structure and reactivity could be studied. It is most important to recognize that the absence of a chemical from Table 3.8 in no way implies that no hazard exists for the unlisted chemical. Chemicals possessing similar structures should be expected to react in a similar fashion. The reader should always consult additional information from the published literature on reactive chemical hazards when dealing with rapidly reacting inorganic chemicals.

REMEDIAL ACTIONS UTILIZING
ELEMENT FIXATION

The published literature contains a few case studies which have depended upon element fixation to remediate soil contamination. For example, oxidation-precipitation was utilized to fix As deposited over approximately 15 km^2 in Manfredonia, Southern Italy[40]. An aerosol mixture containing approximately 12 tons of K_3AsO_3 and H_3AsO_3 escaped from the NH_3 washing column of a chemical factory. A $Ca(OCl)_2$ treatment solution was utilized to convert As(III) to As(V):

$$2K_3AsO_3 + Ca(OCl)_2 \rightleftharpoons 2K_3AsO_4 + CaCl_2 \qquad (3.22)$$

A $Fe_2(SO_4)_3$ treatment solution was utilized to transform the soluble arsenite into the insoluble salt:

$$2K_3AsO_4 + Fe_2(SO_4)_3 \rightleftharpoons 2FeAsO_4(s) + 3K_2SO_4 \qquad (3.23)$$

Potassium permanganate was injected into groundwater through 17 wells and piezometers in the vicinity of a zinc ore smelter near Cologne[41]. Arsenic oxidation resulted in significantly lower As in groundwater, due to the precipitation of As as calcium and iron arsenites.

Due to a major spill of chrome plating solution onto soil at a chromium (Cr) plating facility, soil contained as much as 1,406 ppm Cr[42]. However, the hexavalent Cr originally in the solution was reduced to the more stable, less mobile trivalent Cr by natural soil conditions. As a result, the requirement for the excavation of soil containing Cr was waived by the state environmental regulatory agency.

Zn from a large steel plant's sludge and dredgings piles was believed to have migrated 14 ft. vertically and then several hundred yards laterally to a marshland, resulting in damage to this environment[43]. As much as 3700 mg Zn/kg surface soil was found. However, Zn retention capacity of one acre-ft. of topsoil was estimated to be at least 9300 lbs.; for one acre-ft. of subsoil, at least 1700 lbs. The average amount of Zn present in soil was only 2.3 percent of the Zn retention capacity; the potential for the migration of significant amounts of Zn to groundwater appeared to be very low. Analysis of groundwater from below the piles substantiated this finding.

The restoration of a marshland adjacent to the San Francisco Bay, which was used previously by a ship builder, a construction company, and a manufacturing facility, led to the discovery of a substantial amount of buried fragments of dried paint[43]. As much as 5900 mg Zn/kg soil was found adjacent to the buried paint fragments. Tests revealed that 87 percent of Zn leached from paint fragments would quickly precipitate in soil as fixed Zn. Because the Zn leaching from paint should reside at the site in the precipitated soil mineral form, the remedial action consisted primarily of capping the area with soil.

The disposal of low-level radioactive solid wastes by shallow land burial in humid regions can pose potential groundwater contamination problems due to the mobility of the long-lived fission product ^{90}Sr. However, chemical treatment of soil with sodium carbonate and calcium chloride can fix Sr probably as $Ca(Sr)CO_3$ or $Ca(Sr,Mg)CO_3$[44,45]. In general, crushed limestone was recommended as an effective, low-cost landfill liner aid because it mitigates metal migration via adsorption and by causing metal-carbonate fixation[46,47].

Over 100,000 tons of a titanium manufacturing waste comprised of approximately 21.5 to 50.4 percent TiO_2 and various metals were treated with $CaCO_3$ [48]. Treatment resulted in the generation of a neutral, soil-like material that did not leach significant concentrations of metals.

REFERENCES

1. Bear, F. E. Chemistry of the Soil. New York: Reinhold Publishing Co. (1955).
2. Fairbridge, R. W., and Finkl, C. W. Jr. (eds). The Encyclopedia of Soil Science. Part 1. Stroudsburg, PA: Dowden, Hutchinson, & Ross, Inc. (1979).
3. Polemio, M., Senesi, N., and Bufo, S. A. Soil Contamination by Metals. A Survey in Industrial and Rural Areas of Southern Italy. The Science of the Total Environment 25:71-79 (1982).
4. Allaway, W. H. Agronomic Controls Over the Environmental Cycling of Trace Elements. Advances in Agronomy 20:235-274 (1968).
5. Lisk, D. J. Trace Metals in Soils, Plants, and Animals. Advances in Agronomy 24:267-325 (1972).
6. Page, A. L., Elseewi, A. A., and Straugman, I. Physical and Chemical Properties of Fly Ash From Coal-Fired Power Plants with Reference to Environmental Impacts. Residue Rev. 71:83-120.
7. Hem, J. D. Study and Interpretation of the Chemical Characteristics of Natural Water. Second Edition. U.S. Geological Survey Water-Supply Paper 1473. Washington, D.C.: U.S. Government Printing Office (1970).
8. Durfer, C. N. and Becker, E. Public Water Supplies of the 100 Largest Cities in the United States. U.S. Geological Survey, Water-Supply Paper 1812. Washington, D.C.: U.S. Government Printing Office (1964).
9. Ebens, R. J. and Schacklette, H. T. Geochemistry of Some Rocks, Mine Spoils, Stream Sediments, Soils, Plants, and Waters in the Western Energy Region of the Conterminous United States. U.S. Geological Survey Professional Paper 1237. Washington, D.C.: U.S. Government Printing Office (1982).

10. Helling, C. S., and Dragun, J. Soil Leaching Tests for Toxic Organic Chemicals. *IN* Test Protocols for Environmental Fate and Movement of Toxicants. Arlington, VA: Association of Official Analytical Chemists (1981).
11. Black, C. A. (ed). Methods of Soil Analysis. Part 2. Chemical and Microbiological Properties. Madison, Wi: American Society of Agronomy (1965).
12. Hesse, P. R. A Textbook of Soil Chemical Analysis. New York: Chemistry Publications (1972).
13. Jackson, M. L. Soil Chemical Analysis-Advanced Course. 2nd Edition, 8th Printing. Madison, WI: Dept. of Soil Science, University of Wisconsin (1973).
14. Page, A. L. (ed). Methods of Soil Analysis. Part 2. Chemical and Microbiological Properties. Second Edition. Madison, WI: American Society of Agronomy and the Soil Science Society of America (1982).
15. Greenland, D. J. and Hayes, M. H. B. (eds). The Chemistry of Soil Processes. New York: John Wiley & Sons (1981).
16. Pourbaix, M. Atlas of Electrochemical Equilibria in Aqueous Solutions. Elmsford, NY: Pergamon Press (1966).
17. Kotrly, S. and Sucha, L. Handbook of Chemical Equilibria in Analytical Chemistry. New York: John Wiley & Sons (1985).
18. Smith, R. M. and Martell, A. E. Critical Stability Constants. Volume 4: Inorganic Complexes. New York: Plenum Press (1976).
19. Hem, J. D. Chemical Behavior of Mercury in Aqueous Media. *In* Mercury in the Environment. Professional Paper 713. Washington, D.C.: U.S. Geological Survey (1970).
20. Barney, G. S., Navratil, J. D., and Schulz, W. W. (eds). Geochemical Behavior of Disposed Radioactive Waste. ACS Symposium Series 246. Washington, D.C.: American Chemical Society (1984).
21. Mortvedt, J. J., Giordano, P. M., and Lindsay, W. L. (eds). Micronutrients in Agriculture. Madison, WI: Soil Science Society of America (1972).
22. Cary, E. E., Allaway, W. H., and Olsen, O. E. Control of Chromium Concentrations in Food Plants. 2. Chemistry of Chromium in Soils and its Availability to Plants. Journal of Agricultural and Food Chemistry **25**:305-309 (1977).
23. Santillan-Medrano, J. and Jurinak, J. J. The Chemistry of Lead and Cadmium in Soil: Solid Phase Formation. Soil Science Society of America Proceedings **39**:851-856 (1975).
24. Cavallaro, N. and McBride, M. B. Copper and Cadmium Adsorption Characteristics of Selected Acid and Calcareous Soils. Soil Science Society of America Journal **42**:550-556 (1978).
25. Brookins, D. G. Retention of Transuranic and Actinide Elements and

Bismuth at the Oklo Nuclear Reactor, Gabon: Application of Eh-pH Diagrams. Chemical Geology **23**:309-323 (1978).

26. Brookings, D. G. Eh-pH Diagrams for Elements from Z = 40 to Z = 52; Application to the Oklo Natural Reactor, Gabon. Chemical Geology: **23**:325-342 (1978).

27. Brookings, D. G. Eh-pH Diagrams for Elements of Interest at the Oklo Natural Reactor at 25°C, 1 Bar Pressure and 200°C, 1 Bar Pressure. Report to Los Alamos National Laboratory. CNC-11 (1980).

28. Garrells, R. M. and Christ, C. L. Minerals, Solutions, and Equilibria. New York: Harper and Row (1965).

29. Verink, E. D. Simplified Procedure for Constructing Pourbaix Diagrams. J. Education Modules Mat. Sci. Engg. **1**:535-560 (1979).

30. Clifford, A. F. Inorganic Chemistry of Qualitative Analysis. Englewood Cliffs, N.J.: Prentice-Hall (1961).

31. Brown, T. L., and LeMay, H. E. Jr. Chemistry. 2nd Ed. Englewood Cliffs, NJ: Prentice-Hall, Inc. (1981).

32. Weast, R. C. Handbook of Chemistry and Physics. 52nd Ed. Cleveland, OH: The Chemical Rubber Co. (1971).

33. Santillan-Medrano, J. and Jurinak, J. J. The Chemistry of Lead and Cadmium in Soil: Solid Phase Formation. Soil Science Society of America Proceedings **39**:851-856 (1975).

34. Dragun, J. Copper Availability in Soils and Montmorillonite Suspensions. PhD Thesis. The Pennsylvania State University, University Park, PA (1977).

35. Stumm, W. and Morgan, J. J. Aquatic Chemistry. New York: John Wiley and Sons (1970).

36. Buckman, H. O., and Brady, N. C. The Nature and Properties of Soils. New York: Macmillan (1972).

37. Huang, W. H., and Keller, W. D. Dissolution of Clay Minerals in Dilute Organic Acids at Room Temperature. American Mineralogist **56**:1083-1095 (1971).

38. Huang, W. H., and Keller, W. D. Kinetics and Mechanism of Dissolution of Fithian Illite in Two Complexing Organic Acids. Proceedings of the International Clay Conference, pp. 321-331, Madrid (1973).

39. Bretherick, L. Handbook of Reactive Chemical Hazards. Third Edition. Boston: Butterworths (1985).

40. Liberti, L., and Polemio, M. Arsenic Accidental Soil Contamination Near Manfredonia. A Case History. J. Environ. Sci. Health **A16(3)**:297-314 (1981).

41. Matthess, G. In Situ Treatment of Arsenic Contaminated Groundwater. *In* Glasbergen, P. (ed). Proceedings of an International Symposium on Quality of Groundwater, Noordwijkerhout, The Netherlands, March, 1981. Studies in Environmental Science, Volume 17. Amsterdam, The

Netherlands: Elsevier Scientific Publ. Co. (1981).

42. Thorsen, J. W., and Stensby, D. G. Impact of Chromium Waste Spill in Glacial Till Soils. *In* Proceedings of the 1982 Hazardous Materials Spills Conference, 19-22 April 1982, Milwaukee, WI. Rockville, MD: Government Institutes, Inc. (1982).

43. Dragun, J., Schneiter, R. W., and Erler, T. G. III. Cleanup of Zinc-Contaminated Soil and Groundwater-Soil Chemistry and Engineering Aspects. *In* Saxena, J. (ed). Hazard Assessment of Chemicals. Volume 4. New York: Academic Press (1985).

44. Spalding, B. P. Chemical Treatments of Soil to Decrease Radiostrontium Leachability. Journal of Environmental Quality **10**:42-46 (1981).

45. Browman, M. G., and Spalding, B. P. Reduction of Radiostrontium Mobility in Acid Soils by Carbonate Treatment. Journal of Environmental Quality **13**:166-172 (1984).

46. Fuller, W. H., and Artiola, J. Use of Limestone to Limit Contaminant Movement From Landfills. *In* Land Disposal of Hazardous Wastes. Proceedings of the Fourth Annual Research Symposium, San Antonio, TX, March 6-8, 1978. EPA - 600/9-78-016. Cincinnati, OH: U.S. Environmental Protection Agency (1978).

47. Artiole, J., and Fuller, W. H. Effect of Crushed Limestone Barriers on Chromium Attenuation in Soils. Journal of Environmental Quality **8**:503-510 (1979).

48. Starkey, D. J., and Kroenig, M. H. Land Application/Treatment of Residue Produced in the Manufacture of Titanium Dioxide: A Case History. *In* Proceedings of the 37th Purdue Industrial Waste Conference, May 11-13, 1982. Boston: Ann Arbor Science - Butterworths (1983).

4

Element Adsorption and Mobility in Soil

INTRODUCTION

Water is generally responsible for the mobility of elements in soil systems. If the direction of water movement is known, then the direction of element migration is generally known. However, the rate of element migration usually does not equal the rate of water movement due to fixation and adsorption reactions.

The information presented in Chapter 3 revealed that fixation reactions will remove an element from migrating water and immobilize an element either within the structure of a mineral or at the mineral surface. Adsorption will also remove an element from migrating water. Strictly speaking, adsorption is defined as the accumulation of an element at the surface of soil particles with a decrease in the concentration of the dissolved element in water. The general usage of the term adsorption in the field of environmental science also implies an equilibrium distribution of an element between soil and water.

In this text, the discussions of fixation and adsorption were purposely separated since these reactions are separate and distinct. Although both lead to a reduction in the concentration of an element in water, adsorption is inherently two-dimensional while fixation is inherently three-dimensional. However, the separation of these two reactions during the study of a soil-water system is not an easy task. When no independent data are available upon which one can determine the extent and magnitude of each reaction, the loss of an element to the solid phase is referred to as sorption. The use of this term avoids the implication that either adsorption or fixation has occurred. This text focuses on the magnitude and extent of each reaction because these should be known in order to properly assess potential impacts of the element to human health and the environment.

Chapter 4 will discuss several aspects of element adsorption and mobility. First, this chapter will address cation adsorption mechanisms and how complex formation affects cation adsorption. Then, this chapter will address the mechanisms, extent, and significance of anion adsorption and exclusion. Next, this chapter will discuss how adsorption data is mathematically expressed in terms of equilibrium adsorption models and kinetic adsorption models. Finally, Chapter 4 will discuss approaches to quantifying element mobility in soil that utilize soil physical and chemical properties.

CATION ADSORPTION MECHANISMS

In Chapter 2, the general structure of soil minerals was discussed. Information in Chapter 2 revealed that clay surfaces possess negative charges. Information in Chapter 2 also revealed that these negative charges are the result of isomorphous substitution. These negative charges on soil mineral surfaces are responsible for attracting and accumulating cationic species of elements at soil surfaces.

Humus is also responsible for the accumulation of cationic species of elements at soil surfaces. Humus is the relatively stable fraction of soil organic matter which remains in soil after the chemicals comprising plant and animal residues—amino acids, carbohydrates, fats, waxes, resins, and organic acids—have decomposed. Humus is not an individual chemical but a series of relatively high molecular weight, brown to black colored polymers. Humus is generally subdivided into three general classes: fulvic acid, humic acid, and humin. Fulvic acid is comprised of organic polymers with a minimum molecular weight of 2,000 that are soluble in both acids and bases (see Figure 4.1.). Humic acid is comprised of organic polymers having molecular weights greater than those of fulvic acids, possibly as high as 300,000. Humic acids are soluble in bases but are insoluble in acids. Humin is comprised of very high molecular weight polymers that are insoluble in both acids and bases.

Humus is colloidal in structure. The typical humus colloid is comprised of (a) a central unit containing primarily C and H, and (b) a colloidal surface comprised of carboxylic (COOH) and hydroxyl (OH) functional groups. These groups, when in the dissociated state, possess negative charges. These charges are responsible for accumulating cationic species of elements at soil surfaces.

When an ion in the water phase such as Cu^{2+} is attracted to a soil surface, Cu^{2+} must displace another cation already present at the soil surface. In other words, Cu^{2+} must successfully displace another cation. This competitive process can be described by the following equation:

$$[\text{Soil}]\ Ca^{2+} + Cu^{2+} \rightleftharpoons [\text{Soil}]\ Cu^2 + Ca^{2+} \quad (4.1)$$

The process described by Equation 4.1 is known as exchange phenomena or ion exchange. The term cation exchange specifically refers to the exchange between cations balancing the surface charge on the soil surface and the cations dissolved in water.

The total amount of cations adsorbed by these negative charges on a unit

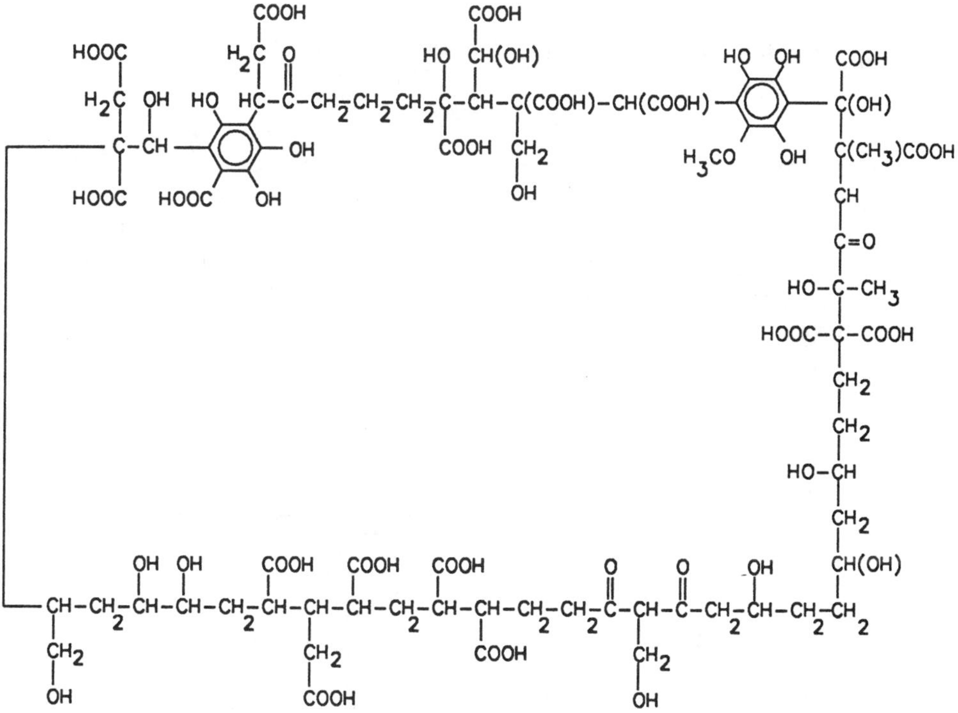

STRUCTURE #1

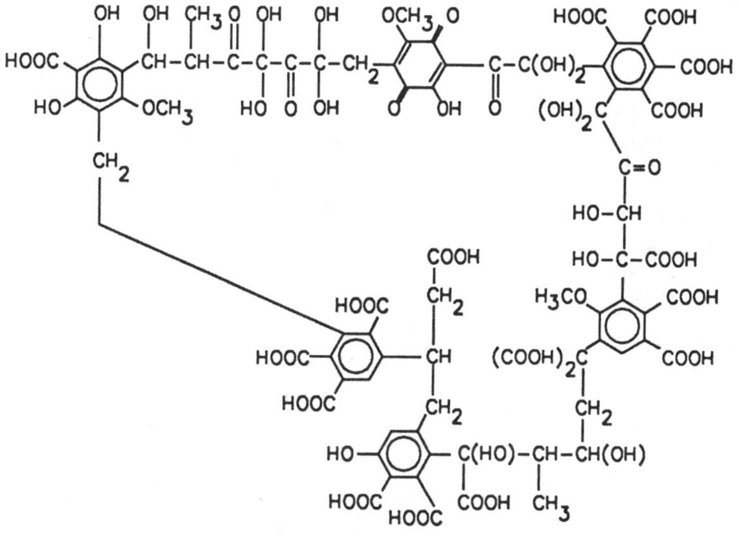

STRUCTURE #2

FIGURE 4.1 Model Fulvic Acid structures containing 15% Aromaticity (#1) and 45% Aromaticity (#2)[1]. (Reprinted with permission of Blackwell Scientific Publications Limited).

mass of soil is defined as the cation exchange capacity of the soil (CEC). The CEC is usually expressed as milliequivalents (meq) per 100 grams of soil. Table 4.1 lists ranges of the CEC for several soils, soil clays, and organic matter. The CEC of a humid region surface soil can be grossly estimated from the percentages of clay and organic matter; one percent of silicate clay should contribute 0.5 meq to the CEC, while one percent of well-humified organic matter should contribute 2.0 meq to the CEC[2].

TABLE 4.1   General Ranges for The Cation Exchange Capacities of Soil Clays, Soil Organic Matter, and Several Soil Types[2,3,4]

	CEC (meq/100 grams)
Soil Clays	
Chlorite	10-40
Illite	10-40
Kaolinite	3-15
Montmorillonite	80-150
Oxides and Oxyhydroxides	2-6
Saponite	80-120
Vermiculite	100-150
Soil Organic Matter	> 200
Soil Type	
Sand	2-7
Sandy Loam	2-18
Loam	8-22
Silt Loam	9-27
Clay Loam	4-32
Clay	5-60

It is important to recognize that, based on an analysis of the information presented in Table 4.1, clays and soils possess widely varying CECs. Even soils within a given soil type can exhibit significant CEC variability. Since CEC represents the capacity of a clay and soil to attract and accumulate cations, wide variations in CEC represent wide variations in a clay's and soil's capacity to adsorb cations. For example, Cecil Clay from Alabama (CEC = 4.8) will not exhibit the same ability to retain certain cations such as Cu^{2+} and Zn^{2+} as Sweeney Clay from California (CEC = 57.5).

The term milliequivalent is defined as one milligram of hydrogen or the amount of any other ion that will displace it[2]. In other words, if a clay has a CEC of 1 meq/100 grams, it is capable of adsorbing 1 mg of hydrogen or its equivalent for every 100 grams of clay. The equivalent amount for Ca would be 20 mg. Each Ca^{2+} has two charges and is equivalent to two H^+. Since Ca has an atomic weight of 40, the amount of Ca^{2+} required to displace 1 mg of adsorbed H^+ is 40/2 or 20 mg. In other words, 20 mg is the weight of 1 meq of Ca^{2+}. If 100 grams of any clay can adsorb 200 mg of Ca^{2+}, the adsorption capacity is 200/20 or 10 meq/100 grams.

An analysis of Equation 4.1 will reveal that the cation exchange process is stoichiometric; in other words, one Cu^{2+} ion can displace only one Ca^{2+} ion. Also, an analysis of Equation 4.1 will reveal that the cation exchange process is reversible. The reversible relationship between a cation in water with an exchangeable cation at equilibrium can be described by the following equation:

$$C_s \underset{k(ads)}{\overset{k(des)}{\rightleftharpoons}} C_e \qquad (4.2)$$

where

C_s = concentration adsorbed on soil surfaces (ug/gram soil)
C_e = concentration in water (ug/ml)
k(des) = desorption rate
k(ads) = adsorption rate

An analysis of Equation 4.2 will reveal that the equilibrium distribution of a cation is governed by two opposing rate processes. The adsorption rate, k(ads), is the rate at which the dissolved cation in water transfers into the adsorbed state. The desorption rate, k(des), is the opposite process: it is the rate at which the cation transfers from the adsorbed state into water. The simplest and most common method for mathematically expressing the extent of adsorption as illustrated in Equation 4.2 is the adsorption coefficient or distribution coefficient, K_d. The K_d is defined as the ratio C_s/C_e. The greater the extent of adsorption, the greater the magnitude of K_d.

The K_ds for several elements are listed in Table 4.2. An analysis of the information presented in Table 4.2 will reveal that the K_ds among the elements are different. Ion exchange selectivity, or the preference of one ion over another at soil adsorption sites, has been the subject of a great deal of study[6]. If ions of equal charge could be treated as point charges, then no preference between ions of equal valence would exist; however, ions do have significantly different hydrated sizes. Since electrostatic forces are involved in the accumulation of ions at soil surfaces, it can be predicted from Coulomb's law that the ion possessing the smallest hydrated radius and largest charge will be preferentially accumulated over ions possessing relatively larger hydrated radii and relatively smaller charge. Numerous published studies have demonstrated this size-charge relationship in soils. The "Lyotropic series" refers to the order of cation preference by soil surfaces. For monovalent cations, the order is generally[6]:

$$Cs^+ > Rb^+ > K^+ > Na^+ > Li^+$$

TABLE 4.2 Ranges for K_ds for Various Elements in Soils and Clays[5] (Reprinted with permission from the American Society of Agronomy, Inc., Crop Science Society of America, Inc., and the Soil Science Society of America, Inc.)

Element	Observed Range (ml/g)	Mean[a]	Standard Deviation[b]
Ag	10 - 1,000	4.7	1.3
Am	1.0 - 47,230	6.7	3.0
As(III)	1.0 - 8.3	1.2	0.6
As(V)	1.9 - 18	1.9	0.5
Ca	1.2 - 9.8	1.4	0.8
Cd	1.3 - 27	1.9	0.9
Ce	58 - 6,000	7.0	1.3
Cm	93 - 51,900	8.1	1.9
Co	0.2 - 3,800	4.0	2.3
Cr(III)	470 - 150,000	7.7	1.2
Cr(VI)	1.2 - 1,800	3.6	2.2
Cs	10 - 52,000	7.0	1.9
Cu	1.4 - 333	3.1	1.1
Fe	1.4 - 1,000	4.0	1.7
K	2.0 - 9.0	1.7	0.5
Mg	1.6 - 13.5	1.7	0.5
Mn	0.2 - 10,000	5.0	2.7
Mo	0.4 - 400	3.0	2.1
Np	0.2 - 929	2.4	2.3
Pb	4.5 - 7,640	4.6	1.7
Po	196 - 1,063	6.3	0.7
Pu	11 - 300,000	7.5	2.3
Ru	48 - 1,000	6.4	1.0
Se(IV)	1.2 - 8.6	1.0	0.7
Sr	0.2 - 3,300	3.3	2.0
Tc	0.003 - 0.28	3.4	1.1
Th	2,000 - 510,000	11.0	1.5
U	11 - 4,400	3.8	1.3
Zn	0.1 - 8,000	2.8	1.9

[a] Mean of the logarithms of the observed values.
[b] Standard deviation of the logarithms of the observed values.

For divalent cations, the order is generally [6]:

$$Ba^{2+} > Sr^{2+} > Ca^{2+} > Mg^{2+}$$

In general, trivalent cations are preferentially adsorbed over divalent cations, which are preferentially adsorbed over monovalent cations.

A further analysis of the information presented in Table 4.2 will reveal that the K_ds measured for an individual element in soils can vary substantially; for example, the ranges of K_ds for Pu and Zn span four orders of magnitude. Because the magnitude of the K_d is dependent upon the size and charge of the cation and upon those soil properties governing the exchange sites on soil surfaces, and because these soil properties vary widely among soils, an element's K_d will also vary widely among soils.

EFFECTS OF COMPLEX FORMATION
ON CATION ADSORPTION

The concentrations of cations that are reported in chemical analyses of groundwater normally represent the total concentrations of each element in solution. The concentration limits specified in drinking-water standards also are expressed in terms of total concentrations of each element or ion of interest. In aqueous systems, however, most cations exist in more than one molecular or ionic form. These forms can have different valences and, therefore, different mobilities owing to different affinities for adsorption and different solubility controls. Knowledge of the distribution of species in solution is sometimes necessary for consideration of the behavior of cations in soil.

In general, the process in which a cation combines with molecules or anions containing free pairs of electrons is known as coordination or complex formation. The cation-anion or cation-molecule combination is known as a complex. The anion(s) or molecule(s) with which the cation forms a complex is usually referred to as a ligand.

To understand the behavior of a complex in soil, the stability of the complex must be ascertained. Stability constants quantify the propensity of the individual elements or molecules to exist together as a complex as opposed to their propensity to dissociate and exist as separate components independent of the complex. At equilibrium, this relationship can be expressed as:

$$A^+ + L^- \rightleftharpoons AL \tag{4.3}$$

where AL is a complex comprised of the association of the cation A^+ with the ligand L^-. The stability of the complex AL is defined as:

$$K = [AL]/[A^+][L^-] \tag{4.4}$$

where the brackets signify the concentration of AL, A^+, and L^- in moles per liter. The larger the value of K, the stability or formation constant, the greater the stability of the complex in solution. In general, the value of K should be greater than approximately 10^7 in order for the complex to be considered a stable one in a soil system.

It is important to note, however, that although a complex may possess a relatively high stability constant, the complex may dissociate due to the presence of another cation that will form a more stable complex with the ligand. For example, the Ni-DTPA complex has a stability constant of $10^{6.1}$. However, this complex will dissociate in the presence of Fe and preferentially form Fe-DTPA, which has a stability constant of $10^{29.2}$. A number of published papers discuss cation competition and complex formation in great detail [7,8,9,10,11].

The effects of ligands on the adsorption of cations by soil surfaces can be classified into five general categories[12]. In the first category, the ligand has a low affinity for both the cation and the soil surface. In this case, the ligand has a negligible effect on cation adsorption. This situation can be described by the following equation:

$$[Soil]\ A^+ \overset{\longrightarrow}{\underset{\longleftarrow}{\rule{0pt}{0pt}}} A^+ + L^- \overset{\rule{1cm}{0pt}}{\underset{\longleftarrow}{\rule{0pt}{0pt}}} AL \qquad (4.5)$$

The ligand will migrate with flowing water; the cation, however, may or may not migrate, depending upon the magnitude of its K_d. A number of naturally occurring inorganic anions fall into this class of ligand such as Cl^-, HCO_3^-, SO_4^{2-}, HSO_4^-, NO_3^-, HS^-, and in some cases CO_3^{2-} and S^{2-}. In general, identification and quantification of these complexes is usually not necessary for an understanding of the behavior of hazardous materials in soil. Therefore, this text will not address this category of complexes further. However, several excellent publications discuss the identification and quantification of these complexes via mathematical computations[7,8,9,10,11].

a soluble complex with the cation, and the complex is adsorbed extensively by the soil. This situation can be described by the following equation:

$$[Soil]\ AL \overset{\rule{1cm}{0pt}}{\underset{\longleftarrow}{\rule{0pt}{0pt}}} AL \qquad (4.6)$$

The magnitude of the arrows indicate the relative magnitude of the flux of the cation A^+, the ligand L^-, or the complex AL out of the water phase and the adsorbed phase. In Equation 4.6, the relative magnitude of the arrows indicates that the soluble complex AL strongly prefers the adsorbed phase.

Probably the most important ligand that affects cation adsorption as illustrated in Equation 4.6 is the hydroxide ion[12]. In general, as the OH^- activity in water increases, the extent of metal adsorption increases. The structures of the major complexes that cations form with OH^- ligands are listed in Table 4.3. Figure 4.2 illustrates the effect of increasing OH^- activity on cation adsorption by soil mineral surfaces.

In the third category, the ligand has a high affinity for the cation and forms a soluble complex with it, but the complex has a low affinity for the soil. This situation can be described by the following equation:

$$[Soil]\ AL \overset{\longrightarrow}{\underset{\longleftarrow}{\rule{0pt}{0pt}}} AL \qquad (4.7)$$

In general, complexes that are not adsorbed can migrate along with water flow. Mobile complexes in silt and clay soils should possess ligands with low

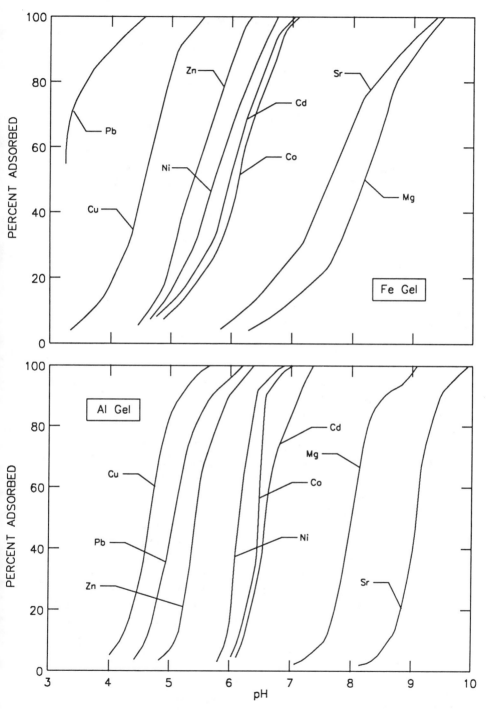

Figure 4.2   The effect of increasing OH⁻ activity on cation
adsorption by two oxide mineral surfaces.[13]
(Reprinted with permission of the Soil Science
Society of America, Inc.).

TABLE 4.3   Structures of Major Cation-Hydroxide Complexes.

Complex[a]	Cation
$M(OH)^+$	Cd, Cu, Pb, Hg, Ni, Zn
$M(OH)_2^+$	Cr, Al, Fe
M_2OH^{3+}	Be^{2+}, Mn^{2+}, Co^{2+}, Ni^{2+}, Zn^{2+}, Cd^{2+}, Hg^{2+}, Pb^{2+}
$M_2(OH)_2^{(2z-2)+}$	Cu^{2+}, Sn^{2+}, UO_2^{2+}, NpO_2^{2+}, PuO_2^{2+}, VO^{2+}, Al^{3+}, Sc^{3+}, Ln^{3+}, Ti^{3+}, Cr^{3+}, Th^{4+}
$M_3(OH)_3^{3+}$	Be^{2+}, Hg^{2+}
$M_3(OH)_4^{(3z-4)+}$	Sn^{2+}, Pb^{2+}, Al^{3+}, Cr^{3+}, Fe^{3+}, In^{3+}
$M_3(OH)_5^{(3z-5)+}$	UO_2^{2+}, NpO_2^{2+}, PuO_2^{2+}, Sc^{3+}, Y^{3+}, Ln^{3+}
$M_4(OH)_4^{4+}$	Mg^{2+}, Co^{2+}, Ni^{2+}, Cd^{2+}, Pb^{2+}
$M_4(OH)_8^{8+}$	Zr^{4+}, Th^{4+}
$M_6(OH)_8^{4+}$	Be^{2+}, Pb^{2+}
$M_6(OH)_{12}^{6+}$	Bi^{3+}

[a] - z indicates the cationic charge; M indicates the metallic cation.

molecular weights, usually less than 300. These may include some naturally occurring organic acids, amino acids, and polyphenols; however, these naturally occurring organic chemicals are present in soils at such low concentrations that they usually do not exert a significant effect on cation migration in most soils.

One example of this type of complex and its effect on element migration involved Co and U complexes with EDTA[14]. Based on the wide range of K_ds for Co and U listed in Table 4.2, one would not necessarily expect these elements to be mobile. Yet, unexpected mobility of these two elements from liquid disposal sites at the Oak Ridge National Laboratory was attributed to the presence of EDTA and possibly dissolved humic substances. At another site, ^{60}Co migrating from a low-level wastewater infiltration pit through sand was complexed with organic ligands[15]; the molecular weight of the complex was approximately 500. Tc(IV)EDTA and Tc(V)DTPA were highly mobile in Hanford soil with only one to six percent adsorption of the Tc-complex[16]; however, citrate had a negligible effect on Tc sorption. EDTA was implicated as the ligand which complexed Pu and Co and caused these radionuclides to migrate in soil adjacent to a waste-filled trench[17]; palmitic and stearic acid type carboxylic acids were also implicated as the ligands which complexed Cs and Sr.

Transition metal organometallic complexes can fall under this third category. Transition metal organometallic complexes are molecules possessing at least one chemical bond between a transition metal cation and a carbon atom. These complexes form because the transition metals, which are seeking electrons to fill their valence orbitals, form bonds to organic molecules that can serve as electron donors. The donation of electron density from the organic molecule to the metal varies in degree according to the structure of both the organic molecule and the transition metal.

Figure 4.3 illustrates the structure of several mononuclear organometallic complexes which can be present in soil due to the release of hazardous materials. The nomenclature describes the bonding geometry: pi indicates that bonding occurs through pi orbitals associated with carbon atoms of various unsaturated molecular fragments. In general, complexes with Co, Cr, Fe, Ir, Mn, Nb, Ni, Os, Pd, Rh, Ru, Ta, Ti, and V have been synthesized and utilized in organic synthesis reactions by organic chemists[18]. Also, these complexes appear to be responsible for the enhanced migration of dibenzylamine into aqueous solutions containing 200 ppm Cu^{2+} and Fe^{3+} from toluene[19]. In addition, these complexes appear to be responsible for the enhanced migration of pyridines, anilines, azanaphthalenes, and azabiphenyls into aqueous solutions containing 200 ppm Fe^{3+} from coal gasification tar[20]. However, no information is available on the formation, stability, transport, and transformations of transition metal organometallic complexes in soil. As a result, the impact of complex formation on predictions of metal and organic chemical migration and transformation at sites where substantial amounts of both are present is not known. Therefore, data inconsistencies and anomalies occurring at sites may be attributable to these complexes, and investigators should at least be aware of their potential existence.

In the fourth category, the ligand is extensively adsorbed by the soil, but the adsorbed ligand has a low affinity for the cation. This situation can be described by the following equation:

$$[Soil]\ L^- \rightleftharpoons A^+ \qquad (4.8)$$

In the fifth category, the ligand is extensively adsorbed by the soil, but the adsorbed ligand has a high affinity for the cation. This situation can be described by the following equation:

$$[Soil]\ LA \rightleftharpoons A^+ \qquad (4.9)$$

The primary difference in the situations illustrated by Equations 4.8 and 4.9 is the mobility of the cation A^+. In the first situation, the cation can migrate along with flowing water. In the second situation, the cation is held at the soil surface and generally cannot migrate along with flowing water.

The primary similarity in the situations illustrated by Equations 4.8 and 4.9 is the extensive adsorption of the ligand. In general, extensive adsorption of organic molecules can be expected in silt and clay soils if the molecular weight exceeds 400. In natural soil systems, the organic chemicals

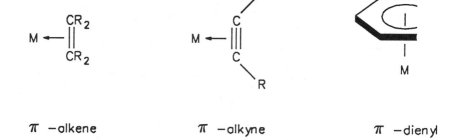

π −alkene π −alkyne π −dienyl

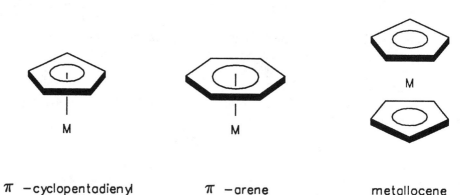

π −cyclopentadienyl π −arene   metallocene

FIGURE 4.3 Structure of selected transition metal
organometallic complexes.

possessing relatively high molecular weights are the chemicals comprising humus. Due to the relatively large molecular surface areas, these molecules are attracted to the surfaces of other colloids via van der Waal's forces. Once the humus molecule is adsorbed, however, only the external surface of the molecule that faces the soil colloid surface is blocked and cannot adsorb cations. The molecule's inner surfaces and the surface facing the water phase can adsorb cations.

Figure 4.4 illustrates the relative affinity of various elements for a fulvic acid. An analysis of Figure 4.4 will reveal that some elements—Fe, Cu—are extensively adsorbed by this fulvic acid; these elements will bind with this particular fulvic acid to form the complex LA illustrated in Equation 4.9. A further analysis of Figure 4.4 will reveal that some elements—Zn, Mn, Ca, Mn—are not extensively adsorbed by this fulvic acid; these elements will react with this adsorbed fulvic acid in a manner illustrated by Equation 4.8. Although fulvic and humic acids will show significant variations in their ability to adsorb cations, the general affinity of cations for humus as a whole should decrease in the order: trivalent cations > divalent cations > monovalent cations.

It is important to note that although fulvic acids are water soluble, cations such as Fe^{3+}, Al^{3+}, Cr^{3+}, Pb^{2+}, Cu^{2+}, Hg^{2+}, Zn^{2+}, Ni^{2+}, Co^{2+}, Cd^{2+}, and Mn^{2+} can form water-insoluble (i.e. precipitated) complexes with fulvic acids[22]. Whether or not an insoluble complex will form depends upon the concentration of the fulvic acid and cations present in the soil system. Fulvic acid/metal weight ratios greater than two favors the formation of soluble complexes[22]. Over the pH range 5-7, trivalent and divalent cations tend to form water insoluble complexes with fulvic acid.

ANION ADSORPTION AND EXCLUSION MECHANISMS

Although most soils possess particles that contain surface charges that are negative, it is important to recognize that soils also possess particles that contain positive surface charges. In general, most soils possess a net negative charge; however, there are tropical soils that can obtain a net positive charge. These positive charges are responsible for accumulating at soil surfaces anionic species of elements such as phosphate, arsenate, molybdate, selenite, sulfate, borate, silicic acid, fluoride, halides, and nitrate[23].

There are three different types of surfaces responsible for attracting and accumulating anions in soil. The first is oxide surfaces. Consider the oxide surface M-OH, where M signifies one of the elements such as Fe or Al that forms oxides and OH is a surface hydroxyl group. M-OH is in equilibrium with soil water or groundwater. Depending upon the pH of the soil system,

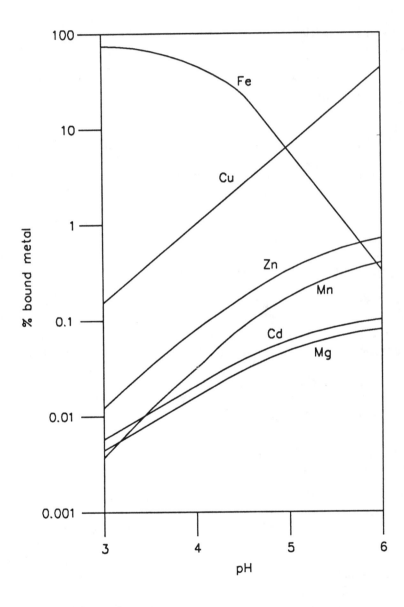

FIGURE 4.4   Relative Affinity of Various Elements for
a Fulvic acid[21]. (Reprinted with permission of Blackwell Scientific Publications
Limited).

M-OH can develop either a positive or negative charge. If the pH is high, then M-OH will develop a negative surface charge due to H^+ dissociation from the OH group:

$$M\text{-}OH + OH^- \rightleftharpoons M\text{-}O^- + H_2O \qquad (4.10)$$

If the pH is low, then M-OH will develop a positive surface charge by attracting H^+:

$$M\text{-}OH + H^+ \rightleftharpoons M\text{-}OH_2^+ \qquad (4.11)$$

An analysis of Equations 4.10 and 4.11 will reveal that the surface charge of oxides is pH dependent. For example, goethite (Alpha-FeOOH) possesses a net negative charge above a pH of approximately 8.5. However, goethite possesses a net positive charge below a pH of approximately 8.5. At approximately pH 8.5, goethite possesses an equal number of positive and negative surface charges. This pH is referred to as the isoelectric point or the zero point of charge[6]. Since the net charge on the surface is dependent upon the pH of the soil system, and since the net surface charge can change as pH changes, this type of surface charge is commonly known as pH dependent charge.

The second type of surface responsible for attracting and accumulating anions is the edges of alumino-silicate clay minerals. The edges of these minerals contain ions that are not fully coordinated. As a result, these edges adsorb H^+ and OH^- to form hydroxylated surfaces and behave in a manner similar to M-OH.

Since all clay minerals have edges, they all participate to some extent in adsorbing anions. However, the 2:1 clay minerals have edges that comprise only about one percent of the total surface charge; therefore, these minerals do not extensively adsorb anions. On the other hand, the 1:1 clay mineral's edges comprise a relatively greater proportion of the total surface charge; as a result, these minerals can adsorb a relatively greater percentage of anions.

The third type of surface responsible for attracting and accumulating anions in soils is organic matter. Soil organic matter possesses functional groups that adsorb H^+ and OH^- to form charged surfaces and behave in a manner similar to M-OH:

$$R\text{-}OH + OH^- \rightleftharpoons R\text{-}O^- \qquad (4.12)$$

$$R\text{-}OH + H^+ \rightleftharpoons R\text{-}OH_2^+ \qquad (4.13)$$

$$R\text{-}NH_2 + H^+ \rightleftharpoons R\text{-}NH_3^+ \qquad (4.14)$$

In general, the adsorption capacity of anions by soil surfaces is relatively smaller than the adsorption capacity of cations by soil surfaces. As a general rule, the anion exchange capacity (AEC) of most U.S. soils should be less than five percent of the CEC. For saline soils, the AEC should be higher, possibly as high as fifteen percent of the CEC. Furthermore, since the AEC is a pH-dependent function, the lower the soil pH, the greater the AEC.

Since soils typically possess a significantly greater number of negative surface charges than positive ones, and since anions possess a net negative charge, repulsion between anions and soil surfaces will occur. This process is known as negative adsorption. The net effect of negative adsorption is the repulsion of anions from the water in the immediate vicinity of the soil surface and from narrow pores where water velocities are slow; as a result, when leaching occurs, anions can migrate at a faster rate than they would if they were evenly dispersed throughout all water present in soil. This repulsion process is known as anion exclusion.

The rate of Cl⁻ migration in soils was two to twenty-five percent greater than the rate of water movement in columns comprised of 14 surface and subsurface soils from southern California[24]. The rate of Cl⁻ movement (as CaCl₂) relative to total water movement in Houston black clay was 37 percent greater at 0.01N, 19 percent greater at 0.1N, and 12 percent greater at 1.0N[25]. In leaching experiments involving 15 soils from across the U.S., the migration rate of Cl⁻ added as 0.01N CaCl₂ was 4 to 67 percent faster relative to total water movement[26]; increased Cl⁻ movement was highly correlated with CEC and surface area.

Soil compaction increases the degree of orientation and aggregation of clay minerals, thereby creating a network of smaller-sized pores. In these pores, ions should be in the vicinity of charged surfaces most of the time, and ion movement should occur primarily by diffusion. As a result, anion exclusion should have a dominant effect on anion migration through these small pores. For example, in six compacted fine-grained soils containing illite, kaolinite, and/or montmorillonite as the dominant clays, the migration rate of Cl⁻ added as brine (30 percent NaCl) was 1.79 to 6.38X faster than if it had been associated with all soil water[27]. These researchers attributed the increased rate of migration to anion exclusion.

The depth of anion movement or penetration in soil cores leached with water can be generally estimated[38]:

$$d = R/(V - V_{ex})$$
(4.15)

where

d = depth, cm

r = amount of water applied, cm

V = volumetric water content of the soil

V_{ex} = volume of water which excludes anions

V_{ex} for Maury silt loam, Pembroke silt loam, and Eden silty clay loam were determined to be 0.082, 0.054, and 0.118, respectively[28]. V_{ex} can be grossly estimated from data presented on a study of anion exclusion in 29 soils[24]:

$$V_{ex} = 0.0085 \, (CEC) \qquad (4.16)$$

where

    CEC = Cation exchange capacity in meq/100 grams,
           where CEC > 10 meq/100 grams.

For a soil containing appreciable amounts of free iron oxides, and consequently some anion exchange capacity, Equation 4.16 will overestimate V_{ex}. For soils with CECs less than 10 meq/100 grams, the subtraction of moisture retained by an air dry soil from the volumetric water content should allow a reasonable estimation of anion exclusion volumes; the volume of moisture retained by air dry soils had a typical range of 2 to 14 percent for some soils utilized in the study[24].

V_{ex} can also be grossly estimated by utilizing data from soil column studies performed on 17 southern California soils[29]:

$$V_{ex} = 0.208 + 0.0063(\text{pH}) - 0.0057(\%\text{Clay}) \qquad (4.17)$$
$$+ \; 0.0063(\text{SP}) + 0.0127(\%\text{OM})$$
$$+ \; 0.0132(\%\text{Fe}_2\text{O}_3) - 0.0135(\text{CEC})$$
$$+ \; 1.2059(\text{SC})$$

where

         pH = Soil pH, ranging from 4.8 to 8.1

  % Clay = percent clay (<2mm) present in soil, ranging from 1.8 to 43.5 percent.

        SP = Saturation percentage, the percent water by weight in a saturated paste, ranging from 22 to 55 percent.

  %OM = percent organic matter present in soil, ranging from 0.13 to 3.22 percent.

  %Fe_2O_3 = percent free iron oxides present in soil, ranging from 0.28 to 1.20 percent.

    CEC = Cation exchange capacity, ranging from 4.2 to 27.4 meq/100 grams.

SC = salt concentration, either 0.01N or 0.3N $CaCl_2$.

It is important to note, however, that some salts can increase soil permeability by causing soil shrinkage and subsequent cracking; as a result, enhanced ion movement may be due to increased soil permeability as well as anion exclusion. For example, aqueous solutions of NaCl, $CaCl_2$, $MgCl_2$, KCl, NH_4Cl, K_2CO_3, $CaSO_4$, $MgSO_4$, K_2SO_4, $NaNO_3$, $FeSO_4$, and FeO increased the permeability of bentonite (see Table 4.4). On the other hand, aqueous solutions of $CaCO_3$, $MgCO_3$, Na_2SO_4, and Na_2SO_3 had a negligible effect on bentonite permeability (see Table 4.4). In addition, aqueous solutions of 500 mg $FeCl_3$/liter and aqueous solutions of 50 and 300,000 mgNi$(NO_3)_2$/liter had a negligible effect on the permeability of White store clay soil, Hoytville clay soil, and Faceville clay soil[31].

TABLE 4.4   Effect of Various Salt Solutions on Bentonite Permeability[30]

Salt	Log Permeability, (cm/sec)								
	0%	0.25%	0.5%	1%	2%	3%	4%	5%	8%
NaCl	7.38	—	—	7.29	7.24	—	—	6.88	—
$CaCl_2$	7.41	6.96	6.79	6.73	6.74	—	—	6.74	—
$MgCl_2$	7.41	7.08	6.85	6.78	6.78	—	—	6.80	—
KCl	7.41	7.46	7.34	7.30	—	6.80	—	6.70	—
NH_4Cl	7.41	7.41	7.29	6.97	6.78	6.54	—	—	—
$CaCO_3$	7.41	—	—	—	7.42	7.38	7.37	7.37	7.43
$MgCO_3$	7.46	—	—	—	7.34	7.30	7.30	7.23	7.21
K_2CO_3	7.41	—	—	7.34	7.20	6.95	6.78	6.71	—
$CaSO_4$	7.46	7.16	6.97	6.96	—	7.03	—	7.01	—
$MgSO_4$	7.46	7.29	7.04	6.94	—	6.90	—	6.92	—
K_2SO_4	7.46	—	—	—	7.27	7.05	6.93	6.84	—
Na_2SO_4	7.46	—	—	—	7.37	7.40	7.40	7.37	—
Na_2SO_3	7.40	—	—	—	7.41	7.43	7.44	7.29	—
$NaNO_3$	7.40	—	—	—	7.35	7.24	7.22	7.10	—
$FeSO_4$	7.33	—	7.24	7.01	6.87	—	6.78	—	—
FeO	7.33	—	7.05	6.93	6.91	—	6.95	—	—

(Copyright ASTM. Reprinted with permission).

EQUILIBRIUM ADSORPTION MODELS

The reversible relationship between C_s and C_e, which is described by Equation 4.2, represents the distribution of an element present at one total concentration, C_t, where $C_t = C_s + C_e$, assuming negligible fixation. Adsorption studies are usually performed over a range of concentrations; data from these studies are presented as a graph commonly known as an adsorption isotherm. Adsorption isotherms for many chemicals have been experimentally derived during the last century. These isotherms have been broadly classified according to initial slope into four general isotherm classes[12,32]: the S-Curve, the L-Curve, the H-Curve, and the C-Curve. Figure 4.5 illustrates the general curve shape of these four isotherm classes.

The slope of the S-Curve is relatively linear near the C_s-C_e intercept, then increases. This suggests that the relative affinity of the soil surface for the chemical at the relatively lower concentration is lower than at higher concentrations. This behavior may result from (a) increasing interactions among adsorbed molecules as the number of adsorbed molecules increases, and/or (b) successful competition by nonadsorbed complexing agents in water. After the complexing capacity for the chemical is exceeded as C_t increases, the soil surfaces gain in the competition and begin to adsorb the chemical.

The shape of the L-Curve is due to soil-chemical interactions of a nature opposite those creating the S-Curve. When C_t is small, the soil surface has a high affinity for the chemical; this results in an initial slope that is steep. As C_t increases, the affinity for each additional increment of C_t decreases. As a result, the slope gradually becomes less steep as C_t increases.

The H-Curve is a special version of the L-Curve. When C_t is relatively small, the chemical is completely adsorbed. As a result, the curve is initially vertical. This condition can be produced by (a) highly specific, strong attractions between the soil surface and the chemical, or by (b) significant van der Waal's interactions occurring between the chemical and the soil surface.

The slope of the C-Curve remains relatively linear at all values of C_t. This condition results primarily from the presence of relatively homogeneous adsorption sites.

Mathematical relationships for two of these Curves—the C- and L-Curves have been developed and extensively used in the published literature. The C-Curve is the simplest of the four and can be mathematically described as:

$$C_s = (K_d)C_e \qquad (4.18)$$

where

C_s = amount of adsorbed chemical

C_e = amount of chemical in water in equilibrium with C_s

K_d = adsorption or distribution coefficient

Equation 4.18 is commonly known as the linear adsorption isotherm model. Use of Equation 4.18 to describe adsorption implies that all adsorption sites present on soil surfaces have a relatively constant affinity for the chemical. This condition generally exists over a limited range of C_t, but usually not over the entire range of all possible values of C_t. Therefore, the use of

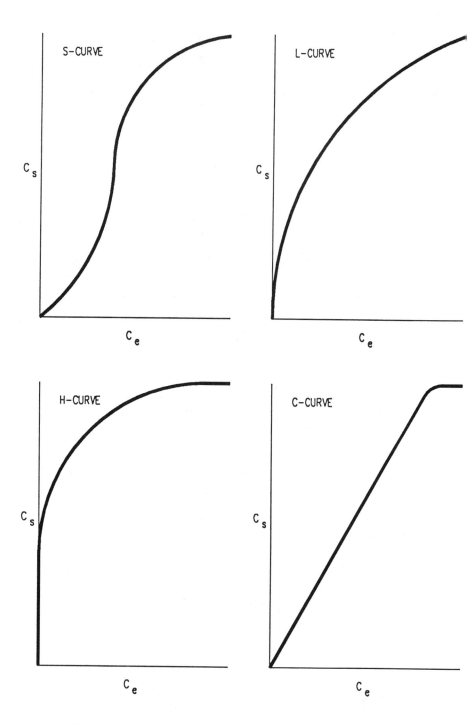

FIGURE 4.5 General shapes of the four classes of
adsorption isotherms.

Equation 4.18 is valid only within this limited range.

The L-Curve is by far the curve most commonly encountered. The L-Curve is primarily described mathematically by either the Bemmelen-Freundlich model (i.e. the Freundlich Equation) or the Langmuir equation. The Freundlich equation is usually expressed as:

$$C_s = KC_e^n \tag{4.19}$$

where

K = constant

n = constant

or by the logarithmic form of Equation 4.19:

$$\log C_s = (1/n)\log C_e + \log K \tag{4.20}$$

The Freundlich equation is a popular expression for the L-Curve because the presence of the constants K and n makes this equation especially applicable for the curve-fitting over a very wide range of C_t values. This feature is useful since the free energy of adsorption in soils typically displays a geometric progression; in other words as C_t increases, the energy of adsorption decreases. However, it is important to note that the use of Equations 4.19 and 4.20 is valid only within the range of C_t values used to determine K and n. Accuracy of the curve beyond this range cannot be guaranteed.

The two standard forms of the Langmuir equation are:

$$C_e/C_s = (1/kb) + (C_e/b) \tag{4.21}$$

and

$$C_s = kbC_e/(1+kC_e) \tag{4.22}$$

where

k = constant

b = constant

The Langmuir equation was developed in 1918 to describe the adsorption of gases by solids. The equation was developed on the basis of assumptions that (a) the free energy of adsorption was constant (b) there are no interactions between the adsorbing molecules, and (c) the surface of a solid possesses a finite number of adsorption sites. While the last assumption generally holds for soils, the first two usually do not. Soils contain a wide array of minerals and organic matter that contribute adsorption sites possessing varying adsorption characteristics. Nonetheless, the Langmuir equation is useful for the empirical "fitting" of adsorption data[33,34], due to the presence of the constants k and b. However, this equation should not be expected to "fit" all adsorption data.

Several researchers have interpreted empirical data and concluded that two different types of surface sites are present in some soils and are responsible for adsorbing certain elements[35]. The adaptation of Langmuir's equation to the adsorption of an element by a soil with two primary adsorbing surfaces is:

$$C_s = k_1 b_1 C_e / (1 + k_1 C_e) + k_2 b_2 C_e / (1 + k_2 C_e) \qquad (4.23)$$

where

$b_1 \& b_2$ = constants, the maximum quantities
of the element that can be adsorbed
by the two surfaces
$k_1 \& k_2$ = constants

Equation 4.23 has been used extensively for P adsorption data and for Zn and Mn[35].

One question usually arises with regard to the use of the Langmuir and Freundlich equations: which is the appropriate one to use? The only real advantage of the Langmuir equation is that it predicts an adsorption maxima, whereas the Freundlich equation does not. In fact, the Freundlich equation can be derived from the Langmuir equation by assuming that the heat of adsorption is a logarithmic function of the surface coverage[35]. In general, the equation to utilize is the one that best fits the adsorption data within the range of C_t values utilized to derive the equation's constant(s).

It is most important to recognize that the Langmuir equation alone cannot be used to determine whether adsorption or mineral precipitation is occurring in a soil system[36]. The same holds generally for the Freundlich equation. Mineral precipitation generally results in a statistically significant, linear correlation of variables in a Langmuir plot when the reacting element's concentrations are much larger than the threshold values needed to initiate mineral precipitation. As a result, one cannot distinguish the contribution of either reaction without additional independent evidence.

KINETIC ADSORPTION MODELS

A review of the scientific literature dealing with input parameters to computer models will reveal that there are two basic approaches that are utilized to mathematically account for adsorption-desorption processes in soil systems, and both involve some form of the parameters identified in Equation 4.2. The first approach involves the direct use of K_d as an input parameter. The principle assumption that is made when K_d is used in modeling chemical transport is that the adsorption and desorption rates k(des) and k(ads) from Equation 4.2 have reached equilibrium conditions quickly. In other words, the change in the solution-phase concentration (C_e) over time is zero. Most researchers have treated the adsorption of organics in environmental systems and models as an environmental process that reaches equilibrium rapidly relative to the movement of water in soil pores. K_d has been successfully utilized as an input parameter in a number of equations and models describing the retardation of chemicals in soils.

The second approach addresses those situations when the adsorption-desorption processes for a chemical may be slow and, therefore, equilibrium may not be attained; the desorption process is the one which is slower and incomplete relative to the adsorption process. To account for this situation, a rate constant(s) is utilized to describe a change in either C_s or C_e with time. Adsorption models utilizing rate constants in this manner are referred to as kinetic adsorption models.

The potential usefulness of kinetic adsorption models can be illustrated by the following conceptual model. At a hazardous waste site, soils containing various levels of a chemical will come into contact initially with chemical-free water percolating through soil pores; this condition can be described by the following equation:

$$C_s(\text{Initial}) \rightleftharpoons C_e(\text{Initial}) \qquad (4.24)$$

The magnitude of the arrows indicate the relative magnitude of the flux of chemical out of each phase which should occur before equilibrium is attained. An analysis of Equation 4.24 will reveal that the flux from the adsorbed phase to the solution phase is the predominant flux. Therefore,

$$C_s(\text{Initial})/C_e(\text{Initial}) >> C_s(\text{equilibrium})/C_e(\text{equilibrium}) \qquad (4.25)$$

and

$$K_d(\text{initial}) >> K_d(\text{equilibrium}) \qquad (4.26)$$

If the process described by Equation 4.24 reaches equilibrium quickly relative to the rate of water movement in soil pores, then the condition described in Equation 4.24 quickly becomes the condition described in Equation 4.2. However, water movement in the pores of soils containing a chemical may occur at a rate such that fluxes between phases do occur but the equilibrium process described in Equation 4.2 is not attained; this nonequilibrium condition (noneq) can be described by the following equation:

$$C_s(\text{noneq}) \rightleftharpoons C_e(\text{noneq}) \qquad (4.27)$$

Equation 4.27 describes a condition observed at times because the desorption process is slower relative to the adsorption process.

Equation 4.25 can be expanded to incorporate the condition described by

Equation 4.27:

$$C_s(\text{initial})/C_e(\text{initial}) > C_s(\text{noneq})/C_e(\text{noneq}) \qquad (4.28)$$

$$> C_s(\text{equilibrium})/C_e(\text{equilibrium})$$

and

$$K_d(\text{initial}) > K_d(\text{noneq}) > K_d(\text{equilibrium}) \qquad (4.29)$$

An analysis of Equation 4.28 and 4.29 will reveal that the magnitude of K_d decreases as equilibrium conditions are approached, and the smallest K_d occurs under equilibrium conditions. If nonequilibrium prevails, then K_d (equilibrium) is an underestimation of the K_d (nonequilibrium); as a result, the calculated chemical travel times and plume velocities should be overestimations of the actual values, and the actual time a chemical would need in order to migrate a certain distance should be significantly greater.

A detailed discussion of kinetic adsorption models, their uses, and recent applications is beyond the scope and intent of this text. The purpose of this discussion was to introduce the concept of these models. A recent review[35], which surveys the more important first-order kinetic sorption-desorption models and their recent applications, should be studied if more information regarding these models is needed.

ELEMENT MOBILITY IN SOIL

An analysis of the information presented in Chapter 3 and in the earlier sections of Chapter 4 will reveal that the mobility of an element in soil is a function of several reactions. Element mobility is directly affected by the extent of fixation, positive and negative adsorption, exclusion, complex formation, and by reaction kinetics. These in turn are influenced by soil physical and chemical properties.

The published literature abounds with studies measuring the migration of elements in soil and identifying those soil physical and chemical properties that affect element migration. In order to briefly highlight the general findings of the published literature, this section will focus on one indepth study involving the migration of eleven elements—As, Be, Cd, Cr, Cu, Hg, Ni, Pb, Se, V and Zn—in ten subsoils representing seven soil orders[37]. The physical and chemical properties of these soils are listed in Table 4.5. One-half to one pore volume of a leachate containing from 70 to 120 ppm of these elements was leached each day for about 25 days through 5 cm x 10 cm soil columns. Figure 4.6 illustrates the results of these tests. An element was classified as "high mobility" if it was detected in the column effluent almost immediately after column leaching was initiated; this condition indicated that the element was weakly retained by the soil. An element was classified as "moderate mobility" if the initial detection of the element in column effluent occurred at some time after column leaching was initiated. An element was classified as "low mobility" if the ratio c/c_0, where c is the concentration of the element in column effluent and c_0 is the initial element concentration in leachate, was generally less than 0.1; in other words, the element was present in leachate in very low concentrations or was absent after extended leaching.

An analysis of the information presented in Figure 4.6 and Table 4.5 will reveal several general trends. First, cations exhibit low mobility in clay and silty clay soils. An analysis of the published literature will reveal that as surface area and the clay content of soil increases, soils' ability to retain cations generally will increase.

TABLE 4.5 Physical and Chemical Properties of Ten Soils Utilized in Determining the Relative Mobilities of Eleven Elements[37]

Soil series	Texture class USDA	pH soil paste	Cation exchange capacity meq/100g	Column bulk density g/cm³	Porosity
Davidson	clay	6.2	9	1.89	0.476
Molokai	clay	6.2	14	1.44	0.429
Nicholson	silty clay	6.7	37	1.53	0.460
Fanno	clay	7.0	33	1.48	0.484
Mohave (Ca)	clay loam	7.8	12	1.54	0.446
Ava	silty clay loam	4.5	19	1.45	0.478
Anthony	sandy loam	7.8	6	2.07	0.360
Mohave	sandy loam	7.3	10	1.78	0.365
Kalkaska	sand	4.7	10	1.53	0.404
Wagram	loamy sand	4.2	2	1.89	0.378

(Reprinted with permission of Williams & Wilkens).

Second, cations usually exhibit moderate and high mobility in sandy, loamy sand, and sandy loam soil. Since the surface area and clay content of these soil types are relatively low, cation retention for these soil types is also expected to be low.

Third, cations can exhibit low, moderate, or high mobility in soils with intermediate textures. The contrast between the two intermediate textured soils used in Figure 4.6 is especially interesting. Ca-Mohave clay loam has a reasonable amount of clay and a relatively high pH of 7.8 (see Table 4.5). As a result, cation fixation and cation-hydroxide complexes are expected to occur; as a result, cation mobility should generally be low. Although Ava silty clay loam also has a reasonable amount of clay, it has a relatively low pH of 4.5. Cation fixation and cation-hydroxide complexes probably will

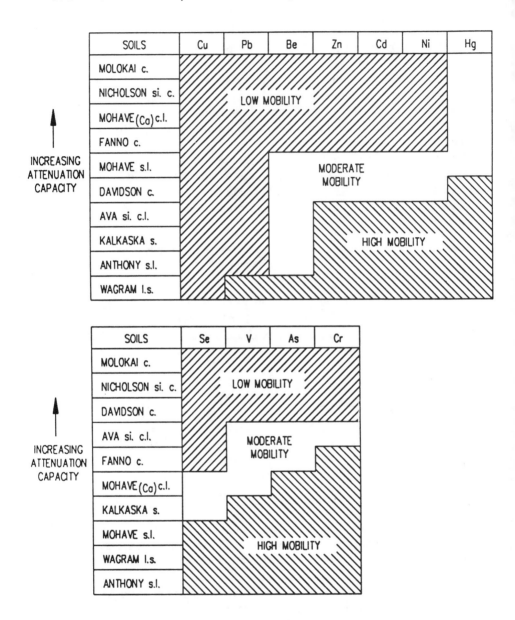

FIGURE 4.6   The relative mobilities of eleven
elements in ten soils. [37] (Reprinted
with permission of Williams & Wilkins).

not occur at this pH for most cations; as a result, cation mobility should be relatively higher in Ava than in Ca-Mohave.

Fourth, anions usually exhibit relatively low mobility in clay and silty clay soils and moderate to high mobility in other soil types. An analysis of the published literature will reveal that as the surface area and clay content of soils increase, the extent of anion adsorption also increases; the net result is a decrease in mobility. Also, the published literature reveals that as the oxide content of soils increase, anion adsorption should increase with a net decrease in mobility.

In summary, general trends in element mobility in different soils can be identified. However, exceptions to the general trend will occur. Exceptions will occur because no one single soil property or element property can be utilized to rank elements and soils with respect to mobility.

PREDICTING ELEMENT MOBILITY

Gross predictions of element mobility and concentrations in soil-water systems can be made via three general approaches. In the first approach, published studies on the migration and concentrations of an element are retrieved. These studies usually involve laboratory, greenhouse, or field studies performed under a diverse variety of environmental conditions and utilizing soils possessing a wide spectrum of physical/chemical properties. The mobility and concentrations of the element in the soil of concern usually parallel those in published studies if soil and environmental conditions in the published studies match those of the site of concern. Due to the vast number of published studies on element mobility in soil, this section will not address this approach. However, an understanding of the information provided in this text should give the basic background needed in order to attempt this approach.

In the second approach, element mobility or concentrations in water are generally estimated via generic mathematical models utilizing a wide spectrum of equations ranging from simple analytic to complex numerical ones. These models are based on assumptions that greatly simplify the computations. However, they do not account for many of the reactions discussed in this text. Therefore, extreme care must always be used when one interprets data on element mobility generated by these models. Two simple examples, the Analytical Leaching Model and the Rapid Assessment Nomograph, are discussed in this section.

In the third approach, element mobility or concentrations are estimated via regression equations. These equations were developed by measuring mobility or concentrations of specific elements in many soils with different physical and chemical properties. The use of these equations require information on the physical and chemical properties of the soil of concern. Although much is known regarding those soil properties that affect element mobility, very little work has been done to develop quantitative regression equations that relate soil properties to element mobility in a manner that is useful in hazardous waste site investigations. Regression equations for Cd, Ni, and Zn are discussed in this section.

An analysis of the models discussed here will reveal that none of these approaches identifies diffusion as an important process governing the migration rate of chemicals. Diffusion is a spontaneous process that results in the movement of a chemical from one region of a soil system to another. Diffusion is caused by the random thermal motions of the chemical and the solvent. It is expressed quantitatively as:

$$J_c = - D_c (dc/dx) \qquad (4.30)$$

where:

J_c = Diffusive flux of chemical c (moles/cm^2 sec).

D_c = Diffusion coefficient for c (cm^2/sec).

dc/dx = Concentration gradient of c in the x direction (moles/cm^4).

A net migration of c occurs by diffusion from a concentrated to a dilute region. This net migration is a statistical phenomenon because there is a greater probability of the chemical moving from a concentrated to a dilute region than visa-versa. In general, the magnitude of D_c for small ions and molecules in water is approximately 10^{-5} cm^2/sec; for a globular protein with a molecular weight of 15,000, D_c is approximately 10^{-6} cm^2/sec.

In general, diffusion is not responsible for the migration of chemicals for significant distances in soil systems. To illustrate this point, the time for a small molecule to diffuse one meter can be estimated:

$$(x_e)^2 = 4D_c t \qquad (4.31)$$

where:

x_e = distance travelled along the x axis.
t = time.

Equation 4.31 quantifies an extremely important and fundamental relationship: the distance a chemical diffuses is proportional to the square root of the diffusion coefficient and time. The diffusion time increases with the square root of the diffusion distance. The time needed for a small molecule to travel one meter by diffusion is $(100)^2/(4)(10^{-5})$ or 2.5×10^8 sec or approximately eight years. The actual diffusion time would be longer in a soil system because some adsorption of the molecule would occur and increase the diffusion time.

Cd, Ni, Zn Regression Equations.   A rare example of simple field-oriented equations for use by soil scientists and environmental engineers was developed for Cd, Ni, and Zn on eleven soils and eight landfill leachates[38]. The physical and chemical properties of the soils utilized in the study are listed in Table 4.6. Soils were packed into 5.2 cm I.D. x 10 cm columns. Cd, Ni, and Zn were added to municipal solid waste leachate at either 100 ppm or 1 millimole liter^{-1} concentrations. Leachates were perfused through the soil columns at a constant rate of 5-10 cm/day darcian velocity (i.e. steady state flow). Regression equations of the following form were developed:

$$V_{c/c_o} = (V/25) (a_1X_1 + a_2X_2 + \cdots + A_nX_n + A_n + 1) \qquad (4.32)$$

where

V_{c/c_o} = velocity of c/c_o, where c is the concentration of the element in effluent and c_o is the initial element concentration, in cm/day.

V = pore water velocity

a_i = the regression coefficient for X_i, listed in Tables 4.7, 4.8, and 4.9.

X_i = the ith independent variable, where i = 1,2,3....N, listed in Tables 4.7, 4.8, and 4.9.

These equations were based on the Lapidus-Amundson model, for which apparent diffusion and forward and backward sorption coefficients were best fitted to the available experimental data. The utilization of Equation 4.32 can best be shown by an application assuming the following soil properties and leachate composition:

Soil: 4% FeO (free iron oxides)
                30% clay
                20% sand
                50% silt
                0.1 meter ^{3}/meter^3 volumetric water content
                0.15 cm/day average darcian velocity

Leachate: 0.11% TSS
                1.09% TOC

This exercise will calculate the time for a soil solution with 3 ppm Cd to migrate to a 5 meter depth, assuming the Cd leachate concentration is 15 ppm and constant water flow; c/c_o and V can be estimated:

$$c/c_o = 3/15 = 0.2$$
$$V = 0.15/0.1 = 1.5 \text{ cm/day}$$

TABLE 4.6 Physical and Chemical Properties of Eleven Soils Utilized to Develop Predictive Equations for Cd, Ni, and Zn Mobility.[38,39]

Soil Series	% FeO	Soil Paste pH	Cation Exch. Capacity meq/100g	Column Bulk Density g/cm³
Davidson clay	17.0	6.4	9	1.40
Nicholson silty clay	5.6	6.7	37	1.53
Fanno clay	3.7	7.0	33	1.48
Mohave(Ca) clay	2.5	7.8	12	1.54
Chalmers silty clay	3.1	6.6	22	1.60
Ava silty clay	4.0	4.5	19	1.45
Anthony silt loam	1.8	7.8	10	1.87
Mohave silt loam	1.7	7.3	10	1.78
Kalkaska sand	1.8	4.7	6	1.53
Wagram sand	0.6	4.2	2	1.89
River sand	0.3	7.2	2	1.73

Substituting the variables and regression coefficients for a c/c_o of 0.2 from Table 4.7 into Equation 4.32 yields:

$$V_{0.2} = [1.5/25] \; [(31.07)(30)^{-1} + \qquad (4.33)$$
$$(-0.227) \; (20) + (0.114)(10^{-2}) \; (20)^2 +$$
$$(0.111)(10^{-1})(4)^2 + (9.385) \; (4)^{-1} +$$
$$(89.19)(0.11) + (-216.4)(0.11)^2 +$$
$$(0.37)(0.09) + (-2.015)]$$

and

$$V_{0.2} = 0.281 \text{ cm/day}$$

The time for the soil solution containing 3 ppm Cd to reach the 5 meter depth would be $(500 \text{ cm})(0.281 \text{ cm/day})^{-1}$ or 1780 days or 4.87 years. The profile for Cd with depth (i.e. a graphical plot of depth as c/c_o) can be produced by calculating $V_{0.1}$ and $V_{0.3}$ through $V_{0.9}$.

Analytical Leaching Model (ALM). The ALM is a model[5] that can estimate the velocity of a chemical in the unsaturated zone, V_c, where:

$$V_c = V_w / (WC \times K) \qquad (4.34)$$

TABLE 4.7 Regression Coefficients for the Velocities of c/c_0 for Cd for Use in Eq. 4.32.[38] (Reprinted with permission from the American Society of Agronomy, Inc., Crop Science Society of America, Inc., and the Soil Science Society of America, Inc.)

Variables*	Relative concentration, c/c_0		
	0.1	0.2	0.3
$(\% \text{ Clay})^{-1}$	32.02	31.07	30.65
% Sand	−0.233	−0.227	−0.223
$(\% \text{ Sand})^2$	$(0.117)(10)^{-2}$	$(0.114)(10)^{-2}$	$(0.112)(10)^{-2}$
$(\% \text{ FeO})^2$	$(0.116)(10)^{-1}$	$(0.111)(10)^{-1}$	$(0.106)(10)^{-1}$
$(\% \text{ FeO})^{-1}$	9.846	9.385	9.057
% TSS	90.83	89.19	87.32
$(\% \text{ TSS})^2$	−218.6	−216.4	−212.3
% TOC	0.322	0.370	0.339
Constant	−2.090	−2.015	−1.923
r^2	0.844	0.843	0.844

Variables*	0.4	0.5	0.6
$(\% \text{ Clay})^{-1}$	29.54	29.91	29.52
% Sand	−0.217	−0.217	−0.214
$(\% \text{ Sand})^2$	$(0.109)(10)^{-2}$	$(0.108)(10)^{-2}$	$(0.107)(10)^{-2}$
$(\% \text{ FeO})^2$	$(0.916)(10)^{-2}$	$(0.101)(10)^{-1}$	$(0.996)(10)^{-2}$
$(\% \text{ FeO})^{-1}$	8.817	8.532	8.293
% TSS	92.20	84.13	83.06
$(\% \text{ TSS})^2$	−234.1	−205.1	−203.0
% TOC	−0.148	0.442	0.494
Constant	−2.069	−1.760	−1.702
r^2	0.845	0.844	0.842

Variables*	0.7	0.8	0.9
$(\% \text{ Clay})^{-1}$	29.18	28.65	28.10
% Sand	−0.212	−0.209	−0.204
$(\% \text{ Sand})^2$	$(0.106)(10)^{-2}$	$(0.105)(10)^{-2}$	$(0.102)(10)^{-2}$
$(\% \text{ FeO})^2$	$(0.970)(10)^{-2}$	$(0.970)(10)^{-2}$	$(0.939)(10)^{-2}$
$(\% \text{ FeO})^{-1}$	8.048	7.761	7.349
% TSS	81.63	79.10	77.81
$(\% \text{ TSS})^2$	−199.8	−194.9	−193.9
% TOC	0.511	0.559	0.426
Constant	−1.585	−1.388	−1.307
r^2	0.840	0.838	0.838

* For clay, sand, and FeO, 1% is equivalent to 10 grams/kg; for TSS and TOC, 1% is equivalent to 10 grams/liter.

TABLE 4.8   Regression Coefficients for the Velocities of c/c_o for Ni for Use in Eq. 4.32.[38] (Reprinted with permission from the American Society of Agronomy, Inc., Crop Science Society of America, Inc., and the Soil Science Society of America, Inc.)

Variables*	Relative concentration, c/c_o		
	0.1	0.2	0.3
(% Clay)$^{-1}$	17.18	15.71	14.72
(% Sand)$^{-1}$	29.29	27.37	26.28
% FeO	0.350	0.333	0.323
(% FeO)$^{-1}$	7.541	7.026	6.775
% TSS	82.19	76.34	73.00
(% TSS)^2	−266.9	−246.5	−234.4
% TOC	−40.13	−37.44	−36.28
(% TOC)^2	132.2	122.9	120.5
Constant	−6.791	−6.298	−6.052
r^2	0.908	0.905	0.907
Variables*	0.4	0.5	0.6
(% Clay)$^{-1}$	14.04	13.460	13.09
(% Sand)$^{-1}$	24.78	23.85	23.06
% FeO	0.312	0.304	0.297
(% FeO)$^{-1}$	6.413	6.200	5.972
% TSS	71.06	69.32	67.59
(% TSS)^2	−228.6	−222.9	−217.3
% TOC	−35.63	−34.71	−34.20
(% TOC)^2	117.5	114.8	113.6
Constant	−5.711	−5.534	−5.350
r^2	0.892	0.885	0.874
Variables*	0.7	0.8	0.9
(% Clay)$^{-1}$	12.70	12.62	12.85
(% Sand)$^{-1}$	21.49	20.89	19.00
% FeO	0.290	0.280	0.263
(% FeO)$^{-1}$	5.734	5.448	5.013
% TSS	65.56	66.12	65.15
(% TSS)^2	−210.0	−212.7	−210.5
% TOC	−32.33	−32.19	−32.57
(% TOC)^2	106.4	112.7	116.6
Constant	−5.195	−5.130	−4.927
r^2	0.861	0.839	0.795

* For clay, sand, and FeO, 1% is equivalent to 10 grams/kg; for TSS and TOC, 1% is equivalent to 10 grams/liter.

TABLE 4.9 Regression Coefficients for the Velocities of c/c_o for Zn for Use in Eq. 4.32.[38] (Reprinted with permission from the American Society of Agronomy, Inc., Crop Science Society of America, Inc., and the Soil Science Society of America, Inc.)

Variables*	Relative concentration, c/c_o		
	0.1	0.2	0.3
% Sand	$(0.210)(10)^{-1}$	$(0.173)(10)^{-1}$	$(0.161)(10)^{-1}$
$(\% Silt)^2$	$(0.657)(10)^{-3}$	$(0.591)(10)^{-3}$	$(0.570)(10)^{-3}$
$(\% FeO)^{-1}$	2.784	2.747	2.697
$(\% FeO)^2$	$(0.373)(10)^{-2}$	$(0.336)(10)^{-2}$	$(0.324)(10)^{-2}$
% TOC	−15.38	−14.92	−14.86
$(TOC)^2$	82.37	80.25	79.18
% TSS	67.11	60.90	58.93
$(\% TSS)^2$	−188.2	−167.5	−162.1
Constant	−6.44	−5.88	−5.68
r^2	0.841	0.844	0.846

Variables*	0.4	0.5	0.6
% Sand	$(0.149)(10)^{-1}$	$(0.136)(10)^{-1}$	$(0.111)(10)^{-1}$
$(\% Silt)^2$	$(0.550)(10)^{-3}$	$(0.526)(10)^{-3}$	$(0.482)(10)^{-3}$
$(\% FeO)^{-1}$	2.660	2.637	2.638
$(\% FeO)^2$	$(0.310)(10)^{-2}$	$(0.288)(10)^{-2}$	$(0.255)(10)^{-2}$
% TOC	−14.72	−14.75	−14.57
$(\% TOC)^2$	78.42	77.79	77.12
% TSS	56.97	54.66	51.12
$(\% TSS)^2$	−156.3	−148.2	−135.9
Constant	−5.48	−5.27	−4.92
r^2	0.848	0.847	0.848

Variables*	0.7	0.8	0.9
% Sand	$(0.954)(10)^{-2}$	$(0.803)(10)^{-2}$	$(0.487)(10)^{-2}$
$(\% Silt)^2$	$(0.466)(10)^{-3}$	$(0.443)(10)^{-3}$	$(0.398)(10)^{-3}$
$(\% FeO)^{-1}$	2.660	2.648	2.698
$(\% FeO)^2$	$(0.254)(10)^{-2}$	$(0.236)(10)^{-2}$	$(0.211)(10)^{-2}$
% TOC	−14.76	−14.96	−15.82
$(\% TOC)^2$	76.71	76.57	78.93
% TSS	51.03	49.37	46.16
$(\% TSS)^2$	−137.1	−132.3	−122.0
Constant	−4.82	−4.61	−4.22
r^2	0.848	0.848	0.846

* For clay, sand, and FeO, 1% is equivalent to 10 grams/kg; for TSS and TOC, 1% is equivalent to 10 grams/liter.

and

V_w = Amount of annual percolating soil water (cm/yr).
WC = Soil moisture content at field capacity (ml/ml).
K = Retardation factor for the chemical.

The retardation factor K relates the movement of the chemical to the amount of annual percolating soil water and can be redefined as:

$$K = 1 + (bK_d / WC) \qquad (4.35)$$

where:

b = Soil bulk density (gram/cm^3).
K_d = Adsorption or distribution coefficient of the chemical (ml/gram).

The travel time, T in years, through the unsaturated zone can be estimated by dividing the depth of the unsaturated soil zone in cm, D, by the chemical's velocity, V_c:

$$T = D / V_c \qquad (4.36)$$

With regard to the accuracy of the ALM, eight comparisons between ALM predictions and leaching constants which were determined directly from field data were made[5]. The leaching constants, L, where:

$$L = 1 / T \qquad (4.37)$$

were calculated by combining Equations 4.34 through 4.37 to obtain:

$$L = V_w(WC)^{-1} (1 + [bK_d/WC])^{-1}D^{-1} \qquad (4.38)$$

In seven comparisons, the ALM-derived leaching constants were within an order of magnitude of the leaching constants determined from five field studies. In the eighth comparison, the ALM output revealed that the K_d for technetium, which was obtained from one published study, could not be an accurate value. It is most important to recognize that the ALM's input parameters for these comparisons were median values obtained from the scientific literature; these were not site-specific values. Significantly closer agreement between calculations and observed leaching constants should occur when the ALM's input parameters are derived from site-specific measurements.

Concentration, Distance, and Time Nomograph. This nomograph (i.e. CDT Nomograph) was developed to provide rapid, approximate solutions for leachate plume movement and concentrations in groundwater at a point directly downgradient from a source[40]. This nomograph involves relatively simple computations which take into account dilution-dispersive mixing and retardation. Although the author has used this approach to a limited extent, he believes it holds much promise as a rapid method for deriving general approximations.

The CDT Nomograph is based upon several assumptions. First, the groundwater flow regime is saturated. Second, groundwater flow is continuous and uniform in direction and velocity. Third, the chemical source is a point in plan view. Fourth, the chemical is evenly distributed over the vertical dimension (aquifer thickness). Fifth, the source supplies a constant mass flow rate. Sixth, there is no dilution of the plume by recharge outside the source area. Seventh, all aquifer properties are homogeneous and isotropic. Eighth, the aquifer is infinite in areal extent.

Three scale factors are utilized in Figure 4.7 to produce a one-dimensional graphical estimation of concentration along the axis in the direction of groundwater flow. The Wilson and Miller Equation was reformulated to derive three scale factors, X_D, T_D, and Q_D, where:

$$X_D = D_x / V \tag{4.39}$$
$$T_D = R_d D_X / V^2 \tag{4.40}$$
$$Q_D = nm(D_X D_Y)^{1/2} \tag{4.41}$$

and

D_X = Longitudinal dispersion coefficient (ft²/day).
D_Y = Transverse dispersion coefficient (ft²/day).
m = Effective aquifer thickness or zone of mixing (ft).
n = Effective porosity of the aquifer or zone of mixing (ft).
R_d = Retardation factor.
V = Velocity of groundwater flow within voids, where:
$$V = KI / n$$
and

K = Hydraulic conductivity.

I = Gradient of groundwater flow.
X_D = Longitudinal dispersivity (ft).

Other parameters identified in Figure 4.7 include:

C_o = Initial concentration at source (mg/liter).
Q = Volume flow rate of source (ft³/day).
QC_o = Mass flow rate of the chemical (ft³/day)(mg/liter).
t = Time from the beginning of source flow to date of sampling (days).
x = Distance along flow pathway from source to sampling point (ft).

The use of the CDT Nomograph will now be illustrated. For the purposes of calculating concentrations, distance, and time, we will assume that:

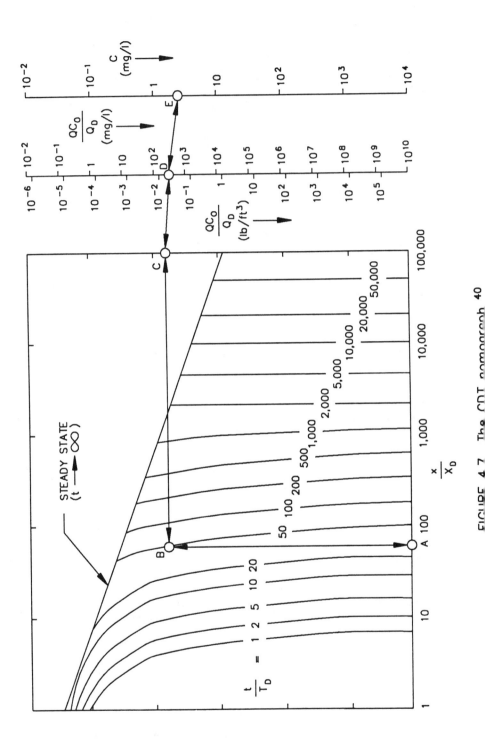

FIGURE 4.7 The CDT nomograph 40

$$
\begin{aligned}
C_o &= 31 \text{ mg/liter} \\
D_X &= 105 \text{ ft}^2/\text{day} \\
D_Y &= 21 \text{ ft}^2/\text{day} \\
m &= 110 \text{ ft} \\
n &= 0.35 \\
Q &= 26{,}800 \text{ ft}^3/\text{day} \\
R_d &= 1 \\
V &= 1.5 \text{ ft/day}
\end{aligned}
$$

The concentration of chemical Z at a distance x of 4200 ft from the source at time t of 2300 days will be estimated. The parameters x/X_D, t/T_D, and QC_o/Q_D are calculated:

$$
\begin{aligned}
x/X_D &= 4200 \text{ ft } / 70 \text{ ft } = 60 \\
t/T_D &= 2300 \text{ days } / 46.7 \text{ days } = 50 \\
QC_o/Q_D &= (26{,}800 \text{ ft}^3/\text{day})(31 \text{ mg/liter}) / (1800 \text{ ft}^3/\text{day}) \\
&= 460 \text{ mg/liter}
\end{aligned}
$$

The quantity x/X_D (i.e. 60) can be found at point A in Figure 4.7. The calculated quantity for t/T_D (i.e. 50) indicates that the t/T_D curve labelled "50" in Figure 4.7 will be used to calculate the concentration of chemical Z. A vertical line from point A is drawn to the point where it intercepts the 50 t/T_D curve at point B; then, a horizontal line from point B is drawn to the point where it intercepts the right edge of the rectangle containing the t/T_D curves at point C.

After point C has been located, the quantity QC_o/Q_D (i.e. 460 mg/liter) is found on the vertical QC_o/Q_D scale adjacent to the rectangle containing the t/T_D curves; this quantity is located at point D. Then, a line which is drawn through points C and D should intersect the vertical concentration scale which is to the right of the vertical QC_o/Q_D scale at point E. Point E identifies the concentration of chemical Z at a distance of 4200 ft from a source after 2300 days.

To use the CDT Nomograph to estimate distance, points B, C, D, and E are located first in order to identify point A; then, the value of x/X_D identified by point A is multiplied by X_D to determine the distance x. To use the CDT Nomograph to estimate time, points A, C, D, and E are located first in order to identify point B; then, the value of the t/T_D curve identified by point B is multiplied by T_D to determine the time t.

The Retardation Equation.   The most commonly accepted and utilized equation for grossly estimating the migration rate of a chemical in a soil-groundwater system is the retardation equation:

$$
V_c = V [1 + K_d (b / P_T)]^{-1} \tag{4.42}
$$

where:

V_c = Velocity of the chemical at the point where $c/c_o = 0.5$.

V = Average linear velocity of groundwater.

K_d = Adsorption or distribution coefficient.

b = Soil bulk density.

P_T = Soil total porosity.

The term $[1 + K_d(b/P_T)]$ is known as the retardation factor. This term should not be confused with the soil TLC R_f (retardation factor) because they are not equivalent parameters.

Equation 4.42 can provide reasonable estimates of the velocity of an inorganic chemical; however, it is important to note that the reactions discussed in Chapter 3 are not directly accounted for in this equation.

Rapid Assessment Nomograph.   This graphical technique was developed as an integrated methodology for assessing chemical movement through both the unsaturated and saturated zone[41]. Although the author has limited exposure to this method, it will be discussed because it appears to be a potentially useful tool. The technique was designed to allow emergency response personnel to make a first-cut, order-of-magnitude estimate of the potential extent of chemical migration/contamination from a waste site or chemical spill within a 24 hour emergency response time frame. The technique was not intended to provide a definitive, in-depth analysis of the complex fate and transport processes of chemicals in the subsurface environment.

This section will not discuss the use of this model or the input parameters needed to utilize the nomograph. However, the reader is encouraged to study and utilize it because it does provide reasonable estimates. The procedures developed in this model are based on a one-dimensional transport equation for flow through a porous medium. The equation considers dispersion, advection, equilibrium adsorption (linear isotherm) and degradation/decay (first order kinetics). Analytical solutions to the transport equation include both continuous (step function) and pulsed inputs of contaminants. The pulse solution was used to simulate short-term releases such as might occur from a spill or tank leak. The nomograph was developed from the continuous input solution but can also be used on pulse problems by subtracting the solutions to two continuous inputs lagged by the pulse duration.

Time and resource limitations expected during an emergency response have required a number of simplifying assumptions in the assessment nomograph; additional simplifications may be needed by the user due to limited data and information available at a particular emergency response site. The major assumptions incorporated into the assessment nomograph are as follows:

1. Homogenous and isotropic properties are assumed for both the unsaturated and saturated zones (or media).
2. Steady and uniform flow is assumed in both the unsaturated and saturated zones.
3. Flow and contaminant movement are considered only in the vertical direction in the unsaturated zone and the horizontal direction in the saturated zone.
4. All contaminants are assumed to be water soluble and exist in concentrations that do not significantly affect water movement.

The nomograph predicts contaminant concentrations as functions of both time and location in either the unsaturated or saturated zone. Separate computations, parameter estimates, and use of the nomograph are required for each zone. The prediction requires evaluation of four dimensionless quantities (A_1, A_2, B_1, and B_2) and subsequent evaluation of the result (C/C_o) through use of the nomograph. The parameters required for this procedure are: initial contaminant concentration, dispersion coefficient, average interstitial pore water velocity, degradation rate coefficient, soil bulk density, soil water content or effective porosity, and partition coefficient. Extensive guidelines for evaluation of these parameters are provided.

REFERENCES

1. Murray, K., and Linder, P. W. Fulvic Acid: Structure and Metal Binding. I. A Random Molecular Model. Journal of Soil Science **34**:511-523 (1983).
2. Buckman, H.O., and Brady, N. C. The Nature and Properties of Soil. New York: The Macmillan Co. (1969).
3. Sethi, R. K., and Chopra, S. L. Adsorption and the Behavior of Pesticides in Soils. Pesticides **11**:15-25 (1977).
4. Greenland, D. J., and Hayes, M. H. B. (eds.) The Chemistry of Soil Processes. New York: John Wiley & Sons (1981).
5. Baes, C. F. III, and Sharp, R. D. A Proposal for Estimation of Soil Leaching and Leaching Constants for Use in Assessment Models. Journal of Environmental Quality **12(1)**:17-28 (1983).
6. Fairbridge, R. W., and Finkl, C. W. Jr. (eds.) The Encyclopedia of Soil Science: Part 1. Stroudsburg, PA: Dowden, Hutchinson, & Ross, Inc. (1979).
7. Stumm, W., and Morgan, J. J. Aquatic Chemistry. New York: John Wiley & Sons, Inc. (1981).
8. Baes, C. F. III, and Mesmer, R. E. The Hydrolysis of Cations. New York: John Wiley & Sons, Inc. (1976).
9. Amacher, M. C. Determination of Ionic Activities in Soil Solutions and Suspensions: Principle Limitations. Soil Science Society of America Journal **48**:519-524 (1984).
10. Baham, J. Prediction of Ion Activities in Soil Solutions: Computer Equilibrium Modeling. Soil Science Society of America Journal **48**:525-531 (1984).

11. Sposito, G. The Future of an Illusion: Ion Activities in Soil Solutions. Soil Science Society of America Journal 48:531-536 (1984).
12. Sposito, G. The Surface Chemistry of Soils. New York: Oxford University Press (1984).
13. Kinniburgh, D. G., Jackson, M. L., and Syers, J. K. Adsorption of Alkaline Earth, Transition, and Heavy Metal Cations by Hydrous Oxide Gels of Iron and Aluminum. Soil Science Society of America Journal 40:796-799 (1976).
14. Means, J. L., Crerar, D. A., and Duguid, J. O. Migration of Radioactive Wastes: Radionuclide Mobilization by Complexing Agents. Science 200:1477-1481 (1978).
15. Killey, R. W., McHugh, J. O., Champ, D. R., Cooper, E. L., and Young, J. L. Subsurface Cobalt-60 Migration From a Low-Level Waste Disposal Site. Environmental Science and Technology 18:148-157 (1984).
16. Martin, L. Y., and Franz, J. A. Effect of Organic Ligands on the Soil Behavior of Technetium-99. Transactions of the American Nuclear Society 35:53 (1980).
17. Toste, A. P., Kirby, L. J., and Pahl, T. R. Role of Organics in the Subsurface Migration of Radionuclides in Ground Water. In Barney, G. S., Navratic, P. D., and Schulz, W. W. (eds). Geochemical Behavior of Disposed Radioactive Waste. ACS Symposium Series 246. Washington, D.C.: American Chemical Society (1984).
18. Lukehart, C. M. Fundamental Transition Metal Organometallic Chemistry. Monterey, CA: Brooks/Cole Publishing Co. (1985).
19. Jorgensen, A. D., Stetter, J. R., and Stamoudis, V. C. Interactions of Aqueous Metal Ions with Organic Compounds Found in Coal Gasification: Model Systems. Environmental Science and Technology 19:919-924 (1985).
20. Stetter, J. R., Stamoudis, V. C., and Jorgensen, A. D. Interactions of Aqueous Metal Ions with Organic Compounds Found in Coal Gasification: Process Condensates. Environmental Science and Technology 19:924-928 (1985).
21. Murray, K., and Linder, P. W. Fulvic Acids: Structure and Metal Binding: II. Predominant Metal Binding Sites. Journal of Soil Science 35:217-222 (1984).
22. Schnitzer, M. and Kerndorff, H. Reactions of Fulvic Acid with Metal Ions. Water, Air, and Soil Pollution 15:97-108 (1981).
23. Parfitt, R. L. Anion Adsorption by Soils and Soil Materials. Advances in Agronomy 30:1-51 (1978).
24. Tullock, R. J., Coleman, N. T., and Pratt, P. F. Rate of Chloride and Water Movement in Southern California Soils. Journal of Environmental Quality 4:127-131 (1975).
25. Thomas, G. W. and Swoboda, A. R. Anion Exclusion Effects on Chloride Movement in Soils. Soil Science 110:163-166 (1970).
26. Smith, S. J. Relative Rate of Chloride Movement in Leaching of Surface Soils. Soil Science 114:259-263 (1972).
27. Ridley, K. J. D., and Bewtra, J. K. Breakthrough of Brine Through Compacted Fine Grained Soils. In Proceedings of the 38th Industrial Waste Conference, May 10-12, 1983, Purdue University, West Lafayette, IN. Boston: Butterworth Publishers (1984).
28. McMahon, M. A., and Thomas, G. W. Chloride and Tritiated Water Flow in

Disturbed and Undisturbed Soil Cores. Soil Science Society of America Proceedings **38**:727-732 (1974).

29. Appelt, H., Holtzclaw, K., and Pratt, P. F. Effect of Anion Exclusion on the Movement of Chloride Through Soils. Soil Science Society of America Proceedings **39**:264-267 (1975).

30. Alther, G., Evans, J. C., Fang, H-Y, and Witmer, K. Influence of Inorganic Permeants Upon the Permeability of Bentonite. *In* Johnson, A. I., Frobel, R. K., Cavalli, N. J., and Peterson, C. B. (eds). Hydraulic Barriers in Soil and Rock. ASTM STP 874. Philadelphia: American Society for Testing and Materials (1985).

31. Peirce, J. J. and Peel, T. A. Effects of Inorganic Leachates on Clay Soil Permeability. *In* Land Disposal of Hazardous Waste. Proceedings of the Eleventh Annual Research Symposium, Cincinnati, OH, April 29-May 1, 1985. EPA-600/9-85-013. Cincinnati, OH: U.S. Environmental Protection Agency (1985).

32. Dixon, J. B., and Weed, S. B. (eds). Minerals in Soil Environments. Madison, WI: Soil Science Society of America (1977).

33. Harter, R. D. Curve-fit Errors in Langmuir Adsorption Maxima. Soil Science Society of America Journal **48**:749-752 (1984).

34. Harter, R. D., and Baker, D. E. Applications and Misapplications of the Langmuir Equation to Soil Adsorption Phenomena. Soil Science Society of America Journal **41**:1077-1080 (1977).

35. Travis, C. C., and Etnier, E. L. A Survey of Sorption Relationships for Reactive Solutes in Soil. Journal of Environmental Quality **10**:8-17 (1981).

36. Veith, J. A. and Sposito, G. On the Use of the Langmuir Equation in the Interpretation of Adsorption Phenomena. Soil Science Society of America Journal **41**:697-702 (1977).

37. Korte, N. E., Skopp, J., Fuller, W. H., Niebla, E. E., and Aleshi, B. A. Trace Element Movement in Soils: Influence of Soil Physical and Chemical Properties. Soil Science: 122:350-359 (1976).

38. Amoozegar-Fard, A., Fuller, W. H., and Warrick, A. W. An Approach to Predicting the Movement of Selected Polluting Metals in Soils. Journal of Environmental Quality **13**:290-297 (1984).

39. Fuller, W. H., Amoozegar-Fard, H., Niebla, E. E., and Boyle, M. Influence of Leachate Quality on Soil Attenuation of Metals. *In* Disposal of Hazardous Waste. Proceedings of the Sixth Annual Research Symposium, Chicago, Ill., March 17-20, 1980. EPA-600/9-80-010. Cincinnati, OH: U.S. Environmental Protection Agency (1980).

40. Pettyjohn, W. A., Prickett, T. L., Kent, D. C., LeGrand, H. E., and Witz, F. E. Predicting Mixing of Leachate Plumes in Groundwater. *In* Land Disposal of Hazardous Waste. Proceedings of the Eighth Annual Research Symposium, March 8-10, 1982, Ft. Mitchell, KY. EPA-600/9-82-002. Cincinnati, OH: U.S. Environmental Protection Agency (1982).

41. Donigian, A., Lo, T. Y. R., and Shanahan, E. Rapid Assessment of Potential Groundwater Contamination Under Emergency Response Conditions. EPA-600/8-83-030. Washington, D.C.: U.S. Environmental Protection Agency (1983).

5

Acids and Bases in Soil

Introduction

A large number of industrial products and waste streams are comprised of acids, while a somewhat smaller number of the products and waste streams are comprised of bases. Spills, leaks, and transportation mishaps may release these acids or bases to soils. All soils are complex acid-base buffer systems that will react with these industrially-derived acids and bases.

This chapter will first discuss the origin of naturally-occurring acids and bases in soil as well as why soil behaves as an acid-base buffer system that resists pH changes. Then, this chapter will discuss how bulk acids and bases, when added to soil, may change soil composition or soil properties. Finally, this chapter will discuss remedial technologies that have been utilized to neutralize the excess acids and bases that have been added to soils.

THE ORIGIN OF HYDROGEN IONS
AND ACIDS IN SOIL

There are several sources for the hydrogen ions and acids that are found in soil. First, ammonium ions are produced naturally in soils during the decomposition of plant residues and humus by heterotrophic microorganisms[1]. Also, ammonium ions are added to agricultural soils in large quantities as fertilizers. Autotrophic bacteria oxidize ammonium ions to form H' and NO_3^-. Arable soils without fertilizer additions usually nitrify between 8 and 40 lb N/acre/yr.

Second, a variety of organic acids are produced through the microbial decomposition of plant residues or through direct release from decaying organic residues[1]. The amount of soil acidity produced in this way varies with local conditions.

Third, the addition of monocalcium phosphate fertilizer to agricultural soil will increase soil acidity[2]. Monocalcium phosphate hydrolyzes in water to yield dicalcium phosphate and phosphoric acid:

$$Ca(H_2PO_4)_2 \longrightarrow CaHPO_4 + H_3PO_4 \qquad (5.1)$$

Fourth, microorganisms that break down plant residue and soil organic matter utilize reduced sulfur (S) in the organic material or sulfate sulfur (SO_4) from the soil system[3]. The organic S is transformed to reduced or incompletely oxidized forms such as sulfides, elemental S, and thiosulfates. These reduced or incompletely oxidized chemicals are then oxidized to (SO_4^{2-}) by aerobic chemautotrophs. The oxidation of the reduced S from the soil organic matter to the sulfate form will inevitably be accompanied by the release of two H^+ ions for each S atom oxidized:

$$H_2S + 2O_2 \longrightarrow 2H^+ + SO_4^{2-} \tag{5.2}$$

$$S + 1.5O_2 + H_2O \longrightarrow 2H^+ + SO_4^{2-} \tag{5.3}$$

If plant uptake of SO_4^{2-} is occurring with plant release of OH^-, then no change in soil acidity will occur; however, if the former process is slow, then SO_4^{2-} may accumulate in soil, and acidity may increase since plants are not releasing OH^- that can neutralize soil H^+.

Fifth, the addition of carbon dioxide (CO_2) to soil water will affect soil acidity[1,2]. CO_2 dissolves in water to produce an acid solution containing H^+, H_2CO_3, HCO_3^-, and CO_3^{2-}. Water that is pure and in equilibrium with atmospheric CO_2 (0.03 percent) will have a pH of approximately 5.6. However, the CO_2 concentration in topsoils is greater because soil organisms are generating CO_2. The CO_2 concentration of soil air varies widely and may occasionally exceed one percent, which would result in a pH of water of approximately 4.9. The total CO_2 produced by topsoils is grossly on the order of $10^{4.4}$ lb/acre/yr.

Sixth, the oxidation of pyrite (FeS_2) in mine spoils, drained tideland soils, estuarine muds, and some argillaceous sediments affects soil pH[1,2]. Pyrite oxidation results in formation of sulfuric acid:

$$2FeS_2 + 7H_2O + 7.5O_2 \rightleftharpoons 4H_2SO_4 + 2Fe(OH)_3 \tag{5.4}$$

Strip mine spoils often contain large amounts of pyrite; as a result, soil pH values can be as low as 2. Acid sulfate soils, which are common in marine floodplains in the temperate and tropical zones of the world, also have very low soil pH values as a result of pyrite oxidation.

Seventh, acidity is deposited from the atmosphere to the soil in all forms of precipitation: rain, mist, fog, snow, hail, and sleet[4]. Sulfur dioxide, nitric oxide, and nitrogen dioxide are primarily responsible for the acidity associated with wet deposition. The average pH of precipitation varies widely across the continental U.S. (see Figure 5.1). Assuming an average precipitation rate of 40 in./yr and an average precipitation pH of 4.2, the acid deposition via precipitation is approximately 0.58 lb H^+/acre/yr. This is a relatively

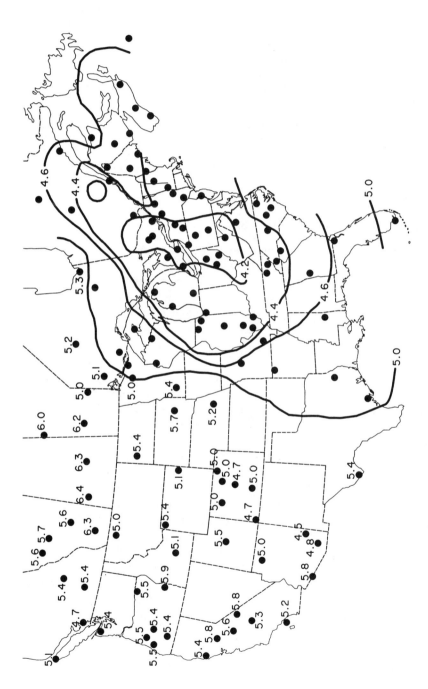

FIGURE 5.1.   The average pH of precipitation in 1981.[4]

modest amount compared to the amount of acidity generated during typical agronomic practices. For example, in the Northeastern U.S. where acid deposition is high, agricultural practice usually necessitates the addition of a few thousand pounds of limestone per acre to neutralize acidity generated by a combination of natural processes and fertilizer amendments[5].

Eighth, acidity is deposited from the atmosphere in the absence of precipitation. This process is commonly known as dry deposition and involves the deposition of particulates and the absorption/adsorption of atmospheric gases by plants, soils, and water[4]. Very little information is available on dry deposition. Some researchers use a gross rule-of-thumb that assumes that dry deposition contributes an amount of acidity equivalent to that contributed by wet deposition[4].

SOIL AS A COMPLEX BUFFER

A buffer is defined as a substance which, when added to a solution, causes a resistance to any change in the amount of H^+ in solution. Soil is a complex buffer that regulates the amount of H^+ in the water phase of a soil system. The components of soil that are responsible for a soil's buffer capacity include organic matter, the cation exchange capacity, aluminum oxide and hydroxide minerals, aluminosilicate minerals, and carbonate minerals[6].

Soil organic matter can buffer the H^+ content of water in a soil system because it possesses carboxyl (RCOOH) and hydroxyl (ROH) functional groups. In neutral and alkaline soils, these functional groups exist in the dissociated state as $RCOO^-$ and RO^-. When H^+ is added to these soils, these dissociated functional groups will preferentially adsorb H^+ over other cations to form RCOOH and ROH, resulting in the removal of H^+ from the water phase of the soil system. In general, these functional groups make organic matter an effective buffer in most soils.

The cation exchange capacity can buffer the H^+ content of water in a soil system. Buffering mechanisms that involve metal cations in solution are affected by the extent of adsorption to cation exchange sites on soil particle surfaces[6]. For example, the extent of Al^{3+} hydrolysis depends upon the Al^{3+} concentration in water, which is governed by the magnitude of its K_d.

Aluminum oxide and hydroxide minerals can buffer the H^+ content of water in a soil system[1]. In neutral soils, solid-phase aluminum occurs in the lattice structure of minerals in the interlayer regions of expanding clay minerals, and in amorphous minerals such as allophane and hydrous oxides. As surface concentrations of H^+ increase, H^+ reacts with the surface of Al minerals:

$$=Al\text{-}OH + H^+ \rightleftharpoons Al^+ + H_2O \tag{5.5}$$

The net result of the reaction described by Equation 5.5 is the creation of positively charged adsorption sites and an increase in the anion exchange capacity of the soil. As surface concentrations of H^+ continue to increase, soluble aluminum polyhydroxy ions and then soluble aluminum ions are produced in addition to H_2O to counteract the addition of H^+:

$$Al(OH)_3 \rightleftharpoons Al_n(OH)_m^{(3n-m)+} \rightleftharpoons Al(OH)_2^+ \rightleftharpoons$$
$$Al(OH)^{2+} \rightleftharpoons Al^{3+} \tag{5.6}$$

Iron and metal oxide and hydroxide minerals should buffer the H^+ content of water in soil systems in a manner similar to Al minerals.

Aluminosilicate minerals can buffer the H^+ content of water in a soil system[1]. In neutral soils, some aluminosilicate minerals such as allophane and the edges of some clay minerals contain dissociated silanol groups ($\equiv SiO^-$). As the surface concentration of H^+ will react with the dissociated silanol groups:

$$\equiv SiO^- + H^+ \rightleftharpoons \equiv SiOH \tag{5.7}$$

The net result is the removal of H^+ from the water phase of the soil system.

Carbonate minerals can buffer the H^+ content of water in a soil system. Calcium carbonates, sodium carbonates, magnesium carbonates, and calcium-magnesium carbonate are present in many soils. As the concentration of H in a neutral or alkaline soil increases, H^+ will react with carbonate minerals:

$$2CaCO_3 + 3H^+ \rightleftharpoons 2Ca^{2+} + HCO_3^- + H_2CO_3 \tag{5.8}$$

The net result is the formation of the carbonate (H_2CO_3) and bicarbonate (HCO_3^-) anions in the water phase of a soil system.

In arid regions where total average annual rainfall is less than 20 in., the amount of rainfall is usually less than the amount of water lost by evaporation and evapotranspiration. Therefore, water does not percolate downward through the unsaturated zone to the groundwater surface. Also, naturally-occurring salts are not dissolved and transported downward but are accumulated in the upper soil horizons. The process of accumulation of soluble salts such as NaCl, Na_2SO_4, $CaCO_3$, and/or $MgCO_3$ is called salinization, while soils with accumulations of these salts are known as saline soils. Saline soils usually possess a soil pH of 8.5 or lower.

The presence of Na salts may result in the saturation of a soil's cation exchange capacity with Na. The process of increasing the soil's exchange capacity with Na is known as sodification, while soils possessing a significant amount of exchangeable Na are known as sodic soils. Sodic soils usually possess a soil pH of 10 or lower. The high pH of most sodic soils is due to a process known as alkalinization. During this process, two events occur. First, hydrolysis of accumulated Na_2CO_3 produces OH^- that causes an increase in soil pH:

$$Na_2CO_3 + 2H_2O \rightleftharpoons 2Na^+ + 2OH^- + H_2CO_3 \qquad (5.9)$$

Second, the Na^+ that saturates the soil's cation exchange capacity hydrolyzes to produce OH^- that causes an increase in soil pH:

$$Soil - Na + H_2O \rightleftharpoons Soil - H + Na^+ + OH^- \qquad (5.10)$$

Likewise, the presence of calcium, magnesium, and potassium salts may result in the saturation of a soil's cation exchange capacity with these cations, with the subsequent production of OH^-:

$$Soil - Ca + 2H_2O \rightleftharpoons Soil - H_2 + Ca^{2+} + 2OH^- \qquad (5.11)$$

$$Soil - Mg + 2H_2O \rightleftharpoons Soil - H_2 + Mg^{2+} + 2OH^- \qquad (5.12)$$

$$Soil - K + H_2O \rightleftharpoons Soil - H + K^+ + OH^- \qquad (5.13)$$

In a Ca, Mg, and K saturated soil, the tendency for these cations to produce OH^- is the factor controlling the amount of H^+ in the water phase of soil. In a soil containing adsorbed H^+ and Al^{3+} as well as these three cations, the effect of OH^- production is not so obvious because it is countered by the effect of adsorbed H^+ and Al^{3+} as discussed earlier. The amount of H^+ in the water phase of a soil-water system will depend upon the relative amounts of adsorbed H^+ and Al^{3+} compared to adsorbed OH^- producing cations. It may be interesting to note that the OH^- producing cations enter soil systems through rainfall, through leaching from plant tissue, through the decay of annual litter fall, as well as through the weathering of soil minerals[7].

MEASUREMENT OF SOIL ACIDITY

Three general types of acidity can be identified in soil. These types of acidity reflect its chemical composition and distribution within the soil system. The first type of acidity is referred to as solution acidity. It is defined as the amount of H^+ present in water which is in equilibrium with soil particles. If water were extracted from a soil by displacement or by pressure extraction, the acidity of the water can be measured and expressed in terms of pH:

$$pH = -\log C_{H^+} \qquad (5.14)$$

where

$$C_{H^+} = \text{hydrogen ion concentration}$$

In general, water possessing pHs less than 7 are considered acidic. Water possessing pHs greater than 7 are considered alkaline or basic. Waters possessing pHs around 7 are considered neutral. Each pH unit represents RT calories of work required to remove the water diluting 1M acid from the standard state to its present pH. Therefore, the pH scale is linear, with each unit pH separated by 1364 calories of work per mole of H^+ at 25 °C. The practical application of this fact deals with averaging pH values[8]. The average pH of 999 samples of water having a pH of 8.0 and the 1000th sample of water with a pH of 3.0 is not obtained by converting pH values to H^+ concentrations, averaging these concentrations, and reconverting the average concentration back to pH (this method would yield an average pH 6.0). The average pH would be obtained by averaging all pH values, resulting in average pH = 8.0.

Soil pH is probably the most commonly measured soil property[2]. The pH of soil usually is measured in a slurry of soil and water. In routine procedures, such as those utilized in agricultural soil testing laboratories, one part of soil is mixed with one or two parts of water, and the pH-reference electrode combination is immersed in the stirred suspension. The description of the degree of acidity or alkalinity of soil on the basis of soil pH is[9]:

	pH
Extremely Acid	<4.5
Very Strongly Acid	4.5 - 5.0
Strongly Acid	5.1 - 5.5
Medium Acid	5.6 - 6.0
Slightly Acid	6.1 - 6.5
Neutral	6.6 - 7.3
Mildly Alkaline	7.4 - 7.8
Moderately Alkaline	7.9 - 8.4
Strongly Alkaline	8.5 - 9.0
Very Strongly Alkaline	>9.1

It is most important to recognize that the soil pH should be regarded as an empirical measurement; it is not possible to define theoretically or to determine experimentally a unique pH value for soil[6]. An example of the latter occurs during the measurement of the pH of a soil-water suspension:

the pH changes if the suspension is stirred or if the position of the reference electrode is changed. The measured pH is a result of (a) solution acidity and (b) the alteration of the liquid junction potential of the reference electrode by electrically charged soil particles. The latter is usually overcome by adding sufficient salt to the solution to increase the ionic strength above 0.005. However, pH also varies with the type and concentration of added salt and with the soil:water ratio of the suspension. Salt additions and the soil:water ratio cause some fraction of the second type of acidity—exchangeable acidity—to be measured along with solution acidity during a typical soil pH measurement.

Exchangeable acidity refers to that portion of total acidity that can be removed from soil cation exchange sites by soil equilibration with a neutral, unbuffered salt such as KCl, $CaCl_2$, or $NaCl^2$. In general, exchangeable acidity is due primarily to monomeric Al^{3+} ions[2]. In soils where organic matter is an important contributor to the cation exchange capacity, a significant proportion of the H^+ released from these organic exchange sites arises from Al hydrolysis.

The third type of soil acidity is known as titratable acidity. It is defined as the amount of acid neutralized via the addition of a base to a soil. The soil components that primarily affect the titratable acidity of partially neutralized soil are (a) the Al, Fe, and metal oxides, hydroxides, and oxyhydroxide minerals, (b) the unionized RCOOH and ROH functional groups of soil organic matter, and (c) the unionized OH groups on soil mineral surfaces[2]. In mineral soils, the most important of these components are the Al and Fe hydroxide minerals having the general formula $[R(OH_x]_n$ where x is less than three[2].

Figure 5.2 shows the titration curves for five Vermont soils. An analysis of these curves will reveal that soils exhibit different buffering behavior. Also, these soils are more buffered at low (<4) and high (>6.8) pH than in the midrange. It is most important to note that although a measured soil pH is directly affected by the amount of titratable acidity, a measured soil pH relates no information regarding the amount of titratable acidity present. In clay soils with appreciable amounts of organic matter, titratable acidity and exchange acidity can be 50,000X to 100,000X the solution acidity.

HOW BULK ACIDS AFFECT SOIL PROPERTIES

Many industrial waste streams are comprised of large volumes of low pH inorganic and organic acid liquids. If these wastes are added to soil, they may alter several soil properties. First, the addition of bulk acids to soil may increase the amount of Al, Fe, and other cations in the water phase of the soil system. H^+ is a very small cation that can successfully compete for

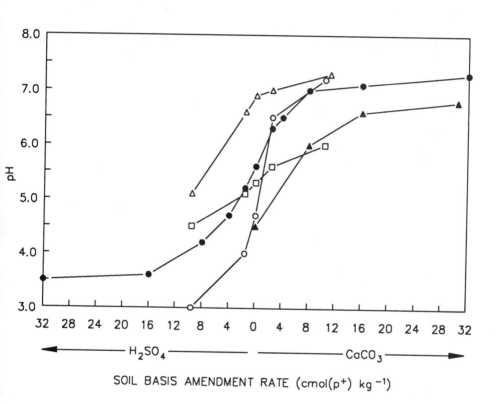

Figure 5.2   Titration curves for five Vermont soils.[10] (Reprinted with permission from the Soil Science Society of America, Inc.)

cation exchange sites and successfully displace other cations such as Ca^{2+}, Mg^{2+}, and the metals into the water phase. Also, the presence of significant amounts of H^+ can dissolve the more acid-soluble soil minerals, releasing cations which were previously fixed in the mineral structure into the water phase. It appears that if sufficient acid waste is added to depress the soil pH below 5, then soils generally lose their ability to retain cations[11]. However, if the soil's buffer capacity is of sufficient magnitude to raise the pH of the liquid waste, then the soil may partially or totally neutralize the waste acidity and attenuate soluble metals present in the waste[12]. A soil's ability to resist pH changes due to acid rain (Table 5.1) can be helpful in grossly estimating its ability to resist pH changes due to additions of waste acids.

TABLE 5.1   Potential for Soils to Neutralize Acidity Deposited From Acid Rain[13,14]

| Soil Factor | Neutralization Potential | | |
	High	Medium	Low
Bedrock/Parent Material	Limestone Dolomite Carbonate/non-carbonate mineral suites	Gabbro Greywacke Mudstone Shale Ultramafic Rocks Volcanic Rocks	Granite Granite Gneiss Orthoquartzite Syenite
% Clay	> 35	10 - 25	< 10
Cation Exchange Capacity (milli-equivalents/grams)	> 20 If free carbonates absent, or any value if free carbonates present or any value if subject to frequent flooding	10 - 20 and free carbonates absent and soil is not subject to frequent flooding	< 10 and free carbonates absent and soil is not subject to frequent flooding
SO_4^{2-} Adsorption Capacity	High SO_4^{2-} adsorption and low organic matter, and high Al_2O_3 and/or Fe_2O_3 + Fe_3O_4 content		Low SO_4^{2-} adsorption and high organic matter, and/or low Al_2O_3 and/or Fe_2O_3 + Fe_3O_4 content
Soil Depth (cm)	> 100	25 - 100	< 25
Soil Drainage	Poor		Rapid
Relief	Level	Rolling	Steep
Soil pH	> 5.0 for clays > 5.5 for loams	4.5 - 5.0 for clays 5.0 - 5.5 for loams > 5.5 for sands	< 4.5 for clays < 5.0 for loams < 5.5 for sands

Second, the addition of bulk acids to soil may cause the dissolution of some of the soil's predominant clay minerals, followed by the formation of other clay minerals. Figures 5.3, 5.4, and 5.5 are mineral stability diagrams or ion activity diagrams that are utilized to estimate the general equilibrium relationship between dissolved ions in a water phase and the predominant minerals comprising the solid phase. Suppose that in the water phase of a soil-water system, $\log a_{SiO_x}$ equalled -2 and $\log[a_{k+}/a_{H+}]^2$ equalled 7. Based on Figure 5.3, montmorillonite should be the predominant clay mineral found in this soil. Suppose that sufficient bulk acid were added to this soil so that significant amounts of SiO_2 dissolved and leached through the soil, resulting

TABLE 5.2 Effect of Bulk Inorganic and Organic Acids on Permeability of Soil and Clay.

Chemical/Waste	Media	Result	Ref.
H_2SO_4 & H_2O, 1N	Kaolinitic clay soil	$1.1 \times$ Perm. incr.	17
2N	Kaolinitic clay soil	$1.13 \times$ Perm. incr.	
15N	Kaolinitic clay soil	$635 \times$ Perm. incr.	
HCl in H_2O, 10 %	Silicate grout	No change	18
HCl in H_2O,	Kaolinite	Negl perm change	19
pH 1.0, 3.0, 5.0	Mg montmorillonite	Negl perm change	
	85% kaolinite -		
	15% montmorillonite	Negl perm change	
H_2SO_4 Waste, pH 1.0	Lacustrine clay	$2 \times$ Perm decr.	20
H_3PO_4 Liquor, pH2.2[a]	Very high plasticity clay	$52 \times$ Perm decr.	21
	High plasticity clay	$35 \times$ Perm decr	
	Medium plasticity clay	$14 \times$ Perm. decr.	
	Low plasticity clayey sand	$9.4 \times$ Perm decr.	
	Nonplastic silty sand		
	+ 29% bentonite	$0.92 \times$ Perm. decr.	
Mill Tailings	Clay soil liners	$2 \times$ to $617 \times$ Perm decr.	22
Leachates, pH			
1.2 - 2.0[b]			
Acetic Acid	6% bentonite-sand	$5 \times$ Perm incr.	23
	Silicate grouted Ottawa sand	Negl perm change	24
	Noncalcareous smectitic clay soil	$10 \times$ Perm decr., then $18 \times$ incr.	25
	Illitic clay soil	$10 \times$ Perm decr., then incr.	25
	Houston black clay soil	$1.9 \times$ Perm decr.	26
	Lufkin clay soil	$1.9 \times$ Perm decr.	26
	Calcareous smectitic clay soil	$100 \times$ Perm decr.	25
	Kaolinitic clay soil	$1,000 \times$ Perm decr.	25

[a] Contained high amounts of Ca, F, Ca oxides, Na, Cl, and SO_4. Leachate pH was 6.8 - 8.0.

[b] Contained high amounts of Al, Ca, Fe, Mg, Si, SO_4.

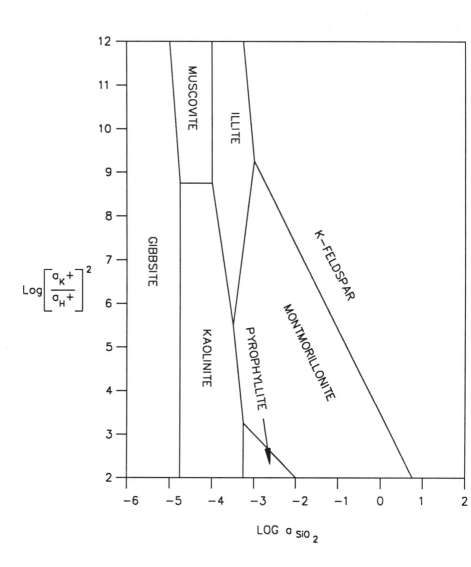

FIGURE 5.3   Mineral stability diagram for the system $Al_2O_3-K_2O-SiO_2-H_2O$.[15] (Reprinted with permission of The Clay Minerals Society).

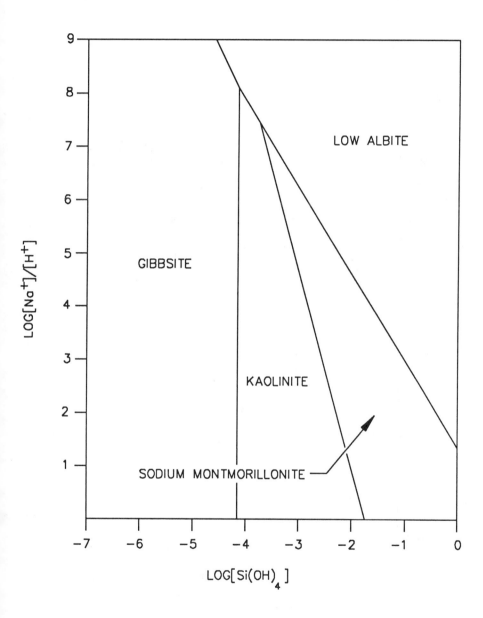

FIGURE 5.4   Mineral stability diagram for the system
$Al_2O_3-Na_2O-SiO_2-H_2O$ at 25°C.[16] (Reprinted
with permission of Freeman Cooper & Co.)

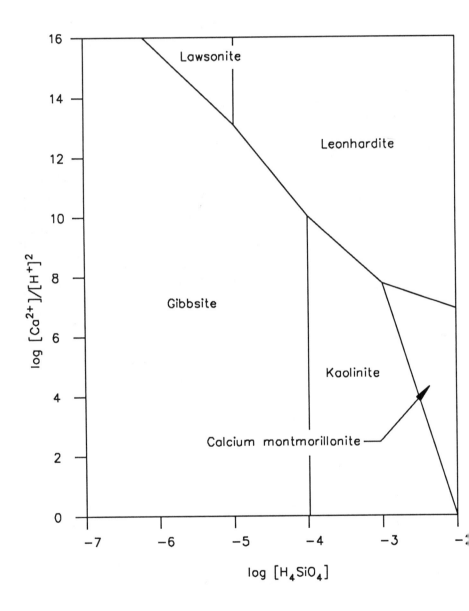

FIGURE 5.5   Mineral stability diagram for the system
$Al_2O_3 - CaO - CO_2 - SiO_2 - H_2O$.[16]
(Reprinted with permission of Freeman
Cooper & Co.)

in the $\log a_{SiO_2}$ at equilibrium equalling -7 and $\log[a_{K^+}]^2$ equalling 3. Based on Figure 5.3, the activities of SiO_2, K^+, and H^+ in the water phase now favor the formation of kaolinite.

Third, the addition of bulk acids to soil may alter soil permeability. Table 5.2 lists the effects of several bulk inorganic and organic acids on the permeability of soil and clay. An analysis of the information presented in Table 5.2 will lead to several interesting conclusions. For inorganic acids comprised of anions such as Cl^- and SO_4^{2-} that do not form sparingly soluble soil minerals, a substantial increase in soil permeability occurred only at a very high acid concentration ($15N\ H_2SO_4$). At lower acid concentrations, permeability changes generally were minimal. For the H_3PO_4 liquor, the buffer capacity of the clay was sufficient to alter the liquor leachate pH from 2.2 to between 6.8 and 8.0; decreased permeability probably resulted from the blocking of clay pores by the precipitation of oxide, sulfate, and phosphate minerals. For the mill tailings leachate, the buffer capacities of the clay soils were sufficient to increase leachate pH, and mineral precipitation in soil pores probably caused decreased permeability. Mineral precipitation should have been aided significantly by the presence of dissolved ions in the H_3PO_4 liquor and in the mill tailings leachate.

Acetic acid appears to be the only organic acid that has been studied to any significant extent to determine quantitatively the effects of organic acids on soil and clay permeability. An analysis of the information on acetic acid presented in Table 5.2 will reveal that the effect of acetic acid on soil permeability varied dramatically. In one soil, a permeability increase occurred. In another soil, a negligible permeability change occurred. In four soils, permeability decreases occurred. In two soils, permeabilities first decreased, then increased; the initial permeability decrease was attributed to partial dissolution and subsequent migration and lodging of soil particles in soil pores, while the subsequent permeability increases were attributed to soil piping which eventually cleared the clogged pores[25]. Additional research is needed in order to understand and predict the effects of acetic acid and other organic acids on soil and clay permeability.

HOW BULK BASES AFFECT SOIL PROPERTIES

Some industrial waste streams are comprised of large volumes of high pH inorganic and organic base liquids. If these wastes are added to soil, they may alter the soil in several ways.

First, the addition of bulk bases to soil may increase the amount of cations in the water phase of the soil system by dissolving the more base-soluble soil minerals. These cations which are released into the water phase were previously fixed in the structure of the more base-soluble minerals. However, this effect should occur if the soil's buffer capacity cannot compensate for the addition of OH^- into the soil-water system and prevent a significant rise in soil pH.

Second, the addition of bulk bases to soil may cause the dissolution of some of the soil's predominant clay minerals, followed by the formation of other clay minerals. This effect is analogous to the one discussed in the previous section. The addition of OH^- to water will depress the H^+ concentration by the formation of H_2O:

$$H^+ + OH^- \rightleftharpoons H_2O \tag{5.15}$$

The depression of H^+ results in an increase in the quantities $\log[a_{k+}/a_{H+}]^2$, $\log[Na^+/H^+]$, and $\log[Ca^{2+}/H^+]^2$ which are found in Figures 5.3, 5.4, and 5.5, respectively. Depending upon the activities of SiO_2, K^+, Na^+, Ca^+, Mg, and H^+, soil conditions may favor the dissolution of one mineral and the formation of another.

Third, the addition of bulk bases to soil may alter soil permeability. Table 5.3 lists the effects of several bulk bases on the permeability and volume occupied by soil and/or clay. An analysis of the information presented in Table 5.3 will lead to several interesting conclusions. In general, substantial permeability decreases were not recorded for permeating bases possessing

TABLE 5.3   Effect of Bulk Bases and Inorganic Salts on the Permeability and Volume of Soil and Clay.

Chemical/Waste	Media	Result	Ref.
$NaOH + H_2O$, pH 9.0	Kaolinite	Negl perm change	19
	Mg montmorillonite	Negl perm change	
	85% kaolinite -		
	15% montmorillonite	Negl perm change	
$NaOH + H_2O$, pH 11.0	kaolinite	Negl perm change	19
	Mg montmorillonite	1.5 × Perm decr.	
	85% kaolinite -		
	15% montmorillonite	Negl perm change	
Hydrazobenzene + H_2O, pH 12.1	Lacustrine clay	4 × Perm decr.	20
$NaOH + H_2O$, pH 13.0	Kaolinite	3 × Perm decr.	19
	Mg montmorillonite	13 × Perm decr.	
	85% kaolinite -		
	15% montmorillonite	2.5 × Perm decr.	
10% $NaOH$[a] in H_2O	Silicate grout	Dissolution	18
10% NH_4OH in H_2O	Silicate grout	80% Shrinkage	18
10% NH_4Cl in H_2O	Silicate grout	41% Shrinkage	18
10% $KCrO_4$ in H_2O	Silicate grout	88% Shrinkage	18

[a] – pH is approximately 14.

pHs below 12. Permeability decreases caused by bases possessing pHs above 12 were probably due to (a) partial dissolution of the more soluble soil minerals followed by the subsequent migration and lodging of soil particles in soil pores, and (b) the formation of new soil minerals.

The 10% NH_4OH solution, which should possess a very high pH, caused a significant shrinkage in the volume occupied by the silicate grout. This was probably caused by the presence of a high concentration of NH_4^+ and not by OH^-. Very small cations such as NH_4^+ and K^+ in soil and clay systems can compete most successfully for cation exchange sites; these small cations can displace other cations, neutralize surface charge, and cause the collapse of the space separating minerals. The result is a reduction in the volume occupied by the soil or clay, as evidenced by the effects of 10 NH_4OH, NH_4Cl, and $KCrO_4$ solutions on the volume occupied by silicate grout (Table 5.3).

REMEDIAL TECHNOLOGIES

The general objective of any remedial technology dealing with the excessively low or high pH of a soil is to achieve neutralization. Neutralization can be defined as the reaction of an acid with a base to form a salt and water [27]. This reaction is represented by the following reaction in which HA and BOH are any acid and base:

$$HA + BOH \longrightarrow BA + H_2O \tag{5.16}$$

Acid   Base   Salt

Bases that are added to soil in order to raise soil pH are usually known as liming materials. Liming materials differ substantially in their ability to neutralize acids [27]. The value of the liming material depends upon the quantity of acid that a unit weight of the liming material will neutralize. This is related to the chemical composition of the liming material and its purity. Pure $CaCO_3$ is the standard against which other liming materials are measured, and its neutralizing value is considered to be 100 percent:

Material	Neutralizing Value %
CaO	179
Ca(OH)$_2$	136
MgCO$_3$	119
CaMg(CO$_3$)$_2$	109
CaCO$_3$	100
CaSiO$_3$	86

The neutralizing value is a relatively simple concept to understand and utilize. Consider the following two reactions involving the neutralization of HCl:

$$CaCO_3 + 2HCl \longrightarrow CaCl_2 + H_2O + CO_2 \qquad (5.17)$$

$$MgCO_3 + 2HCl \longrightarrow MgCl_2 + H_2O + CO_2 \qquad (5.18)$$

In each of these equations, one molecule of carbonate will neutralize two molecules of acid. However the molecular weight of $CaCO_3$ is 100, whereas the molecular weight of $MgCO_3$ is 84. In other words, 84 lbs of $MgCO_3$ will neutralize the same amount of acid as 100 lbs of $CaCO_3$. On an equal weight basis:

$$84/100 = 100/X, \; X = 119 \qquad (5.19)$$

$MgCO_3$ on a weight basis will neutralize 1.19X as much acid as the same weight of $CaCO_3$. Its neutralizing value in relation to $CaCO_3$ is 119 percent.

The degree of fineness is also important, because the rate with which various liming materials will react is dependent on the amount of material surface which is in contact with soil. If the particles comprising the liming material are coarse, the amount of surface area is relatively low and the reaction should be slow; if they are fine, the amount of surface area is relatively high and the reaction should be fast (see Figure 5.6).In general, when rapid results are required, CaO or $Ca(OH)_2$ are usually utilized. The carbonates of Ca and Mg are the most widely utilized liming materials, primarily due to their wide occurrence in nature. Blast-furnace slag, basic slag, electric-furnace slag, fly ash, as well as other process byproducts have been utilized as effective soil liming materials.

In some instances, the liming material surface area can diminish as time progresses due to the formation of crusts. For example, the use of liming materials during the remediation of acid mine drainage occasionally creates crusts of calcium sulfate, iron compounds, and biological slimes; these crusts protect the unreacted material from contact with acids and must be periodically removed.

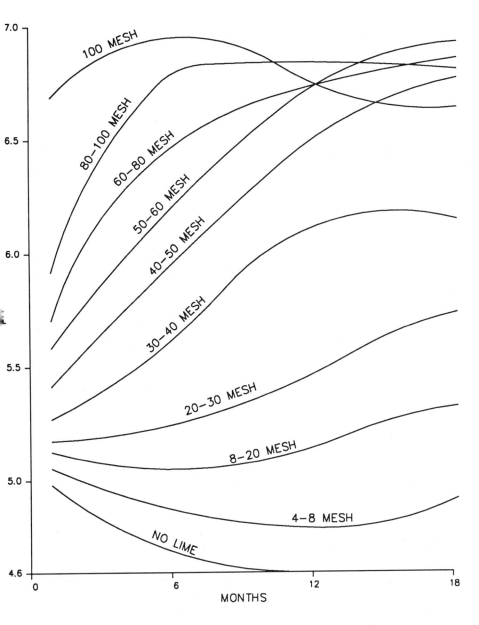

FIGURE 5.6 The effect of dolomitic limestone of different particle sizes on the pH of a Canfield silt loam at various times after application.[28] (Reprinted with permission of Williams & Wilkins).

The amount of liming material required to cause a unit pH change in one soil will in all probability not be the same as the amount required for another soil. In general, soils classified as clays, peats, and mucks have higher buffer capacities and, if acidic, higher lime requirements than coarse-textured soils with little or no organic matter.

The lime requirement for a soil can be estimated by developing a buffer curve or titration curve [27]. First, a known quantity (e.g. 10 grams) of soil is added to a series of small beakers. Second, increasing amounts of acid or base such as $Ca(OH)_2$ or HCl are added to each beaker. Third, water is added to each beaker to equalize the volume of liquid in all beakers. Fourth, the soil-water mixtures are allowed to equilibrate. Fifth, pH measurements are made, and the pH values are plotted against the millequivalents of acid or base added. The pH vs millequivalent acid/base plot is known as a soil buffer or titration curve (see Figure 5.7). According to Figure 5.7, ten grams of soil would require approximately 0.6 milliequivalents of $Ca(OH)_2$ in order to raise its pH from 4.0 to 7.0. Assuming a calcitic limestone will be utilized, the amount required would be 0.6×0.05 or 0.03 gram/10 gram soil. Calculations of the amount of liming material to be applied to soil assume (a) uniform mixing, (b) complete reaction of the liming material, and (c) the use of 100 percent finely divided liming material. Because these assumptions are generally not met, as much as 150 percent of the calculated amount may have to be field applied to soil to achieve the desired pH.

Acidity neutralization has an important role in the remediation of the Gulf Coast Lead (GCL) site in Tampa, Florida [29]. Operations at this facility consisted of recovering Pb from acid batteries. Soil and ground water contamination resulted from the percolation of acid, metal-laden rinsewaters from an unlined surface pond. A subsurface acid reaction barrier, consisting of a deep trench filled with crushed limestone and oyster shells, should intercept and neutralize acidic groundwater which has migrated offsite.

Large areas of land in the United States have been strip mined since 1930. Even larger areas have been affected indirectly by acid mine wash originating from the mined land. Low soil pHs are produced when the products of iron sulfide, $Fe_2(SO_4)_3$, and H_2SO_4 react with soil. The reclamation and revegetation of these soils usually involves soil pH adjustment utilizing various liming materials [30,31,32].

Acidity neutralization played a key role in the remediation of the Ravenfield Tip site in South Yorkshire, England [33]. This site, known as the United Kingdom's Love Canal, was comprised of a group of sandstone quarries which were utilized for the disposal of acid tars, contaminated fuller's earth, slags, foundry sand, chromium waste, fluoride waste, nickel waste, phenolic waste, and other chemicals. Waste was excavated, mixed with quicklime at ratios ranging from 1:4 to 1:10 according to the degree of acidity, deposited,

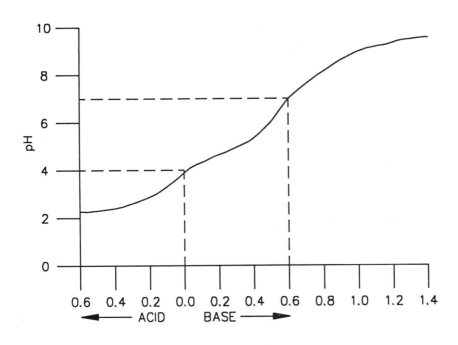

FIGURE 5.7 Hypothetical titration curve for a sandy loam
soil.

and covered with subsoil. The treated wastes were covered with a 15 to 23 cm layer of about 2.5 cm grade limestone chips, which was subsequently covered with subsoil and soil.

Acidity neutralization was utilized to restore ground water quality following the pilot-scale acidic in-situ leaching of a uranium ore body at Nine-Mile Lake near Casper, Wyoming[34]. Groundwater quality in the leached ore zone was restored by diluting and neutralizing extracted groundwater and rein-jecting it into the ore body.

Chemicals that are added to soil in order to lower soil pH are usually known as acidulents. The fundamental soil chemistry of the acidification of soil is the same as that of liming soil[21]. The materials used to reduce soil pH include elemental sulfur, sulfuric acid, aluminum sulfate, iron sulfate, and calcium polysulfide. Also, ammonium sulfate, ammonium phosphate, and similar compounds, primarily nitrogen fertilizers, are also effective in decreasing soil pH under certain soil conditions favoring plant growth.

Elemental sulfur is a most effective soil acidulent in moist, warm, well-aerated soils where it can be converted to sulfuric acid by autotrophic bacteria. Sulfuric acid reacts instantaneously with soil, but it requires specialized equipment. Aluminum sulfate $Al_2(SO_4)_3$ is a popular material among horticulturalists for acidulating the soil in which azaleas, camellias, and similar acid-tolerant ornamental flowers are grown. When added to soil, it hydrolyzes:

$$Al_2(SO_4)_3 + 6H_2O \longrightarrow 2Al(OH)_3 + 3H_2SO_4 \qquad (5.20)$$

In addition to hydrolysis in soil and ground water, Al will replace exchangeable H^+ on the soil particle surface, resulting in a lower pH:

$$Soil\text{-}6H + Al_2(SO_4)_3 \longrightarrow Soil\text{-}2Al + 3H_2SO_4 \qquad (5.21)$$

Iron sulfate, $Fe_2(SO_4)_3$, behaves in a similar manner when applied to soil:

$$Soil\text{-}6H + Fe_2(SO_4)_3 \longrightarrow Soil\text{-}2Fe + 3H_2SO_4 \qquad (5.22)$$

Calcium polysulfide (CaS_5) reacts with oxygen and water to produce sulfuric acid:

$$CaS_5 + 8O_2 + 4H_2O \rightleftharpoons CaSO_4 + 4H_2SO_4 \qquad (5.23)$$

The amount of acidulent required to cause a unit pH change in one soil will in all probability not be the same as the amount required for another soil. The acidulent requirement can be estimated by developing a titration curve in a manner similar to that described above for liming materials.

REFERENCES

1. Hutchinson, T. C. and Havas, M. (eds). Effects of Acid Precipitation on Terrestrial Ecosystems. New York: Plenum Press (1980).
2. Adams, F. (ed.). Soil Acidity and Liming. Second Edition. Madison, WI: American Society of Agronomy (1984).
3. Reuss, J. O. Chemical and Biological Relationships Relevent to the Effect of Acid Rainfall on the Soil-Plant System. Water, Air, and Soil Pollution 7:461-478 (1977).
4. Council for Agricultural Science and Technology. Acid Precipitation in Relation to Agriculture, Forestry, and Aquatic Biology. Report No. 100. Ames, IA: Council for Agricultural Science and Technology (June, 1984).
5. Frink, C. R. and Voigt, G. K. Potential Effects of Acid Precipitation on Soils in the Humid Temperate Zone. Water, Air, and Soil Pollution 7:371-388 (1977).
6. Fairbridge, R. W. and Finkl, C. W. Jr. The Encyclopedia of Soil Science Part 1. Physics, Chemistry, Biology, Fertility, and Technology. Encyclopedia of Earth Sciences Series, Vol. XII. Stroudsburg, PA: Dowden, Hutchinson & Ross, Inc. (1979).
7. McFee, W. W., Kelly, J. M., and Beck, R. H. Acid Precipitation Effects on Soil pH and Base Saturation of Exchange Sites. Water, Air, and Soil Pollution 7:401-408 (1977).
8. Bartlett, R. J. Comments on Arithmetic Average Soil pH Values. Soil Science Society of America Journal 46:444 (1982).
9. U.S. Department of Agriculture. Soil Survey Manual. Washington, D.C.: U.S. Government Printing Office (1951).
10. Magdoff, F. R. and Bartlett, R. J. Soil pH buffering revisited. Soil Science Society of America Journal 49:145-148 (1985).
11. Fuller, W.. and Warrick, A. W. Soils in Waste Treatment and Utilization. Volumes I and II. Boca Raton, FL: CRC Press (1985).
12. Wangen, L. E. and Jones, M. M. The Attenuation of Chemical Elements in Acidic Leachates From Coal Mineral Wastes by Soils. Environ. Geol. Water Sci. 6:161-170 (1984).
13. Bricker, O. P. (ed). Geological Aspects of Acid Deposition. Boston: Butterworth Publishers (1984).
14. Glass, N. R., Arnold, D. E., Galloway, J. N., Hendrey, G. R., Lee, J. J., McFee, W. W., Norton, S. A., Powers, C. F., Rambo, D. L., and Schofield, C. L. Effects of Acid Precipitation. Environmental Science and Technology 16:162A-169A (1982).
15. Garrels, R. M. Montmorillonite/Illite Stability Diagrams. Clays and Clay Minerals 32:161-166 (1984).
16. Helgeson, H. C., Brown, T. H., and Leeper, R. H. 1969. Handbook of Theoretical Activity Diagrams. San Francisco: Freeman Cooper and Co. (1969).
17. Laguros, J. G. and Robertson, J. M. Problems of Interaction Between Industrial Residues and Clays. *IN* Third Annual Conference on Treatment and Disposal of Industrial Wastewaters and Residues, Houston, TX, April 18-20, 1978. Rockville, MD: Information Transfer Inc. (1978).

18. Malone, P. G. and May, J. H. Development of Methods for In-Situ Hazardous Waste Stabilization by Injection Grouting. *In* Land Disposal of Hazardous Waste. Proceedings of the Tenth Annual Research Symposium, April 3-5, 1984, Ft. Mitchell, KY. EPA-600/19-84-007. Cincinnati, OH: U.S. Environmental Protection Agency (1984).

19. Lentz, R. W., Horst, W. D., and Uppot, J. O. The Permeability of Clay to Acidic and Caustic Permeants. *In* Johnson, A. I., Frobel, R. K., Cavalli, N. J., and Petterson, C. B. (eds). Hydraulic Barriers in Soil and Rock. ASTM STP 874. Philadelphia, PA: American Society for Testing and Materials (1985).

20. Everett, E. E. Suitability of the Lacustrine Clays in Bay County, Michigan for Land Disposal of Hazardous Waste. M.S. Thesis, University of Toledo, OH (1977).

21. Gipson, A. H. Jr. Permeability Testing on Clayey Soil and Silty Sand-Bentonite Mixture using Acid Liquor. *In* Johnson, A. I., Frobel, R. K., Cavalli, N. J., and Pettersson, C. B. (eds). Hydraulic Barriers in Soil and Rock. ASTM STP 874. Philadelphia, PA: American Society for Testing and Materials (1985).

22. Peterson, S. R. and Gee, G. W. Interactions Between Acidic Solutions and Clay Liners: Permeability and Neutralization. *In* Johnson, A. I., Frobel, R. K., Cavalli, N. J., and Pettersson, C. B. (eds.). Hydraulic Barriers in Soil and Rock. ASTM STP 874. Philadelphia, PA: American Society for Testing and Materials (1985).

23. Evans, J. C., Fang, H-Y, and Kugelman, I. J. Organic Fluid Effects on the Permeability of Soil-Bentonite Slurry Walls. *In* Proceedings of the National Conference on Hazardous Wastes and Environmental Emergencies, Cincinnati, OH, May 14-16, 1985. Silver Spring, MD: Hazardous Materials Control Research Institute (1985).

24. Lord, A. E. Jr., Weist, F. C., Koerner, R. M., and Arland, F. J. The Hydraulic Conductivity of Silicate Grouted Sands with Various Chemicals. *In* Proceedings of the National Conference on Management of Uncontrolled Hazardous Waste Sites, October 31-November 2, 1983, Washington, D.C. Silver Spring, MD: Hazardous Materials Control Research Institute (1983).

25. Anderson, D. C., Brown, K. W., and Green, J. Organic Leachate Effects on the Permeability of Clay Liners. *In* National Conference on the Management of Uncontrolled Hazardous Waste Sites, Washington, D.C., October 28-30, 1981. Silver Spring, MD: Hazardous Materials Control Research Institute (1981).

26. Anderson, D. and Brown, K. W. Organic Leachate Effects on the Permeability of Clay Liners. *In* Land Disposal: Hazardous Waste. Proceedings of the Seventh Annual Research Symposium, Philadelphia, PA, March 16-18, 1981. EPA-600/9-81-002b. Cincinnati, OH: U.S. Environmental Protection Agency (1981).

27. Tisdale S. L. and Nelson, W. L. Soil Fertility and Fertilizers. New York: Macmillan (1966).

28. Meyer, T. A. and Volk, G. W. Effect of Particle Size of Limestones on Soil Reaction, Exchangeable Cations, and Plant Growth. Soil Science 73:37-52 (1952).

29. Furman, C. and Cochran, S. R. Case Study Investigations of Remedial Response Programs at Uncontrolled Hazardous Waste Sites. *In* Land Disposal of Hazardous Waste. Proceedings of the Eleventh Annual Research Symposium, Cincinnati,

OH, April 29-May 1, 1985. EPA-600/9-85-013. Cincinnati, OH: U.S. Environmental Protection Agency (1985).

30. Blevins, R. L., Bailey, H. H., and Ballard G. E. The Effect of Acid Mine Water on Floodplain Soils in the Western Kentucky Coal Fields. Soil Science **110**:191-196 (1970).

31. Joost, R. E., Olsen, F. J., Jones, J. H., and Stiles, D. Revegetating Acid Coal Mining Sites. Biocycle **24**:47-50 (1983).

32. Johnson, J. R. Reclamation of Abandoned Mine Site. Journal of the Environmental Engineering Division, Proceedings of the American Society of Civil Engineers **105**:597-603 (1979).

33. Khan, A. Q. and Thomas, G. A. Pseudo Love Canal of the United Kingdom: In-Situ Decontamination and Rehabilitation of a Hazardous Chemical Dump Problems of Uncontrolled Disposal Sites. *In* Proceedings of U.S. EPA National Conference on Management of Uncontrolled Hazardous Waste Sites, Washington, D.C., October 15-17, 1980. Silver Spring, MD: Hazardous Materials Control Research Institute (1980).

34. Engelmann, W. H., Phillips, P. E., Tweeton, D. R., Loest, K. W., and Nigbor, M. T. Restoration of Groundwater Quality Following Pilot-Scale Acidic In-Situ Uranium Leaching at Nine-Mile Lake Site Near Casper, Wyoming. Society of Petroleum Engineers 1982: 382-398 (1982).

6

Adsorption and Mobility
of Organic Chemicals

INTRODUCTION

Knowledge concerning the adsorption of organic chemicals to soils is essential for the assessment of the fate of organic chemicals in the environment and for the assessment of potential adverse impacts. If an organic chemical is extensively adsorbed by soil particles, it will not leach through the soil profile. If it remains at the soil surface, deleterious environmental and human health effects may arise due to an increased concentration of the chemical in the zone of plant growth, possibly resulting in the contamination of man's food supply. If the chemical is weakly adsorbed, it may leach through the soil profile and reach groundwater and surface waters; deleterious human health effects may arise if man uses these waters as drinking water.

The degree of adsorption not only affects a chemical's mobility but also other transport and transformation reactions. For example, the rates of volatilization, photolysis, hydrolysis, and biodegradation of many organic chemicals are directly dependent upon the extent of adsorption.

The extent to which an organic chemical is adsorbed is determined by the chemical's structure and soil physical and chemical properties. This chapter discusses these in detail. Several estimation equations which predict the extent of adsorption exist; in general, these equations correlate the adsorption or distribution coefficient with an organic chemical's water solubility or octanol-water partition coefficient. This chapter discusses the proper use and limitations of these equations.

These estimation equations were derived on the basis that water is the primary solvent in the soil system. In many situations, a bulk hydrocarbon is the primary solvent. This chapter discusses how organic chemical adsorption and mobility may change when a bulk hydrocarbon replaces water as the primary solvent in a soil system.

This chapter introduces and discusses a few simple mathematical models that require the adsorption coefficient as an independent variable (input parameter). These models are, in the opinion of the author, simple and reliable equations that can fulfill the need for quick, inexpensive information on the potential migration rate of an organic chemical.

ADSORPTION AND ORGANIC
CHEMICAL STRUCTURE

The extent that any organic chemical is adsorbed onto soil surfaces is directly affected by its structure. Structure determines the magnitude of the organic chemical's gross properties such as molecular volume, water solubility, octanol-water partition coefficient, and vapor pressure that influence the overall adsorptive behavior of the chemical. The chemical structure of any molecule can be subdivided into substructures or fragments. Many gross chemical properties are, in essence, the sums of the fragment contributions to that property. For example, the total lipophilicity or hydrophobicity, P_R, for any chemical structure R can be construed by the simple addition of the proper pi values characterizing the X_n fragments[1]:

$$\log P_R = \sum_n \pi (X_n) \qquad (6.1)$$

However, it is most important to note that one fragment may overwhelmingly influence adsorption relative to the other fragments; its contribution may be relatively large, or it may introduce an interaction that was previously absent. This section will discuss how organic chemical structure and substructure affects its adsorption onto soil particle surfaces. Specifically, this section will discuss six aspects of chemical structure that govern adsorption: molecular size, hydrophobicity, molecular charge, organic molecular fragments that undergo hydrogen bonding, the three-dimensional arrangement and interaction of molecular fragments, and molecular fragments that undergo coordination bonding.

Molecular Size. First, the overall size of the organic molecule significantly influences its adsorption potential. In general, the larger the molecule, the greater its propensity to exist in the adsorbed state. The larger the molecule, the greater the contribution of van der Waal's forces to the adsorption potential. Van der Waal's forces are relatively weak, induced forces existing between otherwise nonattracting atoms and molecules. The *average* distribution of charge on a neutral organic molecule such as octane or decane is symmetrical; as a result, there is no permanent electrostatic polarization or charge separation in terms of positive and negative polar regions or poles on the molecule. However, the electrons within the neutral molecule are not stagnant but mobile so that the electron distribution is somewhat distorted at any instant in time. As a result, small dipoles exist on the molecule's surface. These small, momentary dipoles affect the electron distribution in an adjacent molecule or on the soil surface: the negative regions of the dipole molecule repel electrons in the adjacent soil surface, and the positive regions of the dipole molecule will attract electrons in the adjacent soil surface. The net result of the close proximity of the molecule to the soil surface is the induction of momentary electrostatic polarizations on the soil surface. Although the momentary and induced dipoles are constantly migrating on the molecule's surfaces and the soil surface, the net result is an attraction between the molecule and the soil surface.

In general, van der Waal's forces are weak and of very short range. They act only between molecules and surfaces that are in close proximity. Within a class of organic chemicals, the larger the molecular size, the larger the van der Waal's forces between the molecule and the soil surface, and as a result, the greater the extent of adsorption. For example, even very water soluble organic molecules possessing large molecular volumes such as poly(galacturonic acid)[2], amylose[2], dextran[2], polyvinyl alcohol[3,4], poly(ethylene glycols)[4], poly(vinyl acetate)[4], poly(acrylic acid)[4], poly acrylamide[4], poly(acrylonitrile)[4], antibiotics[5], and water-soluble dyes[6] are adsorbed extensively onto soil surfaces and are extremely difficult to desorb due to multiple van der Waal's forces arising from many points of contact between the surface and the adsorbed molecule. As a general rule, van der Waals forces become the dominating adsorption mechanism in non-sandy soils when the molecular weight of the organic chemical exceeds 400 - 500.

Recent research has shown that a number of organic chemicals—PCBs[7], polynuclear aromatic hydrocarbons[8], bis(2-ethylhexyl)phthalate[8], toluene[9], and cholesterol[10]—as well as a number of pesticides—DDT[8], bromacil[11], triazines[11], diuron[11], chlortoluron[11], and glyphosphate[11]—bind onto dissolved organic macromolecules such as humic acid, fulvic acid, or organic matter. However, it is most important to recognize that in most soil-water systems, these macromolecules are not mobile. They tend to be extensively adsorbed onto soil surfaces due to a large part to van der Waals forces.

Hydrophobicity.   The second factor that affects organic chemical adsorption onto soil surfaces is the molecule's hydrophobicity or lipophilicity. Hydrophobicity refers to the preferential migration to and accumulation of an organic chemical in hydrophobic solvents or on hydrophobic surfaces such as soil organic matter in preference over aqueous solvents or hydrophilic surfaces. In general, molecular fragments comprised of C, H, Br, Cl, and I are hydrophobic fragments, while molecular fragments which contain N, S, O, and P are primarily hydrophilic fragments. The net hydrophobicity of any molecule is governed by the combined effect of hydrophobic and hydrophilic fragments that comprise the molecule.

The preferential accumulation of an organic chemical on hydrophobic soil particle surfaces has been referred to as "hydrophobic bonding" or as "partitioning." The mechanism causing this phenomenon is a subject of considerable controversy. One mechanism proposes that hydrocarbons tend to associate with one another and with hydrophobic surfaces so that restructuring of adsorbed water will not occur[12]. Solution and adsorbed water exists in an ice-like structure with an isotropic network of H_2O-H_2O bonds[13]; because hydrophobic organic chemical-solvent attractive forces should be weak in both the aqueous medium and in a hydrophobic environment, the hydrophobic organic chemical is squeezed out of the aqueous medium due to the strong cohesive forces between water molecules[13]. In thermodynamic terms, this mechanism views hydrophobic bonding as an entropy effect caused by solvent structure.

Another mechanism proposes that hydrophobic attractions between molecules can be entirely ascribed to van der Waal's forces. This position is substantiated in part on thermodynamic calculations[14].

Theoretically and practically, the accumulation of an organic chemical by a soil cannot be defined simply as either "partitioning" or "adsorption." It is a continuum of possible interactions starting with fixed site adsorption and ending with partitioning at both organic and inorganic particles.

Although soil organic matter is usually implicated as the soil solid responsible for accumulating organic chemicals, it is most important to recognize that soil mineral surfaces also possess hydrophobic regions that can preferentially accumulate organic chemicals. An analysis of the information presented in Table 6.1 will reveal that several inorganic particle surfaces, especially clay, have been found to govern the extent of adsorption of various organic chemicals. The organic matter in most soil is intimately bound to clay as a clay-organic complex. As a result, two major types of adsorbing surfaces are available to an organic chemical: clay-organic and clay alone. The relative contribution of organic and inorganic surface areas to adsorption will depend upon the extent to which the clay is coated with organic matter. The influence of clay on organic chemical adsorption can be significant, especially in soils with organic matter contents below one percent; for example, the adsorption of PCBs in a subsurface soil with an organic carbon content of 0.4 percent was greater than PCB adsorption to topsoil with an organic carbon content of 1.2 percent[38].

Molecular Charge.   The third factor that affects organic chemical adsorption onto soil surfaces is the intrinsic positive or negative charge possessed by some organic molecules. A few organic molecules such as the quaternary ammoniums, R_4N^+, oxoniums, R_3O^+, diazoniums, $R\text{-}N \equiv N\text{:}^+$, and some metallo-organics possess functional groups that possess permanent, positive charges. These organic molecules are adsorbed onto the cation exchange sites discussed in Chapter 4.

TABLE 6.1   Soil Properties Governing the Extent of Adsorption of Various Organic Chemicals.

Chemical	Medium	Governing Property	Ref.
Acetochlor	15 U.S. Soils	Clay Organic Matter Soil Surface Area	15
Alachlor	15 U.S. Soils	Clay Organic Matter Soil Surface Area	15
Alkylbenzene Sulfonate	30 U.S. Soils	Iron Oxides Clay	16
Ametryne	34 Puerto Rican Soils	Organic Matter Silt Soil pH	17
Amiben Derivatives	6 Arkansas Soils	Clay Organic Matter	18
Atrazine	25 Missouri Soils	Clay Organic Matter	19
3(4-Chlorophenyl)- 1,1-dimethylurea	12 Soils	Inorganic Minerals Organic Matter	20
o-Cresol	6 U.S. Subsoils	Iron Oxides pH	21
p-Cresol	6 U.S Subsoils	CEC Iron Oxides Surface Area	21
2,4-D	21 Belgian Soils	Exchangeable Al Organic Matter Soil pH	22
DDT	Iowa & Maine Soil	Clay Organic Matter	23
Dicamba	5 Canadian Soils	Organic Matter Soil pH	24
2,4-Dichlorophenol	6 U.S. Subsoils	Iron Oxides pH	21
2,6-Dimethylphenol	6 U.S. Subsoils	Iron Oxides Surface Area	21
Dinoben Derivatives	6 Arkansas Soils	Clay Organic Matter	18
Diphenamid	3 U.S. Soils	Clay Organic Matter	25
Dipropetryn	4 Oklahoma Soils	CEC Clay Organic Matter	26
Diquat	8 Sudanese Soils	CEC	27
Ethylmethyl- phosphonofluoridate	Kaolinite Montmorillonite Organic Matter Sand 6 U.K. Soils	Clay	28

TABLE 6.1　Soil Properties Governing the Extent of Adsorption of Various Organic Chemicals. (cont.)

Chemical	Medium	Governing Property	Ref.
Fluometuron	3 Oklahoma Soils	Clay	29
		Organic Matter	
Lindane	Lake Sediments	Clay	30
		Organic Matter	
Methoxychlor	Iowa & Maine Soil	Clay	23
		Organic Matter	
Metolachlor	15 U.S. Soils	Clay	15
		Organic Matter	
		Surface Area	
Metribuzin	16 Mississippi Soils	Clay	31
		Organic Matter	
Napropamide	3 Oklahoma Soils	Clay	29
		Organic Matter	
p-Nitrophenol	6 U.S. Subsoils	CEC	21
		Clay	
		Iron Oxides	
Paraquat	8 Sudanese Subsoils	CEC	27
Parathion	21 Israeli Soils	Clay	32
		Organic Matter	
Pentachlorophenol	13 Japanese and Korean Soils	Soil pH	33
Phenol	6 U.S. Subsoils	Iron Oxides	21
		Surface Area	
Prometryne	25 Missouri Soils	Clay	19
	5 U.S. Soils	CEC	34
		Organic Matter	
		Surface Area	
	4 Oklahoma Soils	CEC	26
		Clay	
		Organic Matter	
Pichloram	3 Oregon Soils	Clay	35
		Extractable Al	
	5 Canadian Soils	Organic Matter	24
		Soil pH	
Simazine	65 California Soils	CEC	36
		Sand/Clay Ratio	
	25 Missouri Soils	Clay	19
		Organic Matter	
Tebuthiuron	5 U.S. Soils	Clay	37
		Organic Matter	
Terbutryn	3 Oklahoma Soils	Clay	29
		Organic Matter	

Some organic chemicals possess amine functional groups that may or may not possess a positive charge, depending upon the acidity of the system. If sufficient H^+ is present, organoamines will react with H^+, and the amine functional group will acquire a positive charge:

$$R-NH_2 + H^+ \rightleftharpoons R-NH_3^+ \qquad (6.2)$$

The sources of H^+ include H^+ in solution, H_2O associated with adsorbed cations, H^+ from other adsorbed or dissolved organic chemicals, and soil organic matter. After $R-NH_3^+$ is formed, this protonated, charged molecule should be adsorbed by soil cation exchange sites.

An analysis of Equation 6.2 will reveal that the reaction is reversible and will depend upon the acidity of the system. The pK_a of a chemical is a mathematical description of the effect of acidity on the protonation of a molecule such as $R-NH_2$. For any functional group that can accept H^+ or release H^+, the ionization constant K_a can be defined:

$$K_a = [A^m][H^+]/[HA^n] \qquad (6.3)$$

where

$[A^m]$ = Concentration of the dissociated form of the organic molecule A; m equals zero for amines such as $R-NH_2$, and m equals -1 for organic acids such as R-COOH and R-OH.

$[H^+]$ = The concentration of H^+

$[HA^n]$ = Concentration of the complex formed between H^+ and the organic molecule A; n equals $+1$ for amines such as $R-NH_2$, and n equals zero for organic acids such as R-COOH and R-OH.

The relationship between pH and pK_a can be derived by rearranging Equation 6.3 and converting the parameters to negative logarithms:

$$pH - pK_a = \log([A^m]/[HA^n]) \qquad (6.4)$$

Equation 6.4 is an extremely useful equation because the relative ratio of charged to uncharged molecules in the soil-water system can be estimated (see Table 6.2). Once this relative ratio is known, the general existence of a molecular charge that influences the extent of adsorption can be identified.

An example of an organic chemical with a pH dependent charge is 3-amino-1,2,4-triazole. At lower soil pHs, this chemical will become protonated and form an aminotriazolium cation[39]. The cation will adsorb onto soil cation exchange sites and may be displaced by other cations. Other organic chemicals that have been found to behave in a similar fashion include s-tria-zines, substituted ureas, phenylcarbamates, aniline, and anilides[40].

Some organic chemicals possess hydroxyl or carboxyl functional groups that may or may not possess a negative charge, depending upon the acidity of the soil-water system. If the pH is sufficiently high, these organic functional groups will dissociate, and these groups will acquire a negative charge:

$$R\text{--}OH \rightleftharpoons R-O^- + H^+ \tag{6.5}$$

$$R\text{--}COOH \rightleftharpoons R-COO^- + H^+ \tag{6.6}$$

Equation 6.4 can be utilized to estimate the relative ratios of charged to un-charged molecules found in Equations 6.5 and 6.6.

After the negatively charged molecules $R-O^-$ and $R-COO^-$ are formed, they may be adsorbed by oxide surfaces, by the edges of alumino-silicate clay minerals, and by soil organic matter. Information presented in Chapter 4 revealed that these surfaces possess positive charges that can adsorb organic anions to some extent by ion exchange. It is also important to remember that soils typically possess a significantly greater number of negative surface charges than positive ones. Because organic anions possess a negative charge, repulsion between organic anions and soil surfaces will occur. This process was defined in Chapter 4 as negative adsorption. The net result of this process will be decreased adsorption of organic anions at soil surfaces due to charge repulsion. For example, an analysis of organic chemical mobility on soil thin layer chromatography plates revealed that the presence of the carboxylic acid fragment resulted in a statistically significant increase in chemical mobility above that predicted by regression equations based on organic chemical water solubility[41]. The enhanced mobility caused by the presence of the carboxylic acid fragment was attributed to charge repulsion.

Some organic chemicals possess both types of functional groups, i.e. one that can acquire a positive charge and another that can acquire a negative charge. For these organic chemicals, the isoelectric point (IP) is a useful index that aids in predicting the effect of pH on the adsorption of these chemicals. The IP is the pH at which the organic chemical has zero charge. Above the IP, the organic chemical has a net negative charge; below the IP, the organic chemical has a net positive charge. The IP is a general summation of the effects of the pK_as of each functional group in the molecule; it is an empirical value, and no equation that can be utilized for its estimation is available.

Hydrogen Bonding.   The fourth factor that affects organic chemical adsorption onto soil surfaces is hydrogen bonding. Hydrogen bonding occurs whenever a hydrogen atom serves as a bridge between two electronegative

TABLE 6.2 Relationship between pH, pK_a, and the Ratio of $[A^m]/[HA^n]$

pH-pK_a	% A^m	% HA^n	$[A^m]/[HA^n]$
2	99	1	100/1
1	91	9	10/1
0	50	50	1/1
− 1	9	91	1/10
− 2	1	99	1/100

atoms; the hydrogen atom is linked to one electronegative atom by a covalent bond and to the other by an electrostatic bond.

Three distinct types of hydrogen bonding can occur between organic molecules and soil particle surfaces. The first is the formation of a hydrogen bond via the linking of a polar organic molecule to an adsorbed cation through a water molecule in the cation's primary hydration shell:

$$Soil - (M^{n+})OH - H \cdot \cdot \cdot O = C - R_2 \qquad (6.7)$$

This type of bond has been identified for montmorillonite systems containing pyridine, acetone and other ketones, benzoic acid, nitrobenzene, amides, and organic polymers[42,43].

The second type of hydrogen bonding can occur between an adsorbed organic cation and another organic molecule:

$$Soil - (R_3NH^+) \cdot \cdot \cdot O = C - R_2 \qquad (6.8)$$

In the reaction illustrated in Equation 6.8, an adsorbed organic amine will form a hydrogen bond with the oxygen of a carboxyl fragment. The strength of this type of bond depends upon the relative basicity of the neutral organic molecule.

The third type of hydrogen bonding can occur between organic molecule fragments and the oxygen and hydroxyl groups on the surface of soil particles. For example, Raman spectral studies of glycine, lysine, and alanine adsorbed at montmorillonite surfaces revealed that the $R - COO^-$ fragment of these molecules formed hydrogen bonds with metal $- OH$ groups[44]. This type of hydrogen bonding appears to be very weak relative to the first type when relatively small organic molecules are involved. However, this type of hydrogen bonding can be very important when large organic molecules such as polymers are involved: as the number of hydrogen bonds between each functional group and the surface increases, the extent of adsorption of the large organic molecule increases.

Arrangement and Interaction of Molecular Fragments.   The three-dimensional arrangement of the molecular fragments that comprise an organic chemical can have a significant effect on its adsorption potential. The effect can be caused by (a) the intramolecular reaction of adjacent molecular fragments, or (b) the interference with a particular adsorption mechanism caused by the presence of one or more fragments.

The introduction of an OH group into a structure of an organic chemical would normally cause an increase in water solubility of the chemical and a decrease in the extent of adsorption. However, the introduction of an OH group to benzaldehyde to form hydroxybenzaldehyde resulted in increased adsorption onto activated carbon[45]; this effect was probably caused by the formation of intramolecular hydrogen bonding between the CHO and OH groups, which decreased the OH group's ability to form hydrogen bonds with water. The introduction of an OH group to 4-methoxybenzaldehyde to form 4-hydroxy-3-methoxybenzaldehyde also resulted in increased adsorption onto activated carbon[45]; this effect was believed to be due to the creation of steric hindrance by OCH_3 that interfered with the formation of a hydrogen bond between the OH group and water.

The results of adsorption studies of the gamma- and beta-isomers of 1,2,3,4,5,6 - hexachlorocyclohexane (BHC) revealed that the gamma-isomer had greater adsorbability on various soil materials than the beta-isomer[46]. An analysis of the structures of these isomers presented in Figure 6.1 will reveal that the isomers of BHC are not planar; the carbon ring assumes a puckered chair form. In the gamma-isomer, three Cl atoms are equatorial and extend out from what may be considered the plane of the ring; also, three Cl atoms are axial and project above or below the ring. In the beta-isomer, all Cl atoms are equatorial; as a result, the beta-isomer is centrosymmetric and its dipole moment is zero. On the other hand, the dipole moment of gamma-BHC is approximately 2.84 debye units in benzene. Because the beta-isomer possesses a zero dipole moment, the dipole-dipole adsorption mechanism is absent. As a result, the net adsorption forces between the beta-isomer and particle surfaces are relatively less than those between gamma-isomers and soil particle surfaces.

Studies on the adsorption of D and L amino acids onto montmorillonite surfaces revealed that some stereoisomers of organic chemicals possessing identical molecular fragments have significantly different adsorption behavior[47]. L-leucine, L-aspartate, and D-glucose are adsorbed by montmorillonite, while D-leucine, D-aspartate, and L-glucose are adsorbed at a negligible extent.

The three-dimensional arrangement of the molecular fragments of many organic chemicals can change because single bonds can rotate. Fragments will twist and realign in three-dimensional space in an attempt to reach an

beta–BHC   gamma–BHC

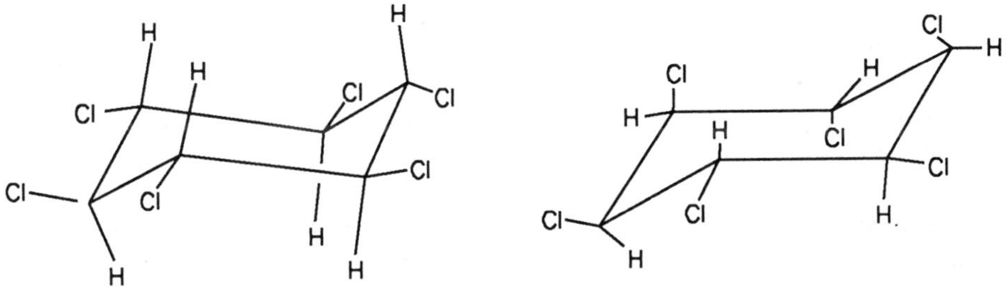

FIGURE 6.1   Structures of beta–BHC and gamma–BHC.

optimum configuration, which in most cases is planar (i.e. flat). The larger the planar surface area, the greater the extent of adsorption. An interesting example of this phenomenon is found with a comparison of the extents of adsorption of DDT, heptachlor, and dieldrin[48]. Based on water solubility, the order of decreasing hydrophobicity is: DDT > dieldrin > heptachlor. However, the order of decreasing adsorption onto the surfaces of illite, kaolinite, and montmorillonite was DDT > heptachlor > dieldrin. An analysis of the three-dimensional arrangement of the molecular structures of these organic chemicals (Figure 6.2) will reveal that DDT can orient into a predominantly planar configuration; heptachlor can orient into a less planar configuration, while dieldrin is least successful in realigning into a planar configuration.

Studies on the mechanism of adsorption of pentachlorophenol (PCP) on soil revealed that molecular fragments on a neutral organic molecule may induce charge polarization within the molecule[49]. Experimental evidence-indicated that a positive charge on the undissociated OH fragment on PCP, induced by electron withdrawing fragments (i.e. Cl), resulted in the PCP molecule's affinity for cation exchange sites on soil surfaces.

Coordination. The sixth factor that affects organic chemical adsorption onto soil surfaces is coordination; coordination is the formation of a weak bond between an organic molecule that is capable of donating electrons and adsorbed cations that are capable of accepting electrons. The net result is a partial overlap of orbitals and a partial exchange of electron density.

Electron donors can be divided into two general groups. The first group is comprised of organic molecules that possess S, N, and O fragments which contain lone pair electrons. These fragments or functional groups include nitro, alcohols, ethers, thiols, thioethers, and amines. For example, infrared studies have revealed that the oxygen of the nitro group on nitrobenzene can displace coordinated water molecules and become directly coordinated to adsorbed K^+ and NH_4^{+}[50]. In another infrared study, octylhydroxamate adsorption onto hematite surfaces was due to coordination[51]. The hydrox-amino acid fragment, $R-C(O)NHOH$, via the carbonyl oxygen and the N, formed coordination complexes with surface Fe^{3+}. In another study, acetone adsorption at bentonite surfaces occurred through coordination between the carbonyl oxygen and adsorbed Na, Ca, Mg, Al, Cu, and NH_4^{+}[52].

The second group of electron donors is comprised of organic molecules that possess electron-rich pi fragments such as alkenes, alkynes, aromatics, and heteroaromatics. These electron-rich fragments can form weak bonds with either adsorbed cations or with soil particle surfaces. For example, specific adsorption of benzene, cyclohexadiene, cyclohexene, 1-butene, cis-2-butene, and 1-hexene onto zeolites occurred through the pi fragments

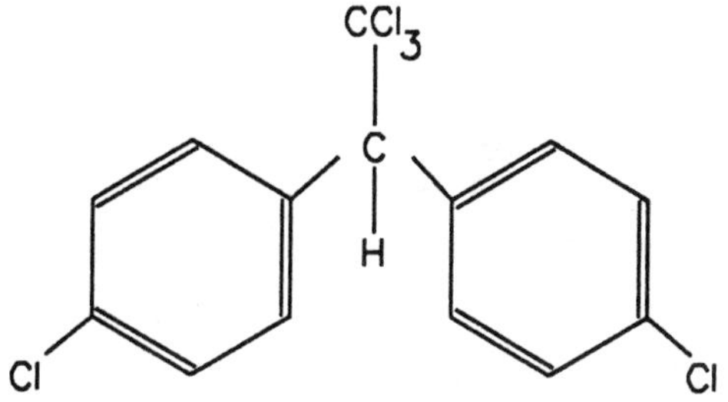

pp'–DDT

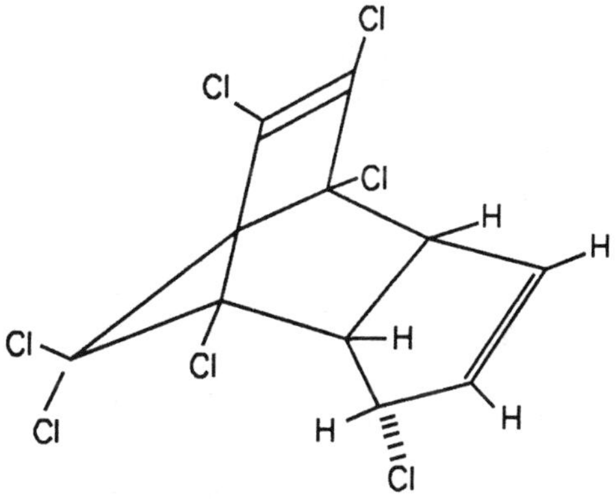

HEPTACHLOR

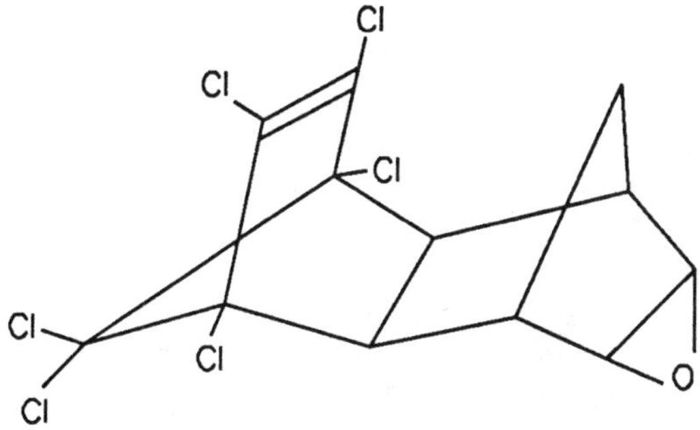

DIELDRIN

FIGURE 6.2 Structures of DDT, Heptachlor, and Dieldrin.

of these organic chemicals[53].

It is important to recognize that coordination can occur between organic chemicals and cations in the water phase of a soil system. By using chromatographic and spectroscopic techniques, aqueous Fe^{3+} and Cu^{2+} were found to extract organic molecules dissolved in toluene and to associate with these organic molecules in the water phase[54].

ESTIMATING ADSORPTION COEFFICIENTS

Chapter Four introduced the concept of adsorption. Adsorption was defined as the accumulation of a chemical at soil surfaces:

$$C_s \underset{k(ads)}{\overset{k(des)}{\rightleftharpoons}} C_e \qquad (6.8a)$$

where

C_s = concentration adsorbed on soil surfaces (ug/gram soil)

C_e = concentration in water (ug/ml)

k(des) = desorption rate

k(ads) = adsorption rate

The adsorption process is usually a reversible one. At equilibrium, the distribution of an organic chemical is governed by two opposing rate processes. The adsorption rate, k(ads), is the rate at which the dissolved organic chemical in water transfers into the adsorbed state. The desorption rate, k(des), is the opposite process: it is the rate at which the organic chemical transfers from the adsorbed state into water. The simplest and most common method for mathematically expressing the distribution of an organic chemical between soil surfaces and water as illustrated in Equation 6.8a is the adsorption or distribution coefficient, K_d. The K_d is defined as the ratio:

$$K_d = C_s/C_e \qquad (6.9)$$

The greater the extent of adsorption (i.e. $C_s > > C_e$), the greater the magnitude of K_d. Several researchers have found that if K_d is normalized on the basis of the soil's organic matter or organic carbon content, much of the variation observed among K_ds over different soils can be eliminated. Normalized K_ds are expressed as either K_{om} or K_{oc}:

$$K_{om} = K_d/om \qquad (6.10)$$

and

$$K_{oc} = K_d/oc \qquad (6.11)$$

where

K_{om} = Soil adsorption coefficient normalized for soil organic matter content.

K_{oc} = Soil adsorption coefficient normalized for the soil organic carbon content.

om = The soil organic matter content (mg organic matter/mg soil).

oc = The soil organic carbon content (mg organic carbon/mg soil).

The relationship between om and oc is assumed to be constant:

$$K_{oc} = 1.724 \ K_{om} \qquad (6.12)$$

K_{oc} and K_{om} values, which have been measured for a wide variety of organic chemicals and soils, are listed in Table 6.3.

In many cases, where it is not feasible to obtain an accurate measurement of the K_d, K_{om}, or K_{oc}, it may be necessary to settle for a reasonable estimate. Several researchers have found that the extent of adsorption of nonionic organic chemicals onto soil particle surfaces was well-correlated with two empirical measurements of organic chemical hydrophobicity: the water solubility, S, and the octanol-water partititon coefficient, K_{ow}. Table 6.4 lists 14 equations which utilize either S or K_{ow} to estimate K_{oc} or K_{om}.

It is most important to recognize that these equations are not universally applicable to all organic chemicals in all soil systems. These estimation equations should not be overextended or used without regard to their limitations. The following guidelines should be utilized whenever one of the equations listed in Table 6.4 is being considered for use in estimating an organic chemical's adsorption behavior in soil.

First, these equations were developed with the assumption that the hydrophobicity of the organic chemical was the only factor that affects the accumulation of organic chemicals onto soil surfaces. These equations neglect

TABLE 6.3 Measured K_{oc} and K_{om} Values for Various Organic Chemicals.

Chemical	K_{oc}	K_{om}	Ref.
acetophenone	35		55
alachlor	190		56
aldrin	410		56
ametryn	392		56
2-aminoanthracene	33,500		57
6-aminochrysene	162,900		57
3-aminonitrobenzene		31	58
4-aminonitrobenzene		44	58
anilide		15	58
aniline		15	58
anisole		20	59
anthracene	26,000		60
anthracene-9-carboxylic acid	517		57
asulam	300		56
atrazine	148		55
	162		55
	216		61
benefin	10,700		56
benzene		18	59
	83		60
alpha-BHC	1995		55
beta-BHC	1995		55
gamma-BHC (lindane)	1995		55
	735		62
2,2′-biquinoline	10,471		55
bromacil	72		55
3-bromoanilide		59	58
4-bromoanilide		51	58
4-bromoaniline		52	58
4-bromonitrobenzene		151	58
3-bromophenylurea		66	58
4-bromophenylurea		76	58
butralin	8200		56
n-butyl-N-phenylcarbamate		105	58
butyranilide		37	63
		30	58
carbaryl	229		55
	310		62
		70	63
carbofuran	105		62
	29		55
carbophenothion	45,400		56
chloramben	21		56
chloramben, methyl ester	507		56
3-chloroacetanilide		50	63

TABLE 6.3 Measured K_{oc} and K_{om} Values for Various Organic Chemicals. (cont.).)

Chemical	K_{oc}	K_{om}	Ref.
2-chloroanilide		22	58
3-chloroanilide		42	58
chlorobenzene		48	59
chlorobromuron	460		56
		219	58
		211	63
2-chlorobiphenyl		1698	59
3-chloro-4-bromonitrobenzene		229	58
3-chloro-4-methoxyacetanilide		54	63
3-chloro-4-methoxyanilide		48	58
3-chloro-4-methoxyaniline		49	58
3-(3-chloro-4-methoxyphenyl)- 1-methylurea		40	58
3-chloro-4-methoxyphenylurea		58	58
3-(3-chloro-4-methylphenyl)- 1-methylurea		72	58
chloroneb	1159		56
3-(3-chlorophenyl)- 1,1-dimethylurea		35	58
3-(3-chlorophenyl)- 1-methylurea		49	58
2-chlorophenylurea		23	58
3-chlorophenylurea		59	58
6-chloropicolinic acid	9		56
chloroxuron	3200		56
chlorppropham	589		55
chlorpyrifos	13,490		55
	6070		62
chloropyrifos-methyl	3300		56
chlorothiamid	107		56
chlortoluron		61	63
		60	58
crotoxyphos	170		56
cyanazine	200		55
	183		61
cycloate	345		56
2,4-D	57		62
	20		56
DBCP	129		56
p, p´-DDT	129		56
	239,883		55
	150,000		62
diallate	1900		56
diamidaphos	32		56
1,2;5,6-dibenzanthracene	2,029,000		64
dibenzothiophene	11,220		55

TABLE 6.3 Measured K_{oc} and K_{om} Values for Various Organic Chemicals. (cont.)

Chemical	K_{oc}	K_{om}	Ref.
dicamba	0.4		56
dichlobenil	235		56
3,4-dichloroacetanilide		153	63
3,4-dichloroanilide		126	58
3,4-dichloroaniline		112	58
1,2-dichlorobenzene	347		55
		186	59
1,3-dichlorobenzene		170	59
1,4-dichlorobenzene		158	59
2,2'-dichlorobiphenyl		4786	59
2,4'-dichlorobiphenyl		7762	59
1,2-dichloroethane	32		55
3,4-dichloronitrobenzene		195	58
3-(3,4-dichlorophenyl)-1-methylurea (DMU)		166	58
		185	63
3,4-dichlorophenylurea		209	63
		179	58
3,6-dichloropicolinic acid	2		56
cis-1,3-dichloropropene	23		56
trans-1,3-dichloropropene	26		56
diflubenzuron	6790		56
7,12-dimethylbenzanthracene	235,700		64
3-(3,5-dimethyl-4-bromophenyl)-1,1-dimethylurea		195	58
3-(3,5-dimethylphenyl)-1,1-dimethylurea		31	58
dinitramine	4000		56
dinoseb	124		56
dipropetryn	1170		56
disulfoton	1780		56
diuron	398		55
	380		55
	387		62
		111	63
		93	58
DMSA	770		56
EPTC	240		56
ethion	15,400		56
ethylbenzene		95	59
ethylenedibromide	44		56
ethyl-N-phenylcarbamate		38	58
fenuron	27		55
	43		55

TABLE 6.3 Measured K_{oc} and K_{om} Values for Various Organic Chemicals. (cont.)

Chemical	K_{oc}	K_{om}	Ref.
fenuron (cont.)		15	63
		8	58
fluchloralin	3600		56
fluometuron	174		55
		38	58
3-fluoroanilide		21	58
4-fluoroanilide		17	58
3-(3-fluorophenyl)-			
1,1-dimethylurea		31	58
3-(4-fluorophenyl)-			
1,1-dimethylurea		15	58
2-fluorophenylurea		12	58
3-fluorophenylurea		34	58
4-fluorophenylurea		32	63
		19	58
glyphosate	2640		56
13H-dibenzo[a,i]carbazole	1,047,129		55
2,2′,4,4′,5,5′-			
hexachlorobiphenyl	416,869		55
2,2′,4,4′,6,6′-			
hexachlorobiphenyl	1,200,000		55
ipazine	1660		55
	813		61
isocil	130		56
isopropalin	75,250		56
isopropyl-N-phenylcarbamate			
(propham)		51	58
leptophos	9300		56
linuron	813		55
	871		55
		133	63
		155	58
malathion	1778		55
methazole	2620		56
methomyl	160		56
4-methoxyacetanilide		23	63
4-methoxyanilide		14	58
methoxychlor	80,000		60
3-(3-methoxyphenyl)-			
1,1-dimethylurea		30	58
3-(4-methoxyphenyl)-			
1,1-dimethylurea		14	58

TABLE 6.3   Measured K_{oc} and K_{om} Values for Various Organic Chemicals. (cont.)

Chemical	K_{oc}	K_{om}	Ref.
2-methoxy-3,5,6-trichloropyridine	920		56
3-methylacetanilide		32	63
3-methylanilide		16	58
3-methylaniline		26	58
4-methylaniline		46	58
9-methylanthracene	65,000		60
3-methyl-4-bromoaniline		105	58
3-methyl-4-bromophenylurea		135	58
methyl-N-(3-chlorophenyl)-N-phenylcarbamate		81	58
3-methylcholanthrene	1,789,000		64
methyl-N-(3,4-dichlorophenyl)-N-phenylcarbamate		316	58
3-methyl-4-fluorophenylurea		36	63
		35	58
methyl isothiocyanate	6		56
2-methylnaphthalene	8500		60
methylparathion	5129		55
	9772		55
methyl-N-phenylcarbamate		31	58
3-(4-methylphenyl)-1,1-dimethylurea		19	58
3-methylphenylurea		21	58
N-methyl-N'-phenylurea		25	63
metobromuron	60		56
		60	58
		58	63
metoxuron		40	63
		32	58
metribuzin	95		56
monolinuron	200		55
	282		55
		38	63
		40	58
monuron	100		55
	182		55
		34	63
		29	58
naphthalene	1300		60
napropamide	680		56
neburon	2300		56
nitralin	960		56
nitrapyrin	458		62
	420		56
3-nitroanilide		50	58

TABLE 6.3 Measured K_{oc} and K_{om} Values for Various Organic Chemicals. (cont.)

Chemical	K_{oc}	K_{om}	Ref.
nitrobenzene		50	58
norfluorazon	1914		56
oxadiazon	3241		56
parathion	4786		55
	10,715		55
pebulate	630		56
2,2',4,5,5'-			
pentachlorobiphenyl	42,500		56
pentachlorophenol	900		56
n-pentyl-N-phenylcarbamate		234	58
permethrin		19300	63
phenanthrene	23,000		60
phenol	27		56
4-phenoxyphenylurea		209	58
3-phenyl-1-cycloheptylurea		135	58
3-phenyl-1-cyclohexylurea		68	58
3-phenyl-1-cyclopentylurea		49	58
3-phenyl-1-cyclopropylurea		30	58
3-phenyl-1-methylurea		11	58
phenylurea		13	58
phorate	3200		56
picloram	17		56
profluralin	8600		56
prometon	350		56
prometryn	48		56
pronamide	200		56
propachlor	265		56
propazine	158		55
	155		55
	363		61
propham	51		56
n-propyl-N-phenylcarbamate		66	58
pyrazon	120		56
pyrene	62,700		64
	84,000		60
pyroxychlor	3000		56
silvex	2600		56
simazine	135		55
	138		55
	215		61
2,4,5-T	53		56
tebuthiuron	620		56
terbacil	51		55

TABLE 6.3   Measured K_{oc} and K_{om} Values for Various Organic Chemicals. (cont.)

Chemical	K_{oc}	K_{om}	Ref.
terbacil (cont.)	41		55
terbutryn	700		56
tetracene	650,000		60
2,3,7,8-tetrachloro-			
dibenzo-p-dioxin	481,340		65
1,1,2,2-tetrachloroethane	79		55
tetrachloroethylene	363		55
thiabendazole	1720		56
triallate	2220		56
1,2,4-trichlorobenzene		501	59
2,4,4′-trichlorobiphenyl		23,988	59
1,1,1-trichloroethane	178		55
3,5,6-trichloro-2-pyridinol	130		56
triclopyr	27		56
trietazine	547		61
3-trifluoromethylanilide		32	58
3-trifluoromethylaniline		132	58
3-trifluoromethylphenylurea		57	63
		52	58
trifluralin	4340		62
	30,550		61
	13,700		56

other factors that may affect organic chemical adsorption such as molecular charge, hydrogen bonding, the arrangement and interaction of molecular fragments, coordination, and reactivity. As a result, one must analyze the structure of the chemical of concern, using the guidelines discussed earlier in this chapter and in other chapters, to determine if these other factors may significantly affect organic chemical adsorption; if no complicating effect is identified, then one of these equations may be used, provided the equation was developed using organic chemicals where structures are very similar to the structure of the chemical of concern. If a similarity between chemical structures is not found, then the equation should not be utilized.

Second, these equations are valid generally for organic chemicals with molecular weights less than about 400. At larger molecular weights, van der Waals forces tend to overwhelm all other effects of chemical structure on adsorption mechanisms; these equations were developed with the assumption that van der Waals forces had a negligible effect on organic chemical adsorption onto soil particle surfaces.

Third, many of these equations were developed using soils with organic carbon contents that were generally above 0.1 percent. On the other hand, subsoils typically possess organic carbon contents at or below this amount.

TABLE 6.4 K_{oc} Predictive Equations.

K_{oc} Equation# 1: $\log K_{oc} = -0.54 \log S + 0.44$ (6.13)

and

S = water solubility (mole fraction)

K_{oc} Equation# 2: $\log K_{oc} = \log K_{ow} - 0.21$ (6.14)

Reference: 60
Soil Percent Organic Carbon Range: 0.09 − 3.29
Organic Chemicals Utilized to Develop K_{oc} Equations #1 and #2:

anthracene
benzene
hexachlorobiphenyl
methoxychlor
9-methylanthracene
2-methylnaphthalene
naphthalene
phenanthrene
pyrene
tetracene

K_{oc} Equation# 3: $\log K_{oc} = -0.557 \log S + 4.277$ (6.15)

and

S = water solubility (micromoles/liter)

Reference: 66
Soil Percent Organic Matter Content (one soil): 1.6
Organic Chemicals Utilized to Develop K_{oc} Equation #3:

DDT
1,2-dibromo-3-chloropropane
1,2-dibromomethane
1,2-dichlorobenzene
2,4′-dichlorobiphenyl
1,2-dichloroethane
1,2-dichloropropane
2,2′,4,4′,5,5′-hexachlorobiphenyl
lindane
parathion
2,2′,5,5′-tetrachlorobiphenyl
1,1,2,2-tetrachloroethane
tetrachloroethene
1,1,1-trichloroethane

K_{oc} Equation# 4: $\log K_{oc} = 0.937 \log K_{ow} - 0.006$ (6.16)

Reference: 61
Sediment Organic Carbon Content (one sediment): 0.033
Organic Chemicals Used to Develop K_{oc} Equation #4:

TABLE 6.4 K_{oc} Predictive Equations. (cont.)

---

atrazine
cyanazine
ipazine
propazine
simazine
trietazine
trifluralin
trifluralin photodegradation products

K_{oc} Equation# 5:   log K_{oc} = 1.029 log K_{ow} − 0.18   (6.17)
Reference: 67
Soil Organic Carbon Range: not specified
Organic Chemicals Used to Develop K_{oc} Equation #5:
   atrazine
   bromacil
   carbofuran
   2,4-D
   DDT
   dicamba
   dichlobenil
   diuron
   lindane
   malathion
   methylparathion
   simazine
   terbacil

K_{oc} Equation# 6: K_{oc} = 0.524 log K_{ow} + 0.855   (6.18)
Reference: 68
Soil Percent Organic Matter Range: 1.0 − 4.0
Organic Chemicals Used to Develop K_{oc} Equation #6:
   3-(3-bromophenyl)urea
   3-(4-bromophenyl)urea
   3-(3-chlorophenyl)urea
   3-(3-chloro-4-methoxyphenyl)urea
   3-(3,4-dichlorophenyl)urea
   1,1-dimethyl-3-(4-chlorophenyl)urea
   1,1-dimethyl-3-(3-chloro-4-methoxyphenyl)urea
   1,1-dimethyl-3-(3,4-dichlorophenyl)urea
   1,1-dimethyl-3-(3-trifluoromethylphenyl)urea
   3-(3-fluorophenyl)urea
   3-(4-fluorophenyl)urea
   3-(3-hydroxyphenyl)urea
   1-methyl-3-(3-chlorophenyl)urea
   1-methyl-3-(3-chloro-4-methoxyphenyl)urea

TABLE 6.4 K_{oc} Predictive Equations. (cont.)

1-methyl-3-(3,4-dichlorophenyl)urea
1-methyl-1-methoxy-3-(4-bromo-3-chlorophenyl)urea
1-methyl-1-methoxy-3-(4-bromophenyl)urea
1-methyl-1-methoxy-3-(4-chlorophenyl)urea
1-methyl-1-methoxy-3-(3,4-dichlorophenyl)urea
phenylurea
3-(4-sulfophenyl)urea
3-(3-trifluoromethylphenyl)urea

K_{oc} Equation# 7: $\log K_{oc} = -0.82 \log S \text{ (in ppm)} + 4.07$ (6.19)
Reference: 64
Soil Percent Organic Carbon Range: 0.11 – 2.38
Organic Chemicals Used to Develop K_{oc} Equation #7:

      dibenzanthracene
      7,12-dimethylbenz[a]anthracene
      3-methylcholanthrene
      pyrene

K_{oc} Equation# 8: $\log K_{oc} = 0.72 \log K_{ow} + 0.49$ (6.20)
Reference: 69
Soil Percent Organic Matter Range: <0.01 – 33.0
Organic Chemicals Used to Develop K_{oc} Equation #8:

      n-butylbenzene
      chlorobenzene
      1,4-dichlorobenzene
      1,4-dimethylbenzene
      1,2,3,4-tetrachlorobenzene
      1,2,4,5-tetrachlorobenzene
      tetrachloroethylene
      1,2,4,5-tetramethylbenzene
      toluene
      1,2,3-trichlorobenzene
      1,2,4-trichlorobenzene
      1,2,3-trimethylbenzene
      1,3,5-trimethylbenzene

K_{oc} Equation# 9: $\log K_{oc} = -0.594 \log S \text{ (mole fraction)} - 0.197$ (6.21)
K_{oc} Equation#10: $\log K_{oc} = 0.989 \log K_{ow} - 0.346$ (6.22)
Reference: 55
Soil Percent Organic Carbon Range: 0.66 – 2.38
Organic Chemicals Used to Develop K_{oc} Equations #9 & #10:

      anthracene
      benzene
      naphthalene
      phenanthrene
      pyrene

TABLE 6.4 K_{oc} Predictive Equations. (cont.)

K_{oc} Equation #11: $\log K_{om} = 0.52 \log K_{ow} + 0.69$ (6.23)

Reference: 70

Soil Percent Organic Matter Range: 0.19 − 6.62

Organic Chemicals Used to Develop K_{oc} Equation #11:

butyranilide
3-chloroacetanilide
3-chloro-4-methoxyacetanilide
3,4-dichloroacetanilide
3,4-dichlorophenylurea
1,1-dimethyl-3-(3,5-dimethylphenyl)urea
1,1-dimethyl-3-(3-chloro-4-methoxyphenyl)urea
1,1-dimethyl-3-(3-chloro-4-methyl)urea
1,1-dimethyl-3-(4-chlorophenyl)urea
1,1-dimethyl-3-(3,4-dichlorophenyl)urea
1,1-dimethyl-3-phenylurea
4-fluorophenylurea
4-methoxyacetanilide
3-methylacetanilide
3-methyl-4-fluorophenylurea
1-methyl-1-methoxy-3-(4-bromophenyl)urea
1-methyl-1-methoxy-3-(4-chlorophenyl)urea
1-methyl-1-methoxy-3-(4-chloro-3-bromophenyl)urea
1-methyl-1-methoxy-3-(3,4-dichlorophenyl)urea
1-methyl-3-(3,4-dichlorophenyl)urea
1-methyl-3-phenylurea
1-naphthalenylmethylcarbamate
permethrin
3-trifluoromethylphenylurea

K_{oc} Equation#12: $\log K_{om} = 0.52 \log K_{ow} + 0.64$ (6.24)

Reference: 58

Soil Percent Organic Matter Range: 1.09 − 5.92

Organic Chemicals Used to Develop K_{oc} Equation #12:

aldicarb
aldicarb sulfone
aldrin
3-aminonitrobenzene
4-aminonitrobenzene
azobenzene
3-bromoanilide
4-bromoanilide
4-bromoaniline
4-bromonitrobenzene
4-bromophenol
3-bromophenylurea
4-bromophenylurea

TABLE 6.4 K_{oc} Predictive Equations. (cont.)

n-butyl N-phenylcarbamate
butyranilide
captafol
captan
carbaryl
chlorfenvinphos
2-chloroanilide
3-chloro-4-bromonitrobenzene
3-chloro-4-methoxyanilide
3-chloro-4-methoxyaniline
3-chloro-4-methoxyphenylurea
2-chlorophenylurea
3-chlorophenylurea
1-cycloheptyl-3-phenylurea
1-cyclohexyl-3-phenylurea
1-cyclopentyl-3-phenylurea
1-cyclopropyl-3-phenylurea
diazinon
3,4-dichloroanilide
3,4-dichloroaniline
3,4-dichloronitrobenzene
3,4-dichlorophenylurea
dieldrin
dimethoate
1,1-dimethyl-3-(3-chloro-4-methylphenyl)urea
1,1-dimethyl-3-(3-chloro-4-methoxyphenyl)urea
1,1-dimethyl-3-(3-chlorophenyl)urea
1,1-dimethyl-3-(4-chlorophenyl)urea
1,1-dimethyl-3-(3,4-dichlorophenyl)urea
1,1-dimethyl-3-(3,5-dimethyl-4-bromophenyl)urea
1,1-dimethyl-3-(3,5-dimethylphenyl)urea
1,1-dimethyl-3-(3-fluorophenyl)urea
1,1-dimethyl-3-(4-fluorophenyl)urea
1,1-dimethyl-3-(3-methoxyphenyl)urea
1,1-dimethyl-3-(4-methoxyphenyl)urea
1,1-dimethyl-3-(4-methylphenyl)urea
1,1-dimethyl-3-phenylurea
1,1-dimethyl-3-(3-trifluoromethylphenyl)urea
diphenylamine
Dowco 275
ethyl N-phenylcarbamate
fenamiphos
3-fluoroanilide
4-fluoroanilide
2-fluorophenylurea
3-fluorophenylurea
4-fluorophenylurea
folpet
hexachlorobenzene
isopropyl N-phenylcarbamate
methiocarb

TABLE 6.4 K_{oc} Predictive Equations. (cont.)

4-methoxyanilide
1-methoxy-1-methyl-3-(4-bromophenyl)urea
1-methoxy-1-methyl-3-(4-chlorophenyl)urea
1-methoxy-1-methyl-3-(3-chloro-4-bromophenyl)urea
1-methoxy-1-methyl-3-(3,4-dichlorophenyl)urea
3-methylanilide
3-methylaniline
4-methylaniline
3-methyl-4-bromoaniline
3-methyl-4-bromophenylurea
1-methyl-3-(3-chloro-4-methoxyphenyl)urea
1-methyl-3-(3-chloro-4-methylphenyl)urea
methyl N-(3-chlorophenyl)carbamate
1-methyl-3-(3-chlorophenyl)urea
methyl N-(3,4-dichlorophenyl)carbamate
1-methyl-3-(3,4-dichloroplhenyl)urea
3-methyl-4-fluorophenylurea
methyl N-phenylcarbamate
3-methylphenylurea
naphthalene
nitrapyrin
3-nitroanilide
oxamil
parathion
pentafluorophenylmethylsulfone
n-pentyl N-phenylcarbamate
phenol
4-phenoxyphenylurea
phorate
n-propyl N-phenylcarbamate
simazine
tetrachlorobenzene
3-trifluoromethylanilide
3-trifluoromethylaniline
3-trifluoromethylphenylurea

K_{oc} Equation#13: $\log K_{oc} = -0.55 \log S + 3.64$ (6.25)
$$\text{and}$$
$$S = \text{water solubility, mg/liter}$$

K_{oc} Equation#14: $\log K_{oc} = 0.544 \log K_{ow} + 1.377$ (6.26)
Reference: 71
Soil Percent Organic Carbon Range: not specified
Organic Chemicals Used to Develop K_{oc} Equations #13 & #14:
    aldrin
    ametryn
    anthracene
    asulam

TABLE 6.4 K_{oc} Predictive Equations. (cont.)

atrazine
benefin
benzene
bromacil
sec-bumeton
butralin
carbaryl
carbophenothion
chloramben
chloramben methyl ester
chlorbromuron
chloroneb
6-chloropicolinic acid
chloroxuron
chlorpropham
chlorpyrifos
chlorpyrifos-methyl
chlorthiamid
crotoxyphos
cyanazine
cycloate
2,4-D acid
DDT
diallate
diamidaphos
dibromochloropropane
dicamba
dichlobenil
3.6-dichloropicolinic acid
cis-1,3-dichloropropene
trans-1,3-dichloropropene
diflubenzuron
dinitramine
dinoseb
dipropetryn
disulfoton
diuron
EPTC
ethion
ethylene bromide
fenuron
fluchloralin
fluometuron
2,2 ',4,4 ',5,5 '-hexachlorobiphenyl
hexachlorobenzene
ipazine
isocil
isopropalin
leptophos
lindane
linuron

TABLE 6.4 K_{oc} Predictive Equations. (cont.)

methazole
metobromuron
methomyl
methoxychlor
2-methoxy-3,5,6-trichloropyridine
9-methylanthracene
methyl isothiocyanate
2-methylnaphthalene
methyl parathion
metribuzin
monolinuron
monuron
naphthalene
napropamide
neburon
nitralin
nitrapyrin
norfluorazon
oxadiazon
parathion
pebulate
2,2′,4,5,5′-pentachlorobiphenyl
pentachlorophenol
phenanthrene
phenol
picloram
phorate
profluralin
prometon
prometryn
pronamide
propachlor
propazine
propham
pyrazon
pyrene
pyroxychlor
silvex
simazine
2,4,5-T
tebuthiuron
terbacil
terbutryn
tetracene
thiabendazole
triallate
3,5,6-trichloro-2-pyridinol
triclopyr
trietazine
trifluralin
urea

The valid use of an equation requires that predictions be made for soils possessing organic carbon contents that fall within the range of those organic carbon contents used to develop the equation.

Fourth, these equations do not take into account the potential contributions of mineral surfaces to organic chemical adsorption. It is most important to note that one should not assume that the contribution of the inorganic soil mineral surfaces to the accumulation of an organic chemical is negligible, and that the organic carbon content is the only governing factor. An analysis of the information presented in Table 6.1 revealed earlier that inorganic mineral surfaces can make a substantial contribution to the accumulation of organic chemicals at soil surfaces. Table 6.5 lists suggested threshold values for significant mineral surface contributions to organic chemical adsorption. If these threshold values are exceeded, then the K_{oc} equations listed in Table 6.4 should not be utilized.

Fifth, these equations do not take into account the fact that soil organic matter varies in chemical structure and composition among soils. It is important to remember that although soil organic matter is usually considered to be a "homogeneous mass," in reality it is a heterogeneous mixture of organic chemicals. And, organic chemicals will be preferentially accumulated by the different organic chemicals which comprise soil organic matter. One excellent example of this phenomenon involved a study utilizing Russell, Chalmers, and Kokomo silt loams[73]; benzidine and alpha-naphthylamine were preferentially and primarily accumulated by humic acid, whereas p-toluidine was primarily accumulated by fulvic acid. This selective accumulation occurs for several reasons. The humic to fulvic acid ratio varies significantly among soils[74]. The aromatic hydrocarbon content and the ratio of polypeptide to polysaccharide products varies significantly among soils. Even the macromolecular structure of the humic fraction of soil organic matter can change as pH, ionic strength, and the concentration of humic material changes[75]. Organic matter is oxidized to varying extents in soil[12]; it generally has a reactivity with respect to water which is similar to that of montmorillonite[76].

Sixth, these equations do not take into account the fact that serious errors may arise by assuming a linear adsorption isotherm occurs when high concentrations of organic chemicals exist in soil systems. Linear adsorption isotherms have usually been assumed to exist for low chemical concentrations because this assumption simplifies computer simulation modeling. However, linear adsorption isotherms may not exist at higher organic chemical concentrations[77]. It is important to recognize that K_{oc} equations are developed primarily at low organic chemical concentrations, and these equations may overestimate the K_{oc} of organic chemicals present at high concentrations.

TABLE 6.5 Threshold Values for Significant Mineral Surface Contributions to Organic Chemical Adsorption[a].

Chemical	cm/oc[b]
Neutral organics with polar functional groups	25-60[c]
Nonpolar organics with less than 10 carbon atoms	>60

[a] Compiled from data presented in Ref. 72.

[b] Ratio of the relative amounts of clay (cm) and organic carbon (oc) in soils with mixed, aged mineral suites; based on the asumption of (a) monolayer coverage of the mineral surface by organic matter, and (b) a 1% carbon content would reduce available mineral surface by 32 m^2/g, which corresponds to a 20% to 30% decrease in surface area for typical swelling field clays.

[c] i.e., if cm/oc < 25, minerals would not be expected to significantly contribute to adsorption.

Seventh, these equations were developed with the assumption that the K_{oc} values for the organic chemicals utilized to develop these equations are single valued (i.e. the K_{oc} is a constant for an organic chemical). Actually, the K_{oc} is not a constant and can vary significantly. An extreme example is alpha-naphthol[78], whose K_{oc} varied from 328 to 15,618. The K_{oc} equations listed in this chapter do not account for the potential variability of K_{oc}s of individual organic chemicals.

These equations do not take into account the salt effect, the change of the adsorption coefficient due to a change in the salt content of water. The dependence of adsorption on salt content was approximated with a derivative of the Setschenow equation[72]:

$$\log[K_d I_2 / K_d I_1] = K_s (I_2 - I_1) \tag{6.27}$$

where

$$I_1, I_2 = \text{Aqueous salt concentrations}$$
$$K_s = \text{Setschenow constant for a given salt (e.g. 0.16 liters/mole for NaCl).}$$

Because K_s is positive for most common salts, an increase in the extent of adsorption of most neutral organic chemicals should occur with increased salt concentration. However, the change in K_d with a change in I usually is not dramatic. For example, a 0.5 molar increase in NaCl would increase K_d by a factor of $10^{0.08}$ or approximately 1.2.

BULK HYDROCARBON EFFECTS ON ORGANIC CHEMICAL ADSORPTION AND MOBILITY

The K_{oc} and K_{om} values presented in Table 6.3 and the K_{oc} equations presented in Table 6.4 can be applied in soil systems whenever water is the primary solvent. However, these equations cannot be utilized in soils whenever a bulk hydrocarbon is the primary solvent. The K_d, K_{oc}, and K_{om} for an organic chemical can be significantly different in a soil-water system than in a soil-bulk hydrocarbon system.

The published literature cites several examples of this phenomenon. Published data are reported as either K_ds or soil thin-layer chromatography R_f values. The TLC R_f value is the ratio of the furthest distance travelled by an organic chemical on a thin-layer chromatography plate to the distance travelled by a solvent front. The TLC R_f is inversely related to the K_d (see Table 6.6); as the TLC R_f increases toward unity, the chemical exhibits greater mobility.

TABLE 6.6 Relative Relationships Between K_d, K_{om}, K_{oc}, and Soil TLC R_f Values[a]

K_d[b]	K_{om}	K_{oc}	Soil TLC	
			R_f	Mobility Class
>10	>200	>2000	0.00 - 0.10	I - Immobile
2 - 10	60 - 200	500 - 2000	0.10 - 0.34	II - Low Mobility
0.5 - 2	20 - 60	150 - 500	0.35 - 0.64	III - Intermediate Mobility
0.1 - 0.5	5 - 20	50 - 150	0.65 - 0.89	IV - Mobile
< 0.1	< 5	< 50	0.90 - 1.00	V - Very Mobile

[a] Based on classification schemes presented in references 58, 62, and 79.
[b] Based on values derived from soil with 2.5% organic matter using a variety of pesticides.

The magnitude of the K_d and the TLC R_f for an organic chemical depends upon whether the organic chemical is miscible in the bulk hydrocarbon. An analysis of the data presented in Table 6.7 will reveal that an organic chemical's mobility generally increases and its extent of adsorption generally decreases as its solubility in the bulk hydrocarbon increases. The further apart the water solubility of the organic chemical is from the water solubility of the bulk hydrocarbon, the less miscible the two will be; the lower the miscibility, the lower the mobility of the organic chemical.

A further analysis of the data in Table 6.7 revealed that undiluted bulk hydrocarbons significantly altered the extent of adsorption and mobility of relatively water-insoluble organic chemicals. Although landfill leachates contain ppm levels of various organic chemicals, its effect on organic chemical K_d and TLC R_f was the same as that of water.

TABLE 6.7 Effect of Various Solvents on Organic Chemical Adsorption and Mobility.

Chemical	Soil	Solvent	K_d	TLC R_f	Ref.
Aroclor 1242	Ava Silty Clay Loam Soil	Water		0.02	80
		Landfill Leachate		0.02	
		Carbon Tetrachloride		1.00	
	Catlin Silt Loam Soil	Water		0.02	80
		Landfill Leachate		0.04	
		Carbon Tetrachloride		1.00	
	Catlin Loam Soil	Water		0.02	80
		Landfill Leachate		0.03	
		Carbon Tetrachloride		1.00	
Aroclor 1254	Ava Silty Clay Loam Soil	Water		0.02	80
		Landfill Leachate		0.02	
		Carbon Tetrachloride		0.96	
	Catlin Silt Loam Soil	Water		0.02	80
		Landfill Leachate		0.04	
		Carbon Tetrachloride		1.00	
	Drummer Silt Loam Soil	Water		0.03	80
		Carbon Tetrachloride		1.00	
Dicamba	Ava Silty Clay Soil	Water		1.00	80
		Landfill Leachate		1.00	
		Carbon Tetrachloride		0.02	
	Catlin Silt Loam Soil	Water		0.85	80
		Landfill Leachate		0.90	
		Carbon Tetrachloride		0.02	
	Flanagan Silt Loam Soil	Water		1.00	80
		Landfill Leachate		1.00	
		Carbon Tetrachloride		0.02	
PCB[b]	Montmorillonitic Sandy Clay and Silty Clay Soils	Water	263		38
		Ethylene Glycol	5.5		
		Kerosene	1.7		
Toxaphene	Cecil B2t Soil	Water	99		81
		Acetonitrile	0.85		
		Ethanol	0.53		
		n-Hexane	0.23		
		Dichloromethane	0.09		
	Cecil A Soil	Water	>99		81
		Acetonitrile	1.04		
		Ethanol	0.61		
		n-Hexane	0.35		
		Dichloromethane	0.18		

[a] Soil thin-layer chromatography R_f value, the ratio of the furthest distance traveled by an organic chemical on a thin-layer chromatography plate with soil as the solid matrix to the distance travelled by a solvent front.

[b] A mixture of polychlorinated biphenyl isomers, with Aroclor 1242 being the predominant mixture present.

PREDICTING ORGANIC CHEMICAL MOBILITY IN SOIL

Several relatively simple mathematical models that require the distribution coefficient as an input parameter have been developed. These models are, in the opinion of the author, simple and reliable equations that can fulfill the need for quick, inexpensive, yet reliable information on the potential migration rate of an organic chemical in a soil system. These equations should be utilized in conjunction with the information provided in this chapter, in other chapters of this book, and with technical common sense.

The Retardation Equation. The most commonly accepted and utilized equation for grossly estimating the migration rate of a chemical in a soil-groundwater system is the retardation equation:

$$V_c = V[1 + K_d(b/P_T)]^{-1} \tag{6.28}$$

where

V_c = The velocity of the chemical at the point where
$c/c_o = 0.5$

V = Average linear velocity of groundwater

K_d = Distribution coefficient

b = Soil bulk density

P_T = Soil total porosity

The term $[1 + K_d(b/P_T)]$ is known as the retardation factor; however, this term should not be confused with the soil TLC R_f (retardation factor) because they are not equivalent terms. Equation 6.28 has been utilized successfully to estimate the velocity of organic chemicals such as PCBs, dioxin, and pesticides within a factor of two to three in relatively unstratified soils.

The retardation equation can be modified for use in estimating the migration rate, V_c, of a chemical in unsaturated zone soil[82]:

$$V_c = V_{sw} (WC + bK_d)^{-1} \tag{6.29}$$

where

V_{sw} = Amount of soil water percolating through the unsaturated zone (inches/year)

WC = Soil water content

b = Soil bulk density

This and derivations of this approach have been utilized successfully to estimate the velocity of numerous pesticides within a factor of two or three in relatively unstratified, unsaturated zone soils.

Convective Mobility Model. This mathematical model was developed to estimate the convective mobility of chemicals that display a linear, equilibrium partitioning between the vapor, liquid, and adsorbed phases of an unsaturated zone soil[83,84]. Convective mobility can be expressed in a variety of ways. One useful index, analogous to chromatography, is to define the convection time t_c to move a distance d:

$$t_c = d(bk_d + WC + P_{sa} K_H)/WF \qquad (6.30)$$

where

$$b = \text{Soil bulk density}$$
$$K_d = \text{Soil-liquid distribution coefficient}$$
$$WC = \text{Soil volumetric water content}$$
$$P_{sa} = \text{Soil volumetric air content}$$
$$K_H = \text{Henry's law constant}$$
$$WF = \text{Water flux}$$

When adsorption is relatively high (i.e. $K_d > 4 \times 10^{-3}$ m³/kg), the WC, P_{sa}, and K_H can be neglected, and t_c should be proportional to K_d and b. Equation 6.30 can be simplified to:

$$t_c = dbK_d/WF \qquad (6.31)$$

Equation 6.31 is a useful equation that should approximately describe the movement of a front or of the peak of a narrow pulse of a chemical. These equations were utilized to classify 35 diverse organic chemicals on the basis of convection time.

The CDT Nomograph. This nomograph was introduced and explained in Chapter 4. It is a graphical approach to predict chemical concentrations in groundwater, based upon a reformulation of the Wilson-Miller analytic equation[85]. This nomograph is based upon principles that are applicable to both inorganic and organic chemicals.

Rapid Assessment Nomograph. This graphical technique was developed as an integrated methodology for assessing chemical movement through both the unsaturated and saturated zone[86]. The technique was designed to allow emergency response personnel to make a first-cut, order-of-magnitude estimate of the potential extent of chemical migration/contamination from a waste site or chemical spill within a 24 hour emergency response time frame. The technique was not intended to provide a definitive, in-depth analysis of the complex fate and transport processes of chemicals in the subsurface environment.

This section will not discuss the use of this model or the input parameters needed to utilize the nomograph. However, the reader is encouraged to study and utilize it because it does provide reasonable estimates.

REFERENCES

1. Hansch, C. and Leo, A. Substituent Constants for Correlation Analysis in Chemistry and Biology. New York: John Wiley (1979).
2. Parfitt, R. L. and Greenland, D. J. Adsorption of Polysaccharides by Montmorillonite. Soil Science Society of America Proceedings 34:862-866 (1970).
3. Stefanson, R. C. The Fate of Polyvinyl Alcohol in Soils. Soil Science 119:426-430 (1975).
4. Greenland, D. J. Interactions Between Organic Polymers and Inorganic Soil Particles. Meded. Fac. Landbouwwetensch. Rijksuniv. Gent 37(3):897-914 (1972).
5. Pinck, L. A., Soulides, D. A., and Allison, F. E. Antibiotics in Soils:Polypeptides and Macrolides. Soil Science 94:129-131 (1962).
6. Miyamoto, S. and Tram, J. Sorption of a Dispersed Dye by Sand and Soil. Journal of Environmental Quality 8:412-416 (1979).
7. Chiou, C. T., Malcolm, R. L., Brinton, T. I., and Kile, D. E. Water Solubility Enhancement of Some Organic Pollutants and Pesticides by Dissolved Humic and Fulvic Acids. Environmental Science and Technology 20:502-508 (1986).
8. Landrum, P. F., Nihart, S. R., Eadle, B. J., and Gardner, W. S. Reverse-phase Separation Method for Determining Pollutant Binding to Aldrich Humic Acid and Dissolved Organic Carbon of Natural Waters. Environmental Science and Technology 18:187-192 (1984).
9. Haas, C. N., and Kaplan, B. M. Toluene-Humic Acid Association Equilibria: Isopiestic Measurements. Environmental Science & Technology 19:643-646 (1985).
10. Hassett, J. P. and Anderson, M. A. Association of Hydrophobic Organic Compounds with Dissolved Organic Matter in Aquatic Systems. Environmental Science and Technology 13:1526-1529 (1979).
11. Madhun, Y. A., Freed, V. H., and Young, J. L. Binding of Ionic and Neutral Herbicides by Soil Humic Acid. Soil Science Society of America Journal 50:319-322 (1986).
12. Goring, C. A. I., and Hamaker, J. W. (eds). Organic Chemicals in the Soil Environment. New York: Marcel Dekker (1972).
13. Tanford, C. The Hydrophobic Effect and the Organization of Living Matter. Science 200:1012-1018 (1978).
14. Van Oss, C. J., Alsolom, D. R., and Neumann, A. W. The Hydrophobic effect: Essentially a van der Waal's Interaction. Colloid and Polymer Science 258:424-427 (1980).
15. Weber, J. B. and Peter, J. C. Adsorption, Bioactivity, and Evaluation of Soil Tests for Alachlor, Acetochlor, and Metachlor. Weed Science 30:14-20 (1982).
16. Fink, D. H., Thomas, G. W., and Meyer, W. J. Adsorption of Anionic Detergents by Soils. Journal of the Water Pollution Control Federation 42:265-271 (1970).
17. Liu, L. C., Cibes-Viade, H., and Koo, F. K. S. Adsorption of Ametryne and Diuron by Soils. Weed Science 18:170-174 (1970).
18. Talbert, R. E., Runyan, R. L., and Baker, H. R. Behavior of Amiben and Dinoben Derivatives in Arkansas Soils. Weed Science 16:10-15 (1968).
19. Talbert, R. E. and Fletchall, O. H. The Adsorption of Some S-Triazines in Soils. Weeds 13:46-52 (1965).

20. Sherburne, H. R. and Freed, V. H. Adsorption of 3(p-Chlorophenyl)-1,1-dimethylurea as a Function of Soil Constituents. Journal of Agricultural and Food Chemistry 2:937-939 (1954).

21. Artiola-Fortung, J. and Fuller, W. H. Adsorption of Some Monohydroxybenzene Derivatives by Soils. Soil Science 133:18-26 (1982).

22. Moreale, A. and Van Bladel, R. Behavior of 2,4-D in Belgian Soils. Journal of Environmental Quality 9:627-633 (1980).

23. Richardson, E. M. and Epstein, E. Retention of Three Insecticides on Different Size Soil Particles Suspended in Water. Soil Science Society of America Proceedings 35:884-887 (1971).

24. Grover, R. Mobility of Dicamba, Pichloram, and 2,4-D in Soil Columns. Weed Science 25:159-162 (1977).

25. Deli, J. and Warren, G. F. Adsorption, Desorption, and Leaching of Diphenamid in Soils. Weed Science 19:67-69 (1971).

26. Murray, D. S., Santelmann, P. W. and Davidson, J. M. Comparative Adsorption, Desorption, and Mobility of Dipropetryn and Prometryn in Soil. Journal of Agricultural and Food Chemistry 23:578-582 (1975).

27. Gamar, Y. and Mustafa, M. A. Adsorption and Desorption of Diquat and Paraquat on Arid-Zone Soils. Soil Science 119:290-295 (1975).

28. Hayes, M. H. B., Lundie, P. R., and Stacey, M. Interactions Between Organophosphorus Compounds and Soil Materials. I. Adsorption of Ethylmethylphosphonofluoridate by Clay and Organic Matter Preparations and by Soils. Pesticide Science 3:619-629 (1972).

29. Wu, C-H, Buehring, N., Davidson, J. M., and Santelmann, P. W. Napropamide Adsorption, Desorption, and Movement in Soils. Weed Science 23:454-457 (1975).

30. Lotse, E. G., Graetz, D. A., Chesters, G., Lee, G. B., and Newland, L. W. Lindane Adsorption by Lake Sediments. Environmental Science and Technology 2:353-357 (1968).

31. Savage, K. E. Adsorption and Mobility of Metribuzin in Soil. Weed Science 24:525-528 (1976).

32. Saltzman, S. and Yaron, B. Parathion Adsorption from Aqueous Solutions as Influenced by Soil Components. International IUPAC Congress of Pesticide Chemistry 2:87-100 (1971).

33. Choi, J. and Aomine S. Adsorption of Pentachlorophenol by Soils. Soil Science and Plant Nutrition 20:135-144 (1974).

34. Doherty P. J. and Warren, G. F. The Adsorption of Four Herbicides by Different Types of Organic Matter and a Bentonite Clay. Weed Research 9:20-26 (1969).

35. Gaynor, J. D. and Volk, V. V. Surfactant Effects on Picloram Adsorption by Soils. Weed Science 24:549-552 (1976).

36. Day, B. E., Jordan, L. S., and Jolliffe, V. A. The Influence of Soil Characteristics on the Adsorption and Phytotoxicity of Simazine. Weed Science 16:209-213 (1968).

37. Chang, S. S. and Stritzke, J. E. Sorption, Movement, and Dissipation of Tebuthiuron in Soils. Weed Science 25:184-187 (1977).

38. Erler, T. G., Dragun, J., and Weiden, D. R. Two Case Studies of Cost Effective Remedial Actions for PCB Contaminated Soil. *In* Proceedings 38th Annual Purdue Industrial Waste Conference. Boston: Butterworth Publishers (1983).
39. Nearpass, D. C. Exchange Adsorption of 3-Amino-1,2,4-Triazole by an Organic Soil. Soil Science Society of America Proceedings **33**:524-528 (1969).
40. Bailey, G. W., White, J. L., and Rothberg, T. Adsorption of Organic Herbicides by Montmorillonite: Role of pH and Chemical Character of Adsorbate. Soil Science Society of America Proceedings **32**:222-234 (1968).
41. Dragun, J. and Helling, C. S. Evaluation of Molecular Modeling Techniques to Estimate the Mobility of Organic Chemicals in Soils: II. Water Solubility and the Molecular Fragment Mobility Coefficient. *In* Disposal of Hazardous Waste: Proceedings of the Seventh Annual Research Symposium, Philadelphia, PA. EPA-600/9-81-002b Cincinnati, OH: U.S. Environmental Protection Agency (1981).
42. Mortland, M. M. Clay-Organic Complexes and Interactions. Advances in Agronomy **22**:75-117 (1970).
43. Farmer, V. C. and Mortland, M. M. An Infrared Study of the Coordination of Pyridine and Water to Exchangeable Cations in Montmorillonite and Saponite. Jour. Chem. Soc. **(A)**:344-351 (1966).
44. Davis, A. R., and Hardin, A. H. Raman Spectral Studies of Organic Molecules Adsorbed at Model Sediment Surfaces. Canadian Journal of Spectroscopy 21 **(5)**:139-143 (1976).
45. Al-Bahrani, K. S. and Martin, R. J. Adsorption Studies Using Gas-Liquid Chromatography - 1. Effect of Molecular Structure. Water Research **10**:731-736 (1976).
46. Mills, A. C. and Biggar, J. W. Solubility-Temperature Effect on the Adsorption of Gamma- and Beta-BHC From Aqueous and Hexane Solutions by Soil Materials. Soil Science Society of America Proceedings **33**:210-216 (1969).
47. Bondy, S. C. and Harrington, M. E. L-Amino Acids and D-Glucose Bind Stereospecifically to a Colloidal Clay. Science **203**:1243-1244 (1979).
48. Huang, J-C and Liao, C-S. Adsorption of Pesticides by Clay Minerals. Journal of the Sanitary Engineering Division, Proceedings of the American Society of Civil Engineers 96(SA5): 1057-1078 (1970).
49. Nose, K. Adsorption of Pentachlorophenol (PCP) on Soil. Bull. Nat. Inst. Agr. Sci. (Japan) Ser C **20**:225-227 (1966).
50. Yariv, S., Russell, J. D., and Farmer, V. C. Infrared Study of the Adsorption of Benzoic Acid and Nitrobenzene in Montmorillonite. Israel Journal of Chemistry **4**:201-213 (1966).
51. Raghavan, S. and Fuerstenau, D. W. The Adsorption of Aqueous Octylhydroxamate on Ferric Oxide. Journal of Colloid and Interface Science **50**:319-330 (1975).
52. Parfitt, R. L. and Mortland, M. M. Ketone Adsorption on Montmorillonite. Soil Science Society of America Proceedings **32**:355-363 (1968).
53. Pfeifer, H., Schirmer, W., and Winkler, H. Nuclear Magnetic Resonance Studies of Molecules Adsorbed on Zeolites A, X, and Y. Advances in Chemistry Series 73 **(121)**:430-440 (1973).
54. Jorgensen, A. D., Stetter, J. R., and Stamoudis, V. C. Interactions of Aqueous Metal Ions with Organic Compounds Found in Coal Gasification: Model Systems.

Environmental Science and Technology **19**:919-924 (1985).

55. Karickhoff, S. W. Semi-Empirical Estimation of Sorption of Hydrophobic Pollutants on Natural Sediments and Soils. Chemosphere **10**:833-846 (1981).

56. Kenega, E. E. and Goring, C. A. I. Relationship Between Water Solubility, Soil Sorption, Octanol-Water Partitioning, and Bioconcentration of Chemicals in Biota. *In* Eaton, J. C., Parrish, P. R., and Hendricks, A. C. (eds). Aquatic Toxicology. ASTM STP 707. Philadelphia, PA: American Society for Testing and Materials (1980).

57. Means, J. C., Wood, S. G., Hassett, J. J. and Banwart, W. L. Sorption of Amino- and Carboxy-Substituted Polynuclear Aromatic Hydrocarbons by Sediments and Soils. Environmental Science & Technology **16**:93-98 (1982).

58. Briggs, G. G. Theoretical and Experimental Relationships Between Soil Adsorption, Octanol-Water Partition Coefficients, Water Solubilities, Bioconcentration Factors, and the Parachor. Journal of Agricultural and Food Chemistry **29**:1050-1059 (1981).

59. Chiou, C. T., Porter, P. E., and Schmedding, D. W. Partition Equilibria of Nonionic Organic Compounds Between Soil Organic Matter and Water. Environmental Science and Technology **17**:227-231 (1983).

60. Karickhoff, S. W., Brown, D. S., and Scott, T. A. Sorption of Hydrophobic Pollutants on Natural Sediments. Water Research **13**:241-248 (1979).

61. Brown, D. S. and Flagg, E. W. Empirical Prediction of Organic Pollutant Adsorption in Natural Sediments. Journal of Environmental Quality **10**:382-386 (1981).

62. McCall, P. J., Swann, R. L., Laskowski, D. A., Unger, S. M., Vrona, S. A., and Dishburger, H. J. Estimation of Chemical Mobility in Soil From Liquid Chromatographic Retention Times. Bulletin of Environmental Contamination and Toxicology **24**:190-195 (1980).

63. Briggs, G. G. Adsorption of Pesticides by Some Australian Soils. Australian Journal of Soil Research **19**:61-68 (1981).

64. Means, J. C., Wood, S. G., Hassett, J. J., and Banwart, W. L. Sorption of Polynuclear Aromatic Hydrocarbons by Sediments and Soils. Environmental Science and Technology **14**:1524-1528 (1980).

65. Jackson, D. R., Roulier, M. H., Grotta, H. M., Rust, S. W., Warner, J. S., Arthur, M. F., and DeRoos, F. L. Leaching Potential of 2,3,7,8-TCDD in Contaminated Soils. *In* Land Disposal of Hazardous Waste. Proceedings of the Eleventh Annual Research Symposium at Cincinnati, OH, April 29-May 1, 1985. EPA - 600/9-85-013. Cincinnati, OH: U.S. Environmental Protection Agency (1985).

66. Chiou, C. T., Peters, L. J., and Freed, V. H. A Physical Concept of Soil-Water Equilibria for Nonionic Organic Compounds. Science **206**:831-832 (1979).

67. Rao, P. S. C. and Davidson, J. M. Estimation of Pesticide Retention and Transformation Parameters Required in Nonpoint Source Pollution Models. *In* Overcash, M. R. and Davidson, J. M. (eds). Environmental Impact of Nonpoint Source Pollution. Ann Arbor, MI: Ann Arbor Science Publishers, Inc. (1980).

68. Briggs, G. G. A Simple Relationship Between Soil Adsorption of Organic Chemicals and their Octanol/Water Partition Coefficients. Proceedings of the 7th British

Insecticide and Fungicide Conference. Volume 1. Nottingham, Great Britain: The Boots Company, Ltd. (1973).

69. Schwarzenbach, R. P. and Westall, J. Transport of Nonpolar Organic Compounds From Surface Water to Groundwater. Laboratory Sorption Studies. Environmental Science and Technology 15:1360-1367 (1981).

70. Briggs, G. G. Adsorption of Pesticides by Some Australian Soils. Australian Journal of Soil Research 19:61-68 (1981).

71. Kenega, E. E. and Goring, C. A. I. Relationship Between Water Solubility, Soil-Sorption, Octanol-Water Partitioning, and Bioconcentration of Chemicals in Biota. *In* Aquatic Toxicology. ASTM STP 707. Philadelphia, PA: American Society for Testing and Materials (1980).

72. Karickhoff, S. W. Organic Pollutant Sorption in Aquatic Systems. Journal of Hydraulic Engineering 110:707-735 (1984).

73. Graveel, J. G., Sommers, L. E., and Nelson, D. W. Sites of Benzidine, Alpha-Naphthylamine, and p-toluidine Retention in Soils. Environmental Toxicology and Chemistry 4:607-613 (1985).

74. Tan, K. H. Variations in Soil Humic Compounds as Related to Regional and Analytical Differences. Soil Science 125:351-358 (1978).

75. Ghosh, K. and Schnitzer, M. Macromolecular Structures of Humic Substances. Soil Science 129:266-276 (1980).

76. Kay, B. D. and Goit, J. B. Thermodynamic Characterization of Water Adsorbed on Peat. Canadian Journal of Soil Science 57:497-501 (1977).

77. Rao, P. S. C. and Davidson, J. M. Adsorption and Movement of Selected Pesticides at High Concentrations in Soils. Water Research 13:375-380 (1979).

78. Hassett, J. J., Banwart, W. L., Wood, S. G. and Means, J. C. Sorption of Alpha-Naphthol: Implications Concerning the Limits of Hydrophobic Sorption. Soil Science Society of America Journal 45:38-42 (1981).

79. Helling, C. S. and Turner, B. C. Pesticide Mobility: Determination by Soil-Thin Layer Chromatography. Science 162:562-563 (1968).

80. Griffin, R. A., Au, A. K., and Chian, E. S. K. Mobility of Polychlorinated Biphenyls and Dicamba in Soil Materials: Determination by Soil Thin-Layer Chromatography. Reprint 1979G. Urbana, IL: Illinois State Geological Survey (1979).

81. LaFleur, K. S. Toxaphene-Soil-Solvent Interactions. Soil Science 117:205-210 (1974).

82. Baes C. F. III and Sharp, R. D. A Proposal for Estimation of Soil Leaching Constants for Use in Assessment Models. Journal of Environmental Quality 12:17-28 (1983).

83. Jury, W. A., Farmer, W. J. and Spencer, W. F. Behavior Assessment Model for Trace Organics in Soil: II. Chemical Classification and Parameter Sensitivity. Journal of Environmental Quality 13:567-572 (1984).

84. Jury, W. A., Spencer, W. F., and Farmer, W. J. Behavior Assessment Model for Trace Organics in Soil: I. Model Description. Journal of Environmental Quality 12:558-564. (1983.)

85. Kent, D. C., Pettyjohn, W. A., and Prickett, T. A. Analytical Methods for the Prediction of Leachate Plume Migration. Ground Water Monitoring Review

5(2):46-59 (1985).
86. Donigian, A., Lo, T. Y. R., and Shanahan, E. Rapid Assessment of Potential Groundwater Contamination Under Emergency Response Conditions. EPA-600/8-83-030. Washington, D.C.: U.S. Environmental Protection Agency (1983).

7

Organic Chemical Diffusion and Volatilization

INTRODUCTION

The transport of an organic chemical in soil air or from the soil surface to the atmosphere can be an important mechanism that must be addressed in order to protect human health and the environment. Soil chemists, soil physicists, civil and environmental engineers study how soils affect the transport of organic chemicals in soil air and from the soil surface to the atmosphere.

In this text, volatilization is defined as the loss of a chemical from the soil surface into the atmosphere. The rate at which an organic chemical will move through soil air and volatilize into the atmosphere is controlled by the following equilibria:

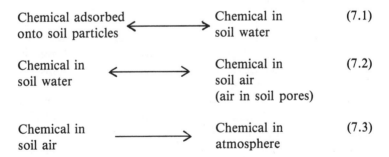

$$\text{Chemical adsorbed onto soil particles} \longleftrightarrow \text{Chemical in soil water} \tag{7.1}$$

$$\text{Chemical in soil water} \longleftrightarrow \text{Chemical in soil air (air in soil pores)} \tag{7.2}$$

$$\text{Chemical in soil air} \longrightarrow \text{Chemical in atmosphere} \tag{7.3}$$

This chapter will focus on the diffusion and volatilization of organic chemicals as depicted in Equations 7.1, 7.2, and 7.3. An analysis of Equations 7.1, 7.2, and 7.3 will reveal that in order for a chemical to volatilize from a soil surface, two events must occur. First, the chemical within the soil must move to the soil surface. Second, the chemical must move away from the soil surface into the atmosphere above the soil. These two events directly depend upon diffusion.

Diffusion in this text is defined as the average rate of migration or velocity of a chemical in air. Diffusion can be caused by three different gradients. First, diffusion can be caused by a concentration gradient. Chemicals should migrate from regions of higher concentration to regions of lower concentration; the greater the difference, the greater the diffusion rate. Diffusion which is primarily dependent upon a concentration gradient is known as concentration diffusion or ordinary diffusion.

Second, diffusion can be caused by a total pressure gradient. For example, helium and other light elements will concentrate in the upper portions of the earth's atmosphere due to diffusion driven by pressure differences in the atmosphere; these gases respond to the pressure gradient created by gravity, but they do not respond directly to gravity. Diffusion which is primarily dependent upon a pressure gradient is known as pressure diffusion.

Third, diffusion can be caused by temperature differences. In a gas mixture, the lighter components will concentrate near the warmer zone. If the temperature gradient is maintained, the gases will continue to separate until concentration diffusion balances thermal diffusion. Diffusion which is primarily dependent upon thermal gradients is known as thermal diffusion.

Concentration diffusion, pressure diffusion, and thermal diffusion affect the velocity of a chemical in the atmosphere. However, it is important to note that, under normal circumstances in soil systems, only concentration diffusion is important[1].

ORGANIC CHEMICAL PROPERTIES AFFECTING DIFFUSION AND VOLATILIZATION

The structure of an organic chemical will determine how the chemical behaves in air because structure determines (a) the inherent behavior of a molecule as well as (b) how the molecule will behave when in the immediate presence of other molecules. Structure determines the nature of the intermolecular forces between molecules.

With regard to (a), the chemical structure determines the molecular weight, which is directly linked to the diffusion coefficient of a chemical. In general, the rates at which different nonreactive chemicals diffuse at a given temperature are inversely proportional to the square roots of their molecular weights:

$$D_{A1}/D_{A2} = (MW_1/MW_2)^{1/2} \qquad (7.4)$$

where

D_{A1} = diffusion coefficient or diffusivity of chemical 1 in air

D_{A2} = diffusion coefficient of chemical 2 in air

MW_1 = molecular weight of chemical 1

MW_2 = molecular weight of chemical 2

This relationship is due to the fact that different chemicals possess the same mean kinetic energy, so their mean square speeds are an inverse ratio of their masses. Table 7.1 lists D_A values for several chemicals in air.

TABLE 7.1 Diffusion Coefficients for Selected Chemicals in Air[2,3,4]

Chemical	$D_A{}^*$	T^{**}	Chemical	$D_A{}^*$	T
Acetic Acid	0.133	25	Ethyl Acetate	0.089	30
Ammonia	0.28	25	Ethyl Alcohol	0.119	25
				0.145	42
n-Amyl Alcohol	0.07	25	Ethyl benzene	0.077	25
sec-Amyl Alcohol	0.072	30	Ethylene dibromide	0.070	0
Aniline	0.072	25	Ethyl ether	0.093	25
	0.075	30			
Benzene	0.088	25	Formic Acid	0.159	25
Biphenyl	0.061	0	Hexane	0.080	21
n-Butyl Alcohol	0.097	25.9	Hexyl Alcohol	0.059	25
Butylamine	0.101	25	Hydrogen	0.410	25
i-Butyric Acid	0.081	25	Methyl Alcohol	0.159	25
i-Caproic Acid	0.06	25	n-Octane	0.060	25
Carbon Dioxide	0.164	25	Oxygen	0.178	0
	0.177	44		0.206	25
Carbon Disulfide	0.107	25	n-Pentane	0.071	21
Chlorobenzene	0.075	30	Propionic Acid	0.099	25
Chloropicrin	0.088	25	n-Propyl Alcohol	0.100	25
Chlorotoluene	0.065	25	n-Propyl benzene	0.059	25
Cyclohexane	0.086	45	Toluene	0.088	30
Diethylamine	0.105	25	Water	0.220	0
Diphenyl	0.068	25		0.256	25
			Xylene	0.071	25

* in cm^2/sec
** in °C

With regard to (b), the intermolecular forces that affect the diffusion and volatilization of organic chemicals are van der Waal's forces and dipole-dipole interactions. Information presented in Chapter 6 revealed that van der Waal's forces are relatively weak, induced forces existing between otherwise non-attracting atoms and molecules. The *average* distribution of charge on a neutral organic molecule such as propane or butane is symmetrical; as a result, there is no permanent electrostatic polarization or charge separation in terms of positive and negative polar regions or poles on the molecule. However, the electrons within the neutral molecule are not stagnant but mobile so that the electron distribution is somewhat distorted at any instant in time. As a result, small dipoles exist on the molecule's surface. This momentary dipole affects the electron distribution in an adjacent molecule: the negative regions of the dipole molecule will repel electrons in the adjacent molecule, and the positive regions of the dipole molecule will attract electrons in the adjacent molecule. The net result of the close proximity of two molecules is the induction of momentary electrostatic polarizations on the adjacent molecule's surfaces. Although the momentary and induced dipoles are constantly migrating on the molecule's surfaces, the net result is an attraction between the molecules.

In general, van der Waal's forces are weak and of very short range. They act only between the surfaces of molecules that are in close proximity. Within a class of organic chemicals such as alkanes, the larger the surface area, the larger the van der Waal's forces between surfaces and, as a result, the lesser the vapor pressure and volatility.

Dipole-dipole interactions are the second kind of intermolecular force affecting diffusion and volatilization. A dipole-dipole interaction is the attraction of the permanent positive pole at one end of a polar molecule to the negative pole on an adjacent molecule. As a result of dipole-dipole interactions, polar molecules usually exhibit lower vapor pressures and volatility than non-polar, neutral molecules of comparable molecular size.

The vapor pressure of a chemical is the pressure exerted by its vapor when in equilibrium with the chemical in a solid or liquid form. If a chemical in liquid form is placed in a closed container, molecules of the chemical that possess relatively high kinetic energy migrate to the liquid surface, evaporate into the atmosphere within the container, and diffuse throughout the atmosphere. As evaporation continues, the chemical's concentration in the atmosphere increases. However, some of these molecules collide with the liquid surface and are captured by the liquid; this process is known as condensation. As the concentration of molecules increases in the gaseous phase, the rate of condensation increases. Eventually, the rate of condensation equals the rate of evaporation, and at this time the system is in equilibrium. At equilibrium, the molecules in the container atmosphere exert a pressure on the container walls; this pressure is defined as the equilibrium vapor pressure or simply the vapor pressure.

It is important to note that the vapor pressure is a measure of the tenacity with which the molecules of the chemical are attracted to each other. If a chemical has high intermolecular forces acting between its molecules, then only the very few molecules near the surface that possess very high kinetic energy will escape into the gas phase; the rate of evaporation will be low and will be equaled by a low rate of condensation. Therefore, a low vapor pressure indicates the presence of high intermolecular attractions.

The boiling point is also a guide to the strength of intermolecular attractions. The temperature at which the vapor pressure of a chemical equals the vapor pressure of the atmosphere is defined as the boiling point. If the boiling point of a chemical is high, a large amount of energy, usually in the form of heat, must be applied to the chemical in order to elevate the chemical's kinetic energy and overcome intermolecular attractions. A high boiling point indicates the presence of high intermolecular attractions. A relatively low boiling point indicates the presence of a high vapor pressure.

The vapor pressure and boiling point are properties of a chemical that describe its propensity to exist in the atmosphere as opposed to existing as a pure chemical in liquid or solid form. In general, chemicals with vapor pressures less than 10^{-7}mm Hg should be present in the atmosphere or soil air in negligible amounts. Also, chemicals with vapor pressures greater than 10^{-2}mm Hg should be present primarily in the atmosphere or soil air. Those chemicals possessing vapor pressures between these two values may have a tendency to exist in both the atmosphere or in soil air as well as in liquid or solid form.

The magnitude of van der Waal's forces and dipole-dipole interactions determines the solubility of a chemical in water and in organic solvents. The solubility of a chemical will affect the diffusion and volatilization of a chemical from water into air. The greater the water solubility of a chemical, the greater the propensity for that chemical to exist in the water phase instead of in air which is in equilibrium with it.

The distribution of a chemical between water and air can be mathematically described in at least two ways. First, the equilibrium constant, K_{wa}, where:

$$K_{wa} = (ug\ chemical/ml\ water)/(ug\ chemical/ml\ air) \qquad (7.5)$$

can be grossly utilized to identify the probable mechanism of chemical movement in soil systems[5,6]. If K_{wa} equals unity, the chemical movement should be almost entirely via diffusion in the air phase. If K_{wa} is approximately 100, then chemical movement should occur primarily via diffusion in the air phase. If K_{wa} is approximately 10,000, then chemical movement should occur by both diffusion and mass flow in both air and water-phases. If K_{wa} is approximately 10^6, then chemical movement should occur mainly by mass flow in the water phase. If K_{wa} is approximately 10^8, then chemical movement should occur almost entirely by mass flow in the water phase.

The propensity for a chemical to volatilize from an aqueous phase and exist in the atmosphere or in soil air can be grossly estimated via Henry's Law. Henry's Law states that when a solution becomes very dilute (i.e. mole fraction less than 0.001), the vapor pressure of a chemical should be proportional to its concentration:

$$V_p = K_H C \qquad (7.6)$$

where

V_p = vapor pressure of the chemical

$$C = \text{concentration of the chemical in water}$$

$$K_H = \text{Henry's Law constant}$$

The determination of Henry's Law constants is a very costly laboratory procedure. If Henry's Law constants are not available for a chemical from the published literature or from experimental data, an estimate can be derived using the following equation:

$$K_H = (V_p)(MW)/760(S) \tag{7.7}$$

where

$$V_p = \text{vapor pressure of the chemical (mmHg)}$$

$$MW = \text{molecular weight of the chemical}$$

$$S = \text{solubility (mg/l or grams/m}^3)$$

$$K_H = \text{Henry's Law constant (atm·m}^3/\text{mole)}$$

It is important to note that Henry's Law constants are reported in various dimensions in the published literature such as atm·m³/mole, atm·cm³/gram, or it can be dimensionless. Therefore, numerical values of K_H which are obtained from the published literature should be used with care and caution since the dimensions of K_H depend upon the dimensions of V_p, C, or S.

In general, chemicals having K_H values less than 5×10^{-6} atm·m³/mole encounter resistance to mass transfer primarily in the gas phase; in other words, these chemicals should be present only in negligible amounts in the atmosphere or in soil air. Also, chemicals having K_H values greater than 5×10^{-3} atm·m³/mole encounter resistance to mass transfer primarily in the aqueous phase; in other words, volatilization and diffusion in soil air should dominate. Those chemicals possessing K_Hs between these two values encounter resistance to mass transfer in both gas and aqueous phases.

Laboratory-derived K_Hs are available for many chemicals with relatively high vapor pressures and water solubilities. In general, these values are in agreement with predicted values. On the other hand, laboratory-derived K_Hs are generally not available for chemicals with relatively low vapor pressures and water solubilities due to the difficulties involved in measuring minute concentrations of these chemicals. As a result, comparisons between predicted and experimental values are rare. In one such comparison, the average error for predicted K_Hs for polychlorinated biphenyl cogeners relative to measured K_Hs was a factor of five[7]. Deviations from Henry's Law may occur at finite concentrations if (a) the concentration of the chemical exceeds three mole percent[8], or (b) if the vapor pressure is sufficiently high that it departs measurably from the fugacity[9], or (c) if the chemical structure is very large[9], or (d) if the chemical possesses a charge[9]. An analysis of conditions (a), (c), and (d) will reveal that Henry's Law should apply as long as molecules of the chemical never approach one another closely enough to interact significantly. If significant interactions occur between molecules of the chemical, then Henry's Law may not apply; likewise, if significant interactions occur between molecules of the chemical and other dissolved or precipitated chemicals, then Henry's Law may not apply.

The extent of volatilization and diffusion of an organic chemical is significantly affected by the chemical composition of the liquid or waste it is in. An interesting experimental study addressing the effect of solvent mixtures and solvent-waste mixtures exemplifies the effect[10]. The evaporation rates of acetone, isopropanol, perchloroethylene, white spirit, acetone-water, isopropanol-water, and perchloroethylene-oil from a one square meter shallow pool and from pulverized domestic waste 0.5 meters deep and one meter square were measured (see Table 7.2). The conclusions derived from this study should be useful when assessing the general evaporative behavior of solvents

TABLE 7.2 Evaporation Rates for Various Solvent, Water, and Waste Mixtures[a]

Solvent/Mixture	Evaporation Rate $(kg/meter^2/hr)$	Mean Wind Speed (Meter/Sec)
Acetone	4.92	6.7
Acetone(43wt%)-Water	6.80 (Initial 10 min)	4.0
Acetone(11wt%)-Waste	8.70 (Initial 10 min)	6.7
	0.11 (200-440 min)	6.7
Acetone(3.0wt%)-Water (4.0wt%)-Waste	3.70 (Initial 10 min)	4.0
	0.47 (200-440 min)	4.0
Isopropanol	1.10	0.5
Isopropanol (42wt%)-Water	1.54 (Initial 10 min)	4.5
Isopropanol (10wt%)-Waste	3.22 (Initial 10 min)	0.5
	0.15 (200-440 min)	0.5
Isopropanol (3.8wt%)-Water (5.2wt%)-Waste	1.90 (Initial 10 min)	4.5
	0.65 (200-440 min)	4.5
Perchloroethylene	3.13	10.0
Perchloroethylene (68wt%)-Oil	1.44 (Initial 20 min)	2.5
Perchloroethylene (16wt%)-Waste	6.09 (Initial 10 min)	10.0
	0.36 (200-440 min)	10.0
Perchloroethylene (8.6wt%)-Oil (3.8wt%)-Waste	5.64 (Initial 10 min)	2.5
	0.18 (200-440 min)	2.5
White Spirit	0.29	5.0
White Spirit (11wt%)-Waste	6.93 (Initial 20 min)	5.0
	0.11 (200-440 min)	5.0

[a] compiled from data from reference 10.

and solvent mixtures. First, the evaporation rate of volatile solvents dissolved in water is lower from waste than from a pool of liquid over relatively long periods of time. The same holds for less volatile water miscible solvents (i.e. isopropanol). Second, volatile solvents dissolved in oil (i.e. perchloroethylene) appear to have a lower evaporation rate relative to the pure solvent. Third, the presence of a solid phase (i.e. domestic waste) significantly reduces the evaporation rate over longer time periods.

In conclusion, it is important to remember that although Henry's Law is based upon the principles of chemical thermodynamics, it is beyond the scope of thermodynamics to predict the deviations from the laws of dilute solutions on the basis of the molecular properties of solutes and solvents[9].

SOIL PROPERTIES AFFECTING DIFFUSION AND VOLATILIZATION

Because soil is a porous medium, the diffusion of a chemical through soil air will be somewhat modified relative to diffusion through air alone. This section will discuss several ways in which soil affects the diffusion of a chemical through soil air.

The most commonly utilized relationship describing diffusion is Fick's Law:

$$\text{Flux} = dq/dt = -D_A(dc/dx) \tag{7.8}$$

where

$$dq/dt = \text{rate that a chemical moves past a given point per unit cross-sectional area}$$

$$c = \text{concentration of the chemical}$$

$$x = \text{distance}$$

$$D_A = \text{diffusion coefficient in air}$$

$$dc/dx = \text{the chemical's concentration gradient in the x direction and perpendicular to the cross-sectional area.}$$

D_A is always positive; the negative sign in Equation 7.8 indicates that diffusion occurs in the opposite direction from the concentration gradient, i.e., from higher to lower concentrations.

Fick's Law is strictly an empirical relation, when applied to gases[1], that was borrowed from studies with solutes and shown to agree well with observed diffusional processes in air. The comparison of the Stefan-Maxwell equations and Fick's Law for diffusion of gas through vapor shows that only under certain circumstances is D_A a constant independent of the mole fraction of the gas and the diffusion flux of other gases. These cases are (a) the diffusion of a trace amount of gas into a multicomponent gas mixture, (b) equimolar, counter-current diffusion in a binary gas mixture, and (c) equimolar, counter-current diffusion of two gases in a ternary system with a stagnant third gas. Measured values of D_A can only be accurately extrapolated to other systems when all the above circumstances are similar.

The presence of soil affects the diffusion of a chemical through soil air in several ways. First, the diffusion coefficient in soil air, D_{SA}, should be smaller than the diffusion coefficient of a chemical in unobstructed air, D_A, because the soil particles will interfere with the flux of chemical[11]. D_{SA} is usually some fraction of D_A due to the fact that (a) cross-sectional area in which diffusion can occur has been decreased by the presence of soil particles, (b) the diffusion distance is significantly increased since the diffusion pathway is not straight but tortuous, and (c) part of the cross-sectional area not occupied by soil particles represents blocked pores that do not contribute to the diffusion pathway. The smaller the soil particles, the smaller D_{SA} will be.

In general, as bulk density increases, porosity decreases. The general overall effect of increasing soil bulk density is to decrease D_{SA}[6]. The equation developed to estimate the effect of varying soil porosity on the D_{SA} of hexachlorobenzene[12] can be utilized to derive a general understanding of the variation of bulk density on D_{SA}:

$$D_{SA} = D_A (P_{SA})^{10/3}/(P_T)^2 \qquad (7.9)$$

where

D_{SA} = diffusion coefficient of a chemical in soil air (cm^2/day)

D_A = diffusion coefficient of a chemical in air (cm^2/day)

P_{SA} = soil air-filled porosity (cm^3/cm^3)

P_T = total soil porosity (cm^3/cm^3)

Second, the soil moisture content also affects D_{SA}. At higher soil moisture contents, the cross-sectional area where diffusion can occur will be relatively smaller. As a result, D_{SA} must also decrease. Fick's first law for diffusion can be modified to account for these two factors:

$$\text{Flux} = -D_{SA}(dc/dx) = -A(P_{SA} - P_B)D_A(dc/dx) \qquad (7.10)$$

and

$$-D_{SA} = -A(P_{SA} - P_B)D_A \qquad (7.11)$$

where

D_A = diffusion coefficient in unobstructed air

D_{SA} = diffusion coefficient in soil air

A = correction factor for soil tortuosity

P_{SA} = soil porosity, the fraction of open volume in soil

P_B = volume of blocked pores

Experimental values for A are quite varied; suggested values[2] range from 0.6 to 0.7 with a median value of 0.66. Experimental values for P_B vary from 0.0 to 0.25 with an apparent median value of 0.1[2]. The relationships identified in Equations 7.10 and 7.11 are valid over a range of P_{SA} of 0.2 to 0.7[2].

Third, D_{SA} is affected by the adsorption of the chemical onto the surfaces of soil particles. Organic matter and the clay mineral content both significantly influence volatile organic chemical emissions from soil. The emission concentrations of n-pentane were reduced by one percent for every percent incorporation into sand of a silt plus clay fraction containing organic matter[13]; also, n-pentane emission concentrations were reduced by four percent for every percent incorporation into sand of 400°C fired silt plus clay fraction (i.e. organic matter removed)[13].

Many adsorption sites exist in soil which preferentially adsorb water, and these will adsorb some organic chemicals only when soil moisture content is low[2,6]. In some relatively dry soils, soil surfaces may hold organic chemicals that would volatilize under relatively wet soil conditions[2,14].

Fourth, the presence of dissolved higher molecular weight organics in soil-water may affect the flux of a chemical. The presence of these organics significantly enhanced the water solubility of several low molecular weight chlorinated hydrocarbons[15]; this resulted in a lower concentration of these chlorinated hydrocarbons in the vapor phase, which in turn affected the magnitude of dc/dx in Equation 7.10.

Fifth, the nature of the soil surface significantly affects chemical diffusion and water evaporation. Under high radiation conditions, the soil surface may develop a crust within a few hours after rainfall or irrigation; the dry soil crust acts as a capillary break and greatly reduces the evaporation rate of water and the movement of organic chemicals. Snow cover acts in a manner similar to a soil crust.

CLIMATIC FACTORS AFFECTING DIFFUSION AND VOLATILIZATION

The climatic factors affecting volatility are different from those affecting diffusion. In general, chemical volatilization is significantly affected by total pressure gradients, atmospheric turbulence, and air temperatures. In general, these factors are not major factors influencing chemical diffusion in soil. The reasons why these are not major factors are discussed in greater detail below.

Chemicals in the vapor state move in response to the gradient in total pressure[16]. Air will flow into the soil as atmospheric pressure rises, and out as it falls, until pressure equilibrium is restored between soil and atmosphere. However, the mean atmospheric pressure changes slowly and so for practical purposes, the attainment of equilibrium can be regarded as almost instantaneous. As a result, the exchange of gas can be estimated via Boyle's Law[16]. In a soil freely permeable to air and uniformly porous to a depth of 1 meter, where an impervious layer exists, an increase in barometric pressure of 33 mbar/1 bar would compress the existing soil air into the lower 1000/1033 of its original volume, admitting fresh air to a depth of 33/1033 meters or 0.032 meters. This pressure change is about the maximum pressure change experienced over one day. As a result, the depth reached by fresh air is only a small fraction of the total depth. Consequently, this mechanism should not be a significant one affecting air-phase movement of most chemicals in subsoils.

Atmospheric turbulence or wind results in a succession of relatively small, transient pressure changes. However, these pressure changes make negligible direct contributions to chemical movement in soil air relative to atmospheric pressure gradients[16]. On the other hand, the rate of air flow can influence chemical volatilization directly and indirectly[6]. If the relative humidity of air is not 100 percent, then increases in air flow will hasten the drying of soil. This indirect effect lowers the soil-water content and, therefore, significantly affects volatilization. The direct effect of increased air flow involves a more rapid removal of the chemical from the soil surface which leads to an increased movement of the chemical to the soil surface.

It may be interesting to note that in the very thin layer next to the soil surface, volatile chemical movement occurs via molecular diffusion. Above this layer, however, atmospheric turbulence far exceeds molecular diffusion in affecting volatile chemical movement. At a height of one meter, diffusivity due to atmospheric turbulence may be as much as 5000X the contribution of molecular diffusion.

Soil and air temperature usually do not appear as variables in equations quantifying diffusion and volatilization. However, several parameters such as D_A, vapor pressure, vapor density, solubility, and adsorption vary as temperature varies; also, their intricate interactions due to temperature variations in soil are not completely understood. The general overall effect of increasing temperature is an increase in the chemical's diffusion rate[6]. In general, the following equation[12] which was derived for hexachlorobenzene can be used to obtain a gross estimate of how temperature affects D_A:

$$D_{A2} = D_{A1}(T_2/T_1)^{1/2} \qquad\qquad (7.12)$$

where

D_{A1}, D_{A2} = diffusion coefficients

T_1, T_2 = temperature (°K)

In general, as soil temperature rises, air increases in pressure and flows to re-establish equilibrium. Temperature changes, whether diurnal or seasonal, are greatest at the soil surface and decrease exponentially with depth, so that the effects of significant daily temperature changes on gas exchange is limited to the upper few inches of soil[7]. Therefore, it is extremely unlikely that this mechanism will affect chemical movement at any significant soil depth.

If climatic conditions are favorable for the evaporation of soil water, chemical volatilization could be enhanced by the "wick" effect[6]. Soil water containing a dissolved chemical moves toward the soil surface by capillary action in response to water evaporation from the soil surface. The net result, known as the wick effect, is an enhancement of chemical movement to the soil surface for subsequent volatilization. The degree of enhancement is related to the water evaporation rate. In dry soils, water evaporation rates are very low, and the extremely slow water movement to the surface is primarily via vapor phase diffusion. In moist soils, the chemical volatilizes along with the water; the chemical:water ratio which volatilizes is generally the same as the chemical:water ratio present in soil water[17].

ESTIMATING DIFFUSION AND VOLATILIZATION RATES

A reasonable amount of research has been directed toward predicting the behavior of volatile chemicals. Addressing and correcting problems associated with the discharge of chemicals or hazardous materials may require the estimation of (a) the rate of volatilization of a pure chemical or hazardous materials in a pool of pure chemical at the soil surface, (b) the rate of volatilization of a chemical or hazardous materials dissolved in a pool of water at the soil surface, (c) the rate of volatilization of a chemical or hazardous material diffusing through soil to the soil surface, or (d) the migration rate of chemicals or hazardous materials through soil air. In cases where the chemical or waste has been discharged or placed on the soil surface, the rate of vapor generation of the pure chemical for steady state conditions can be estimated[18]:

$$E = 2 \, P_v W_A (L_A D_A V / (3.1416) f)^{1/2} (W_c / W) \qquad (7.13)$$

where

$$E = \text{emission rate (cm}^3/\text{sec)}$$

$$P_v = \text{equivalent vapor pressure (\%), where}$$

$$P_v = [\text{vapor pressure (mmHg)}]/760$$

$$W_A = \text{width of area occupied by the chemical/waste (cm)}$$

$$L_A = \text{length of area occupied by chemical/waste (cm)}$$

$$D_A = \text{diffusion coefficient of the chemical in air (cm}^2/\text{sec)}$$

$$V = \text{wind speed (cm/sec)}$$

$$W_c/W = \text{weight fraction of the chemical in contaminated soil/waste (gram/gram)}$$

$$f = \text{correction factor, where}$$

$$f = (0.985 - 0.00775 \, P_v) \text{ where the range of } P_v \text{ is 0 - 80\%.}$$

The volumetric emission rate E can be converted into a mass emission rate:

$$Q = E \, (MW)/G \qquad (7.14)$$

where

$$Q = \text{mass emission rate (gram/sec)}$$

$$E = \text{volumetric emission rate (cm}^3/\text{sec)}$$

$$MW = \text{molecular weight (gram/mole)}$$

$$G = 24{,}860 \text{ cm}^3/\text{mole}$$

An approach to estimating emissions from the landspreading of a chemical involves the use of the evaporation rate of a model compound to predict the unknown evaporation rate of the chemical in question[19]:

$$E_a = E_b \, [P_{va}(MW_a)^{1/2}] / [P_{vb}(MW_b)^{1/2}] \qquad (7.15)$$

where

$$E_a, E_b = \text{emission rates of chemicals a and b} \\ (\text{gram/cm}^2 \cdot \text{month})$$

P_{va}, P_{vb} = vapor pressures of chemicals a and b (mmHg)

MW_a, MW_b= molecular weights of chemicals a and b (gram/mole)

The potential mass emission rate can be estimated by using water evaporation rates as the model compound flux and by factoring into Equation 7.15 the area of the disposal site and the weight fraction of the chemical in question:

$$E_a = E_w(1-RH)^{-1}P_{va}(MW_a)^{1/2}[P_{vw}(MW_w)^{1/2}]^{-1}A(W_a/W) \qquad (7.16)$$

where

E_w = water evaporation rate (gram/cm²·month)

RH = relative humidity

MW_w = molecular weight of water (gram/mole)

P_{vw} = vapor pressure of water (mmHg)

A = surface area of site (cm²)

(W_a/W) = weight fraction of chemical a in the deposited waste (gram/gram)

E_w can be easily obtained by multiplying the approximate water evaporation rate (cm/month) by the density of water (1 gram/cm³).

Liss and Slater developed a formula to predict the volatilization of compounds from the water phase[20]:

$$Kv = [dMW^{1/2}(3.1407 \times 10^{-4}) + (2.0413 \times 10^{-4}\ TS/P_v)]^{-1} \quad (7.17)$$

d = depth of water (cm)

MW = molecular weight of the chemical

T = temperature (°K)

S = Solubility of the chemical in water (mole/liter)

P_v = vapor pressure of the chemical (Torr)

The volatility of lindane from soil which has been uniformly treated can be generally estimated[21]:

$$Q = D_{sa}C_o/(3.1416\ Dt)^{1/2} \qquad (7.18)$$

where

Q = chemical flux (gram/cm^2/sec)

D_{sa} = diffusion coefficient of the chemical in soil air (cm^2/sec)

C_o = chemical concentration in soil (gram/cm^3 total volume)

t = time (sec)

Equation 7.18 is identical with the solution for heat flow in an infinite solid. The boundary conditions used to drive this equation should not be violated provided:

$$\exp\left(-L^2/4D_{sa}t\right) < 0.01 \tag{7.19}$$

or

$$t < L^2/18.4D_{sa} \tag{7.20}$$

where

L = depth of soil layer treated uniformly with lindane (cm)

Equation 7.18 is applicable on finite soil systems in the region 0 to L as long as the chemical concentration at the lower boundary of the soil layer is not decreased by chemical moving upward or downward. For cases where diffusion can occur downward across the lower boundary:

$$Q = [D_{sa}C_o/(3.1416\ D_{sa}t)^{1/2}]\ [1 - \exp\left(-L^2/4D_{sa}t\right)] \tag{7.21}$$

Note that Equation 7.21 reduces to Equation 7.18 for large values of $L^2/4D_{sa}t$.

An approach to estimating the emission rate of a volatile chemical from soil beneath a clean soil cover is[11]:

$$E = fD_aC_{sv}AP_T^{4/3}(L)^{-1}(Wc/W) \tag{7.22}$$

where

E = emission rate of the chemical (gram/sec)

f = emission rate enhancement factor, where

f = 6 - 7 if sweep gas is being produced, and f = 1 under normal conditions

D_A = diffusion coefficient of the chemical in air (cm^2/sec).

A = exposed area (cm^2)

P_T = soil total porosity (dimensionless fraction)

L = effective depth of soil cover (less than or equal to 35 cm)

W_c/W = weight fraction of the chemical in contaminated soil/waste beneath the cover (gram/gram or ppm)

C_{sv} = saturated vapor concentration (gram/cm^3),

where C_{sv} = V_p (MW)/RT

and

V_p = vapor pressure (mmHg)

MW = Molecular weight (gram/mole)

R = 6.23×10^4 mmHg cm^3/mole $^\circ$K

T = temperature ($^\circ$K)

This equation was originally derived to estimate hexachlorobenzene emissions through the soil cover of a landfill. An analysis of Equation 7.22 will reveal that diffusion is assumed to be the only transport process operating, and that a given concentration of the chemical will volatilize, resulting in a saturated vapor concentration. The vapor then diffuses through the soil cover at a rate determined by the diffusion coefficient, the cover depth, and the porosity.

Jury et al.[23,24,25,26] have proposed several simple methods to estimate diffusion rates and volatilization losses. Equations 7.23 through 7.32 address their methods, which are based on boundary layer limits. The soil and atmosphere are connected by a stagnant air boundary layer through which water vapor and chemical vapor are assumed to move by diffusion. The extent to which this boundary layer limits the volatilization flux may be used as a criterion for classifying volatile organics into general categories, similar to the volatilization groups used to classify chemical losses from water bodies. To achieve this, a distinction is made between processes where no water flow (E) is occurring (i.e. E = O) and processes where both volatilization and evaporation are occurring.

A review of the experimental evidence supporting these simple methods[26] will reveal that reasonable agreement was found between published experimental data and simulation data for DDT, diazinon, dieldrin, dimethoate, lindane, parathion, and triallate. An analysis of the information presented on these simple methods[23,24,25,26] indicates that Equations 7.23 through 7.32 can be utilized to generate gross estimates of diffusion and volatilization rates for organic chemical with structures similar to those chemicals analyzed by these researchers (see Table 7.3). These equations were tested using chemi-

TABLE 7.3 Chemicals Utilized in the Development and Analysis of Models to Predict Soil
Diffusion Rates and Volatilization Losses[a]

Atrazine	Dieldrin	Nitrobenzene
Benzene	Diuron	n-Octane
Biphenyl	EPTC	Parathion
Bromacil	Ethoprophos	Phenanthrene
Bromobenzene	Ethylene dibromide	Phenol
Carbofuran	Lindane	Phorate
Carbon tetrachloride	Mercury	Prometryne
Chlorobenzene	Methyl bromide	Simazine
Chloroform	Methyl parathion	Triallate
2,4-D	Monuron	Trifluralin
DDT	Napropamide	Vinyl chloride
Diazinon	Naphthalene	

[a] compiled from data from references 17,18,19, and 20.

cals with (a) vapor pressures ranging from $10^{-5.22}$ to $10^{4.3}$ mg/l, (b) water
solubilities ranging from $10^{-2.52}$ to $10^{4.91}$ mg/l, and (c) K_{oc}s ranging from
$10^{-1.70}$ to $10^{2.38}$ m^3/kg where the organic C fraction was assumed to be
0.0125.

When mass flow by convection is small or negligible, a chemical is able
to move through the soil only by liquid or vapor diffusion. A characteristic
diffusion time for a chemical was expressed as[24]:

$$t_D = l^2/D_E \tag{7.23}$$

where

t_D = diffusion time (days)

l = distance transversed by the chemical (meters)

D_E = effective soil diffusion coefficient (meter^2/day)

and

$$D_E = [(D_A K_H P_{sa}^{10/3}/P_T^2) + (D_w(WC)^{10/3}/P_T^2)] \cdot$$
$$(P_b K_d + WC + P_{sa} K_H)^{-1} \tag{7.24}$$

where

D_A = diffusion coefficient of the chemical in air (m^2/day)

K_H = Henry's Law constant (dimensionless)

P_{sa} = soil air content (meter^3/meter^3)

P_T = soil total porosity

D_w = diffusion coefficient of the chemical in water
(meter^2/day)

WC = soil water content (meter^3/meter^3)

P$_b$ = soil bulk density (kg/meter^3)

K$_d$ = adsorption or distribution coefficient (meter^3/kg)

Chemicals that move predominantly in the vapor phase should have a relatively small t_D. For these chemicals, Equation 7.23 can be rewritten:

$$t_D \cong (P_b f_{oc} K_{oc} + WC + P_{sa} K_H) P_T^2 l^2 / D_{sa} K_H P_{sa}^{10/3} \qquad (7.25)$$

where

f$_{oc}$ = soil organic carbon fraction (dimensionless)

K$_{oc}$ = soil organic carbon partition coefficient (meter^3/kg)

If K$_d$ is known, it can be substituted into Equation 7.25 in place of $f_{oc} K_{oc}$.

In the case of no water evaporation, when a chemical is initially uniformly incorporated in the soil, the maximum volatilization flux rate through the soil surface to the atmosphere that could occur is:

$$Q_1 = C_o [D_E/(3.1416)t]^{1/2} \qquad (7.26)$$

where

Q$_1$ = maximum volatilization flux rate

C$_o$ = total concentration (gram/meter^2)

D$_E$ = effective soil diffusion coefficient (meter^2/day) as defined in Equation 7.24

t = time (days)

This flux rate is that which would occur with no boundary layer resistance in the air or equivalently when the surface concentration is held at zero for at >0.

When a boundary layer of thickness d is present, the maximum flux that can move through the boundary layer occurs when no soil resistance is present and the gas concentration C$_G$ at the soil surface is held at its initial value C$_G$(o) = C$_o$/R$_G$:

$$Q_2 = D_A C_o / R_G d \qquad (7.27)$$

where

Q$_2$ = maximum flux through the boundary layer

D$_A$ = diffusion coefficient of the chemical in air (meter^2/day)

C_o = total concentration (gram/meter^3)

d = boundary layer thickness (meter)

R_G = ratio of the total chemical concentration to the chemical's concentration in soil air

and

$$R_G = R_L/K_H = (P_bK_d + WC + P_{sa}K_H)/K_H \qquad (7.28)$$

where

R_L = ratio of the total concentration to the chemical's concentration in soil water

K_H = Henry's Law constant (dimensionless)

P_b = soil bulk density (kg/meter^3)

K_d = adsorption or distribution coefficient (meter^3/kg)

WC = soil water content (meter^3/meter^3)

P_{sa} = soil air content (meter^3/meter^3)

Equation 7.27 assumes that the concentration of the chemical in the free air above the boundary layer is zero.

When a boundary layer is present, it will act to restrict volatilization fluxes only if the maximum flux through the boundary layer Q_2 is small compared with Q_1, the rate at which the chemical moves to the soil surface. A benchmark criterion[24] for boundary layer influence when volatilization occurs without water evaporation is:

$$(K_H)^2/K_d < < 9x10^{-8}(kg/meter^3) \qquad (7.29)$$

In the case of water evaporation, if upward water flow carries an insignificant amount of chemical compared with upward diffusion, then Equations 7.27 – 7.29 should be utilized. However, if the upward chemical convection is dominant, as it will be if the solution concentration of the chemical is high or if evaporation and volatilization both occur for a long time period, then the upward flux of chemical toward the boundary layer is approximately:

$$Q_3 = C_LE = C_oE/R_L \qquad (7.30)$$

where

Q_3 = upward flux of the chemical

C_L = solution concentration

Co = initial soil concentration (mg/cm^3)

R_L = ratio of the total concentration to the concentration of the chemical in soil water

E = evaporation rate (meter/day)

If it is assumed that water evaporation is regulated by the boundary layer, then the rate of water vapor diffusion across the boundary layer is:

$$E = D_{wa} P_{wv}(1 - RH)/2P_{wl}d \qquad (7.31)$$

where D_{wa} = binary diffusion coefficient for water vapor in air (2 meter^2/day)

P_{wv} = saturated water vapor density

RH = atmospheric relative humidity

P_{wl} = liquid water density

The criterion for boundary layer restriction on volatilization in this case should occur generally when:

$$K_H < < 2.5 \times 10^{-5} \qquad (7.32)$$

APPLICATIONS AND LIMITATIONS OF SOIL GAS ANALYSIS

The traditional approaches to detect the presence of volatile organic chemicals (VOCs) in groundwater involve the sampling and analysis of groundwater obtained from soil borings or monitoring wells. Recently, a new approach—soil gas analysis—has been developed. This approach involves the collection of a known volume of soil air (i.e. soil gas) obtained from soil pores within a few feet of the soil surface. The soil gas is analyzed by a portable gas chromatograph to determine VOC concentrations. Based on Henry's Law, the concentration of a VOC in soil air should be directly proportional to the VOC concentration in groundwater at equilibrium; under ideal equilibrium conditions, the relationship between these two concentrations should be quantitatively described by the Henry's Law constant.

Soil gas analysis has been successfully utilized during several subsurface investigations. A linear correlation of greater than 95 percent significance existed between groundwater and soil gas concentrations of chloroform obtained at a depth of four feet[27]. The hydrology of this soil-groundwater system was a relatively simple one: unconfined groundwater occurred at a depth of seven to fifteen feet in calcified, unconsolidated alluvium overlying a clay aquiclude in Pittman, NV. At another site, shallow soil gas sampling was utilized to identify potential sources of volatile organic chemicals and to identify potential migration pathways in shallow groundwater (15-20 ft depth to the groundwater surface) in a sand and gravel alluvial deposit[28]. At another site, a static trapping device containing a charcoal absorbent placed immediately below the soil surface was utilized to delineate the extent of trichloroethylene contamination in shallow groundwater in a gravelly, clayey, sand alluvium[29]. The trapping device remained below the soil surface for seven to thirty days to ensure a time-integrated collection of soil-gas.

Soil gas analysis can provide useful information regarding the source and migration of volatile organic chemicals in soil-groundwater systems. However, it is most important to recognize that certain soil properties, chemical properties, and atmospheric conditions can seriously limit the applicability of soil gas analysis. Therefore, it is important that data derived from soil gas analyses be interpreted with the following considerations in mind.

First, an analysis of Equations 7.10 and 7.11 will reveal that the flux of a chemical in soil air is caused by a concentration gradient. Concentration gradients can exist upward toward the soil surface, downward toward the groundwater surface, and horizontally (i.e. parallel to the groundwater surface). As a result, one should not assume that the source of a chemical detected in soil air is immediately above or below a sampling point; the potential for horizontal migration should be assessed.

Second, the absence of a chemical in soil air does not necessarily mean that the chemical is not present in groundwater immediately below the sampling point. An analysis of Equations 7.10 and 7.11 will reveal that blocked soil pores impede the flux of a chemical in soil air. Percolating and capillary water retained by a soil horizon or lens can act as an effective barrier to the vertical migration of a chemical in soil air. Because sandy soils generally have a much smaller water retention capacity than silt soils, which retain less water than clay soils, the presence of silt and clay horizons and lenses impedes the flux of a chemical to the soil gas sampling point.

Third, the inability to detect in soil air a chemical which is present in groundwater may be due to the downward migration of a plume into an aquifer[28]. The downward vertical flow component of a plume in an aquifer will create a layer of clean groundwater above the plume. The chemical must diffuse through this layer before it can diffuse through soil air. Because the diffusion coefficients of most chemicals in water are an order of magnitude smaller than their diffusion coefficients in air, a layer of water above a contaminated plume can impede the flux of a chemical to the soil gas sampling point.

Fourth, the inability to detect in soil air a chemical which is present in groundwater may be due to increasing atmospheric pressure. For example, an increase in barometric pressure of 33 mbar would allow fresh air to penetrate the upper three feet of a 100 ft unsaturated zone. This may cause a volatile chemical to migrate away from a shallow soil gas sampling point.

Fifth, the inability to detect in soil air a chemical present in a light bulk hydrocarbon pancake may be due to masking by water[30]. Percolating soil water may accumulate atop a pancake and impede the flux of a chemical to the soil gas sampling point.

Sixth, the inability to detect in soil a chemical present in groundwater may be due to adsorption. It is important to recognize that soil acts like a gas chromatographic column with respect to the diffusion of organic chemicals through soil air. If several organic chemicals are present, they will not necessarily reach a sampling point simultaneously. One should conclude that sufficient time for the chemical to migrate to the sampling point has elapsed before concluding that the chemical may not be present.

Seventh, the inability to detect in soil a chemical present in groundwater may be due to the biodegradation of the chemical in the unsaturated zone. Biodegradation is extensively discussed in Chapter 9 and will not be covered here.

Eighth, the source of some organic chemicals in surface soils is plant or microorganism metabolism and not industrial sources. Seasonal fluctuations of the concentrations of these chemicals should be anticipated.

REMEDIAL TECHNOLOGIES

Remedial technologies addressing the ability of volatile organic and inorganic chemicals to diffuse through soil air and volatilize into the atmosphere can be divided into two general classes. The first class involves the enhancement or the mitigation of volatile chemical movement through soil or waste. Technologies in this class include soil venting and soil covers. The second class involves remedial technologies that treat soil as a physical structure upon which a waste or hazardous material is placed in order to facilitate volatilization into the atmosphere. Technologies in this class include landspreading and overland flow land treatment systems.

Soil venting can be an effective technology for removing volatile organic chemicals from soil. Two principle types of soil venting systems are available. Positive differential pressure systems induce soil air and volatile organic chemical flow away from the control points, while negative differential pressure systems induce flow toward the control points. Control points are usually slotted or screened casing, placed in or near the zone of subsurface contamination.

Soil venting systems may be either passive or active. Passive methods utilize naturally occurring differences in air and chemical pressures to induce flow. Active methods require the artificial generation of differential pressures to accomplish soil air and chemical movement. Practical experience has demonstrated that the active generation of negative differential pressures typically provides the most favorable field results for removing volatile chemicals from soil.

Several studies attest to the usefulness and applicability of soil venting. During 11 days of continuous venting of unsaturated river sand, 57 percent of the original 75 gallons of gasoline was removed by venting[31]. Carbon dioxide production indicated that microbes may have consumed as much as 2 percent. Venting was most effective in removing lower molecular weight components, especially the paraffins and olefins. Negative differential pressure systems were utilized at three different sites to remove subsurface gasoline vapors, while positive differential pressure systems were utilized in some buildings to prevent vapor movement into basements[32].

Soil covers can be an effective technology for either mitigating or eliminating the movement of volatile chemicals. Properly designed landfill covers reduce gas generation and retard the movement of gases to the soil surface. Soil covers also are used extensively to control radon (Rn) gas emissions.

Radium (Ra) in trace amounts is found in all soils and water. The ^{226}Ra content in uranium mill tailings is usually several orders of magnitude greater than the average background concentration found in soil and water. ^{226}Ra decays to form radioactive ^{222}Rn gas, which poses a serious environmental and health risk. The reduction and control of ^{222}Rn emission from uranium mill tailings disposal sites is an area of major concern in the mining and milling sections of the nuclear industry.

Current practice calls for earthen or other cover materials to control ^{222}Rn emissions. In general, earthen covers about three or more meters thick usually reduce emissions below federal emission standards. Compacted clay/gravel layers approximately 20 cm or less in depth have been relatively effective in reducing ^{222}Rn emissions[33]. Also, compacted barriers comprised of bentonite/gravel, bentonite/lime/gravel, and lime/alum gelatin reduced ^{222}Rn tailing emissions by 96 to 99 percent[33].

Landspreading of soil and waste is also an effective means of decontamination. The spreading of contaminated soil or waste over the land surface allows the volatile components to diffuse through the soil or waste quickly and volatilize into the atmosphere. During the cleanup of two abandoned gravel pits in which hazardous waste was disposed, over 3,000 yd^3 of soil was landspread and decontaminated due to volatilization[34]. The biodegradation of organic chemicals during landspreading will be discussed in great detail in Chapter 9.

Overland flow land treatment can be an effective method of removing volatile organic chemicals from water. This treatment method is usually employed in areas where soils possess relatively slow water infiltration rates. In general, water is applied to the top of a vegetated, gently sloping soil. Water is applied generally at rates of 6.5 to 40 cm/week and is allowed to flow in a thin sheet over the surface. The runoff is collected in ditches and discharged.

The efficiency of removing 13 trace organics from wastewater was studied on an outdoor, prototype overland flow land treatment system[35]. More than 94 percent of each chemical was removed at an application rate of 0.4 cm/hr at 0.12 m³/hr/meter of width. Chemicals included: bromoform, chlorobenzene, chloroform, dibromochloromethane, diethylphthalate, 2,4-dinitrophenol, m-nitrotoluene, naphthalene, aroclor 1242, pentachlorophenol, phenanthrene, and toluene.

REFERENCES

1. Jaynes, D. B. and Rogowski, A. S. Applicability of Fick's Law to Gas Diffusion. Soil Science Society of America Journal **47**:425-430 (1983).
2. Goring, C. A. I. and Hamaker, J. W. (eds). Organic Chemicals in the Soil Environment. New York: Marcel Dekker, Inc. (1972).
3. Jost, W. Diffusion. Third edition. New York: Academic Press (1960).
4. Thibodeaux, L. Chemodynamics. New York: John Wiley & Sons (1979).
5. Riley, D. Physical Loss and Redistribution of Pesticides in the Liquid Phase. *In* British Crop Protection Council Symposium Proceedings (1976).
6. Guenzi, W. D. (ed). Pesticides in Soil and Water. Madison, WI: Soil Science Society of America, Inc. (1974).
7. Burkhard, L. P., Armstrong, D. E., and Andrew, A. W. Henry's Law Constants for the Polychlorinated Biphenyls. Environmental Science and Technology **19(7)**:590-596 (1985).
8. Prausnitz, J. M. Molecular Thermodynamics of Fluid-Phase Equilibria. Englewood Cliffs, NJ: Prentice Hall (1969).
9. Lewis, G. N., Randall, M., Pitzer, K. S., and Brewer, L. Thermodynamics. New York: McGraw-Hill (1961).
10. Jones, C. J. and McGugan, P. J. An Investigation of the Evaporation of Some Volatile Solvents from Domestic Waste. Journal of Hazardous Materials **2**:235-251 (1977/78).
11. Penman, H. L. Gas and Vapor Movements in the Soil: 1. The Diffusion of Vapors Through Porous Solids. Journal of Agricultural Science **30**:437-462 (1940).
12. Farmer, W. J., Yang, M. S., Letey, J., Spencer, W. F., and Roulier,

M. H. Land Disposal of Hexachlorobenzene Wastes: Controlling Vapor Movement in Soils. *In* Proceedings of the Fourth Annual Research Symposium, March 6-8, 1978, San Antonio, TX. EPA-600/9-78-016. Cincinnati, OH: U.S. Environmental Protection Agency (1978).

13. Manos, C. G. Jr., Williams, K. R., Balfour, W. D., and Williamson, S. J. Effects of Clay Mineral-Organic Matter Complexes on Gaseous Hydrocarbon Emissions From Soils. Proceedings of the NWWA/API Conference on Petroleum Hydrocarbons and Organic Chemicals in Groundwater, Houston, TX, Nov. 13-15, 1985. Dublin, OH: National Water Well Assocation (1985).

14. Bohn, H. L., Prososki, G. K., and Eckhardt, J. G. Hydrocarbon Adsorption by Soils as the Stationary Phase of Gas-Solid Chromatography. Journal of Environmental Quality 9(4):563-565 (1980).

15. Callaway, J. Y., Gabbita, K. V., and Vilker, V. L. Reduction of Low Molecular Weight Halocarbons in the Vapor Phase Above Concentrated Humic Acid Solutions. Environmental Science and Technology 18(11): 890-893 (1984).

16. Currie, J. A. Movement of Gases in Soil Respiration. *In* Sorption and Transport Processes in Soil. S.C.I. Monograph No. 37. London: Society of Chemical Industry (1970).

17. Spencer, W. F. and Cliath, M. M. Pesticide Volatilization as Related to Water Loss From Soil. Journal of Environmental Quality 2(2):284-289 (1973).

18. Shen, T. T. Estimating Hazardous Air Emissions From Disposal Sites. Pollution Engineering 13:31-34 (1981).

19. Hartley, G. S. Evaporation of Pesticides. *In* Advances in Chemistry Series 86. Washington, DC: American Chemical Society (1969).

20. Liss, P. S. and Slater, P. G. Flux of Gases Across the Air-Sea Interface. Nature 247:181-184 (1974).

21. Mayer, R., Letey, J., and Farmer, W. J. Models for Predicting Volatilization of Soil-Incorporated Pesticides. Soil Science Society of America Proceedings 38:563-568 (1974).

22. Baker, L. W. and Mackay, K. P. Screening Models for Estimating Toxic Air Pollution Near a Hazardous Waste Landfill. Journal of the Air Pollution Control Association 35(11):1190-1195 (1985).

23. Jury, W. A., Spencer, W. F., and Farmer, W. J. Behavior Assessment Model for Trace Organics in Soil: I. Model Description. Journal of Environmental Quality 12(4):558-564 (1983).

24. Jury, W. A., Farmer, W. J., and Spencer, W. F. Behavior Assessment Model for Trace Organics in Soil: II. Chemical Classification and Parameter Sensitivity. Journal of Environmental Quality 13(4): 567-572 (1984).

25. Jury, W. A., Spencer, W. F., and Farmer, W. J. Behavior Assessment Model for Trace Organics in Soils III. Application of Screening Model. Journal of Environmental Quality **13(4)**:573-579 (1984).
26. Jury, W. A., Spencer, W. F., and Farmer, W. J. Behavior Assessment Model for Trace Organics in Soil: IV. Review of Experimental Evidence. Journal of Environmental Quality **13(4)**: 580-586 (1984).
27. Kerfoot, H. B., Kohout, J. A., and Amick, E. N. Detection and Measurement of Groundwater Contamination by Soil-Gas Analysis. Proceedings of the National Conference on Hazardous Wastes and Hazardous Materials, Atlanta, GA, March 4-6, 1986. Silver Spring, MD: Hazardous Materials Control Research Institute (1986).
28. Wittmann, S. G., Quinn, K. J., and Lee, R. D. Use of Soil Gas Sampling Techniques for Assessment of Groundwater Contamination. *In* Proceedings of the NWWA/API Conference on Petroleum Hydrocarbons and Organic Chemicals in Groundwater, Houston, TX, Nov 13-15, 1985. Dublin, OH: National Water Well Assocation (1985).
29. Malley, M. J., Bath, W. W., and Bongers, L. H. A Case History: Surface Static Collection and Analysis of Chlorinated Hydrocarbons From Contaminated Groundwater. *In* Proceedings of the NWWA/API Conference on Petroleum Hydrocarbons and Organic Chemicals in Groundwater, Houston, TX, Nov. 13-15, 1985. Dublin, OH: National Water Well Association (1985).
30. Yaniga, P. M. Hydrocarbon Retrieval and Apparent Hydrocarbon Thickness: Interrelationships to Recharging/Discharging Aquifer Conditions. *In* Proceedings of the NWWA/API Conference on Petroleum Hydrocarbons and Organic Chemicals in Groundwater, Houston, TX, Nov. 5-7, 1984. Worthington, OH: National Water Well Association (1984).
31. Thornton, J. S. and Wootan, W. L. Jr. Venting for the Removal of Hydrocarbon Vapors From Gasoline Contaminated Soil. J. Environ. Sci. Health **A17(1)**:31-44 (1982).
32. O'Conner, M. J., Agar, J. G., and King, R. D. Practical Experience in the Management of Hydrocarbon Vapors in the Subsurface. *In* Proceedings of the NWWA/API Conference on Petroleum Hydrocarbons and Organic Chemicals in Groundwater, Houston, TX, November 5-7, 1984. Worthington, OH: National Water Well Association (1984).
33. Opitz, B. E., Martin, J. W., and Sherwood, D. R. Gelatinous Soil Barrier for Reducing Contaminant Emissions at Waste Disposal Sites. *In* Proceedings of the National Conference on Management of Uncontrolled Hazardous Waste Sites, Washington, D.C., Nov 29-Dec 1, 1982. Silver Spring, MD: Hazardous Materials Control Research Institute (1982).

34. Davey, J. R. The Abatement of Uncontrolled Hazardous Waste Sites—Freetown and Dartmouth, MA. *In* Proceedings of the U.S. EPA National Conference on Management of Uncontrolled Hazardous Waste Sites, Washington, D.C., October 15-17, 1980. Silver Spring, MD: Hazardous Materials Control Research Institute (1980).
35. Jenkins, T. F., Leggett, D. C., Parker, V. C., and Oliphant, J. L. Toxic Organics Removal Kinetics in Overland Flow Land Treatment. Water Research **19(6)**:707-718 (1985).

8

Organic Chemical Reactions in Soil

INTRODUCTION

Many organic chemicals in soil systems can be degraded through one or more chemical (i.e. abiotic) reactions. In this text, the terms chemical and abiotic reactions refer to those reactions that do not involve (a) metabolically active microorganisms, (b) extracellular enzymes, or (c) metabolic intermediates such as NADH, NADPH, flavins, flavoproteins, hemoprotein, iron porphyrins, chlorophylls, cytochromes, and glutathiones. The ability of soil microorganisms and extracellular enzymes to degrade organic chemicals will be discussed in detail in the next chapter.

In general, five organic chemical reactions are known to occur in soil systems: hydrolysis, substitution, elimination, oxidation, and reduction. These will be defined and discussed in this chapter.

Aluminosilicate minerals which catalytically transform organic chemicals have been used for decades in petroleum refining and in the chemical and petrochemical industries. These mineral catalysts are important in many high temperature, high pressure reactions such as ammonia synthesis, hydrocarbon conversion to gas, dehydrocyclization, the Fisher-Tropsch synthesis, paraffin and cycloalkane isomerization, olefin hydroisomerization, diene hydroisomerization, aromatic hydroisomerization, ethylbenzene isomerization to xylenes, oxidation, and reduction. Many of these aluminosilicate minerals are present in soil. These minerals can catalyze the chemical reactions of organic chemicals at ambient pressure and temperature in soil. The specific role that a mineral may play will depend upon the type of reaction; this will be discussed in this chapter.

Certain organic chemicals can hydrolyze, oxidize, or reduce very quickly upon contact with water in soil. If these chemicals are present in bulk quantities, the reaction may be violent. This chapter lists several classes of organic chemicals that can react rapidly and violently with water in soil.

The soil system can be manipulated to initiate or accelerate the degradation rate of many organic chemicals. It is most unfortunate that this area has received little attention. This chapter will discuss a few case studies, which have been reported in the published literature, addressing remedial technologies for initiating or accelerating in-situ organic chemical reactions in soil.

HYDROLYSIS

Reaction Mechanisms.   Hydrolysis is a chemical reaction in which an organic chemical reacts with water or a hydroxide ion:

$$R-X + H_2O \longrightarrow R-OH + H^+ + X^- \tag{8.1}$$

$$R-X + OH^- \longrightarrow R-OH + X^- \tag{8.2}$$

In Equations 8.1 and 8.2, a nucleophile (H_2O or OH^-) attacks an electrophile (RX) and displaces a leaving group (X). During these reactions, a carbon-leaving group bond is cleaved and a carbon-oxygen bond is formed. Hydrolysis reactions are generally classified as nucleophilic displacement reactions.

There are two general types of nucleophilic displacement reaction mechanisms: S_N1 and S_N2. The S_N1 (substitution, nucleophilic, unimolecular) reaction mechanism involves two separate reactions:

$$R-X \rightleftharpoons R^+ + X^- \tag{8.3}$$

$$R^+ + H_2O \longrightarrow R-OH + H^+ \tag{8.4}$$

First the molecule R-X must ionize to produce the carbonium ion R^+ and a dissociated leaving group X^-. This is usually the slower or rate limiting reaction. Then, the carbonium ion R^+ undergoes a relatively rapid nucleophilic attack by OH^- to form the reaction product R-OH. The overall rate of product formation is governed by the rate of dissociation of R-X (Equation 8.3), which is affected by electron donating and withdrawing fragments comprising the molecule R-X. Also, the dissociation is always assisted by the solvent; in soil systems, water is both the solvent and the nucleophile. The overall rate of formation of the product R-OH is not dependent upon the concentration of R-X.

The S_N2 (substitution, nucleophilic, bimolecular) reaction mechanism involves two molecules (i.e. bimolecular) in a single reaction:

$$H_2O + R-X \longrightarrow H_2O--R--X \longrightarrow H^+ + HO-R + X^- \tag{8.5}$$

The nucleophilic H_2O approaches $R-X$ from a position 180° away from the leaving group X (i.e. the backside of the molecule), as illustrated in Equation 8.5 by $H_2--R--X$. The group X must leave as the group OH comes in. This is a one-step reaction mechanism with no intermediate chemical complexes forming. The energy needed to break the $R-X$ bond is supplied by the simultaneous formation of the $R-OH$ bond.

Reaction Rates. A review of data published in the scientific literature reveals that the hydrolysis of organic chemicals in water is generally first-order with respect to the organic chemical's concentration. As a result, rate constants can be calculated using the following equation:

$$k = (2.303/t) \log (a/[a-x]) \qquad (8.6)$$

where:

$\quad$ k = rate constant, 1/time

$\quad$ t = time

$\quad$ a = initial concentration, ppm

$a-x$ = concentration at time t, ppm

The half-life, the time needed for half of the concentration to react, can be estimated if k is known and by assigning a and x the values of 1 and 0.5, respectively, into Equation 8.6:

$$t_{1/2} = 0.693/k \qquad (8.7)$$

where:

$\quad$ $t_{1/2}$ = half-life

Because hydrolysis for organic chemicals in water is generally first-order with respect to concentration, the k and $t_{1/2}$ obtained at relatively high concentrations can be extrapolated to low concentrations, provided other factors affecting reactivity are constant.

Table 8.1 lists hydrolysis half-lives for several organic chemicals. Half-lives vary from seconds to tens of thousands of years, depending on the presence of molecular fragments susceptible to hydrolysis. In general, the fragments susceptible to hydrolysis include alkyl halides, chlorinated amides, carbamates, esters, epoxides, phosphonic acid esters, phosphoric acid esters, and sulfones. Fragments that generally are not susceptible to hydrolysis include aldehyde, alkane, alkene, alkyne, aliphatic amide, amine, carboxy, and nitro fragments.

It is most important to recognize that Equation 8.6 is an oversimplification for most organic hydrolysis reactions. The total hydrolysis rate constant, k_T, can be expressed as:

$$k_T = k_H[H^+] + k_N + k_{OH}[OH^-] \qquad (8.8)$$

where:

$\quad$ k_H = rate constant for acid-catalyzed hydrolysis

$[H^+]$ = hydrogen ion concentration

k_N = rate constant for neutral hydrolysis

k_{OH} = rate constant for base-catalyzed hydrolysis

$[OH^-]$ = hydroxyl ion concentration

Equation 8.8 can be rewritten to eliminate the term $[OH^-]$ from the equation:

$$k_w = [H^+][OH^-] \tag{8.9}$$

$$k_H = k_H[H^+] + k_N + k_{OH}k_w/[H^+] \tag{8.10}$$

An analysis of Equations 8.8 through 8.10 reveals several important facts. First, the total rate of hydrolysis for many organic chemicals, k_T, is dependent upon the acidity of the soil system. For these chemicals, the reaction rate is proportional to the concentration of acid $[H^+]$ present. Acids can catalyze hydrolysis reactions by converting an organic chemical (e.g. R_1-C=O) to an ion (R_1-C$^+$-OH) that possesses a positive charge on the carbon atom:

$$R_1 - C = O + H^+ \longrightarrow R_1 - C^+ - OH \tag{8.11}$$

The ion is significantly more susceptible to nucleophilic attack, as discussed earlier.

Acid catalysis may also occur with organic chemicals that do not form a complex with H^+ but can form hydrogen bonds:

$$R_2 - C = O + H_2O \longrightarrow R_2 - C^+ = O \cdots H_2O \tag{8.12}$$

Again, the positive charge on the C atom makes the complex more susceptible to hydrolysis than the uncharged organic molecule.

Second, the total rate of hydrolysis for many organic chemicals is dependent upon the alkalinity of the soil system. For these chemicals, the reaction rate is proportional to the concentration of base $[OH^-]$ present: as the concentration of OH^- in the reaction illustrated by Equation 8.2 increases, the total rate of hydrolysis increases.

Third, the total rate of hydrolysis is dependent upon the magnitudes of three rate constants—k_H, k_N, and k_{OH}—in addition to pH. The magnitudes of these three rate constants vary with organic chemical structure.

Because the total hydrolysis rate is dependent upon four parameters, the pH-rate profiles for organic chemicals that react via acid and base catalyzed hydrolysis may be either U-shaped or V-shaped, depending upon the magnitude of k_N (see Figure 8.1). For organic chemicals that react by water-acid

TABLE 8.1 Hydrolysis Half-Lives for Various Organic Chemicals.

Chemical	half-life[a,b]
acetamide	3950 y
atrazine	2.5 h
	0.3-5.0 y[c]
	0.4-1.0y[e]
	3.0-5.0 y[f]
aziridine	154 d
benzoyl chloride	16 s
benzyl bromide	1.32 h
benzyl chloride	15 h
benzyl ethanoate	1.1 y
benzylidene chloride	0.1 h
bis-4-nitrophenyl phosphate	7 d[g]
bromoacetamide	21,200 y
bromochloromethane	44 y
bromodichloromethane	137 y
bromoethane	30 d
1-bromohexane	40 d
3-bromohexane	12 d
bromomethane	20 d
bromomethylepoxyethane	16 d
1-bromo-3-phenylpropane	290 d
1-bromopropane	26 d
3-bromopropene	12 h
chloroacetamide	1.46 y
chlorodibromomethane	274 y
chloroethane	38 d
1-(2-chloroethyl) N-phenylcarbamate	140 y
chlorofluoroiodomethane	1.0 y
chloromethane	339 d
chloromethylepoxyethane	8.2 d
1-chloromethyl-1- methylepoxyethane	16 d
2-chloro-2-methylpropane	23 s
1-(3-chlorophenyl) N-phenylcarbamate	1.1 h
2-chloropropane	2.9 d
3-chloropropene	69 d
cyclohexene oxide	4.7 d[i]
cyclopentanecarboxamide	5500 y
diazinon	9.5 d[i]
dibromomethane	183 y
1,3-dibromopropane	48 d
dichloroacetamide	0.73 y
1-(2,2-dichloroethyl)- N-phenylcarbamate	4.4 y

TABLE 8.1 Hydrolysis Half-Lives for Various Organic Chemicals. (cont.)

Chemical	half-life[a,b]
dichloroiodomethane	275 y
dichloromethane	704 y
dichloromethyl ether	25 s
diethyl methylphosphonate	990 y
diethyl phenylphosphonate	440 y
diisopropyl ethylphosphonate	5.5 y
diisopropyl methylphosphonate	663,000 y
dimethoxysulfone	1.2 m
1,2-dimethylepoxyethane	15.7 d
1,1-dimethylepoxyethane	4.4 d
1,1-dimethylethyl ethanoate	140 y
1,1-dimethylethyl trifluoroethanoate	<8.9 m
dimethyl methylphosphonate	88 y
2,4-dinitrophenyl ethanoate	9.4 h
diphenyl phosphate	20.6 d[g]
epxoyethane	12 d
3,4-epoxy-1,5-cyclohexadiene	8 m
3,4-epoxycyclohexene	6 m
3,4-epoxycyclooctane	52 m
1,3-epoxy-1-oxopropane	3.5 m
ethion	9.9 d[i]
N-ethylacetamide	70,000 y
ethyl acetate	136 d[i]
ethyl butanoate	5.8 y
ethyl trans-2-butenoate	17 y
ethyl difluoroethanoate	23 m
ethyl dimethylethanoate	9.6 y
ethyl ethanoate	2 y
1-ethyl N-methyl-N-phenylcarbamate	44,000 y
ethyl methylthioethanoate	87 d
ethyl methylsulfinylethanoate	6.2 d
ethyl (4-nitrophenyl) methylphosphonate	5.5 y
ethyl phenylmethanoate	7.3 y
ethyl propanoate	2.5 y
ethyl propenoate	3.5 y
ethyl propynoate	17 d
ethyl pyridylmethanoate	0.41 y
fluoromethane	30 y
2-fluoro-2-methylpropane	50 d
hydroxymethylepoxyethane	28 d
1-hydroxymethyl-1-methylepoxyethane	16 d
iodoethane	49 d
iodomethane	110 d

TABLE 8.1 Hydrolysis Half-Lives for Various Organic Chemicals. (cont.)

Chemical	half-life[a,b]
2-iodopropane	2.9 d
3-iodopropene	2.0 d
isobutyramide	7700 y
isopropyl bromide	2.0 d
isopropyl ethanoate	8.4 y
isopropyl phenylmethanoate	35 y
malathion	8.1 d[i]
methoxyacetamide	500 y
1-(4-methoxyphenyl) N-phenylcarbamate	3.2 d
N-methylacetamide	38,000 y
methyl chloroethanoate	14 h
methyl dichloroethanoate	38 m
methylepoxyethane	14.6 d
N,N-methylethylacetamide	18,500 y
methyl parathion	10.9 d[i]
1-methyl N-phenylcarbamate	4000 y
methyl trichloroethanoate	<15 m
methyl trifluoroethanoate	<3.6 m
monomethyl phosphate	1.0 d[g]
1-naphthyl N,N-dimethylcarbamate	1,200 y
1-naphthyl N-methylcarbamate	8.5 d
1-(4-nitrophenyl) N-methyl-N-phenylcarbamate	2,700 y
1-(4-nitrophenyl) N-phenylcarbamate	26 s
parathion	17 d[i]
phenyl dichloroethanoate	3.7 m
phenyl ethanoate	38 d
1-phenyl N-methyl-N-phenylcarbamate	5,200 y
1-phenyl N-phenylcarbamate	1.5 d
phosphonitrilic hexaamide	46 d[g]
propadienyl ethanoate	110 d
ronnel	1.6 d[i]
tetrachloromethane	7000 y (1 ppm) 7 y (1000 ppm)
tribromomethane	686 y
trichloroacetamide	0.23 y
1-(2,2,2-trichloroethyl) N-phenylcarbamate	252 d
trichloromethane	3500 y
trichloromethylbenzene	19 s
triethylphosphate	5.5 y
tri(ethylthio)phosphate	8.5 y
1-(2,2,2-trifluoroethyl) N-phenylcarbamate	2.2 y

TABLE 8.1   Hydrolysis Half-Lives for Various Organic Chemicals. (cont.)

Chemical	half-life[a,b]
trimethylphosphate	1.2 y
tri(4-nitrophenyl)phosphate	11 m
triphenoxyphosphate	1.3 y
valeramide	11300 y
p-xylyl bromide	4.3 m
p-xylyl chloride	0.43 h

[a] d = days, h = hours, m = minutes, s = seconds, y = years.
[b] in water at pH 7, from references 1,2,3,4, and 5, unless specified otherwise.
[c] in Ella silt loam soil, pH = 4, reference 6.
[e] in Polygan silt soil, pH = 7, reference 6.
[f] in Kewaunee clay soil, pH = 8, reference 6.
[g] average obtained using Clarion loam soil (pH = 4.6), Nicollet clay loam soil (pH = 6.2), and Harps clay loam soil (pH = 7.6), reference 7.
[i] in water, pH = 6.0, reference 3.

catalysis or by water-base catalysis, the pH-rate profiles are somewhat different than those illustrated in Figure 8.1. Figure 8.2 illustrates pH-rate profiles for three organochlorides—allyl chloride, methyl chloride, and chloroform—that hydrolyze under water and base catalysis conditions and two organochlorides—t-butyl chloride and benzyl chloride—that are not subject to acid-and base-catalyzed hydrolysis at typical environmental pHs.

Soil-Catalyzed Hydrolysis.   Soil can profoundly affect the hydrolysis half-lives of some organic chemicals. This effect is caused by four different soil factors: the pH at soil particle surfaces, the presence of metals in soil, the adsorption of the organic chemical, and the soil water content.

The pH at soil particle surfaces in acid soils can be as much as 4 pH units lower than the pH of the water phase of a soil system. The strength of acid sites on a dried kaolinite was measured and found to be comparable to 90 percent sulfuric acid[9]. One generally accepted explanation is that soil particles possess negative charges that attract H^+. Another generally accepted explanation is that an insufficient number of water molecules are present to satisfy the hydration requirements of exchangeable cations and exposed Al ions at crystal edges. The residual H_2O molecules become highly polarized and more readily dissociate in the presence of proton acceptors. This causes the hydration deficient-exchangeable cations and edge Al ions to become active acid sites. The net result of the above processes is the greater accumulation of H^+ at soil particle surfaces relative to the water phase. An organic chemical susceptible to acid-catalyzed hydrolysis may have a shorter half-life at or near particle surfaces than in the water phase of a soil system.

In soil, metal ions may be important catalysts for some organic chemi-

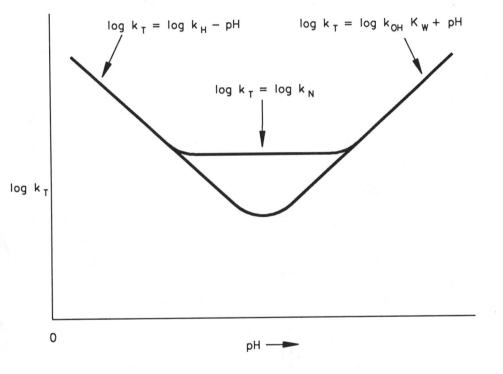

FIGURE 8.1   pH—rate profile for organic chemicals undergoing acid— and base—catalyzed hydrolysis.

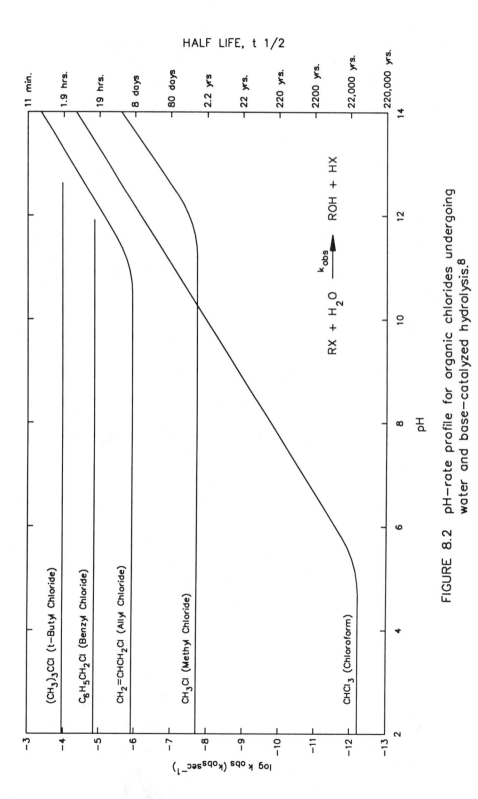

FIGURE 8.2 pH-rate profile for organic chlorides undergoing water and base-catalyzed hydrolysis.[8]

cals. The metal M^+ may act like a Lewis acid that directly polarizes the organic chemical:

$$R-C=O \ + \ M^+ \longrightarrow R-C^+-O\cdots M \qquad (8.13)$$

which, as a result, would enhance substitution reactions. Polarization can also occur by coordination of the metal to the leaving group:

$$R\text{-}X \ + \ M^+ \longrightarrow R^+ \ + \ XM \qquad (8.14)$$

The fragment R^+ is now susceptible to hydrolysis by H_2O and OH^-. For example, the catalysis of organophosphate hydrolysis by Cu^{2+} was related to the ability of the organophosphate to form Cu^{2+} complexes[10]:

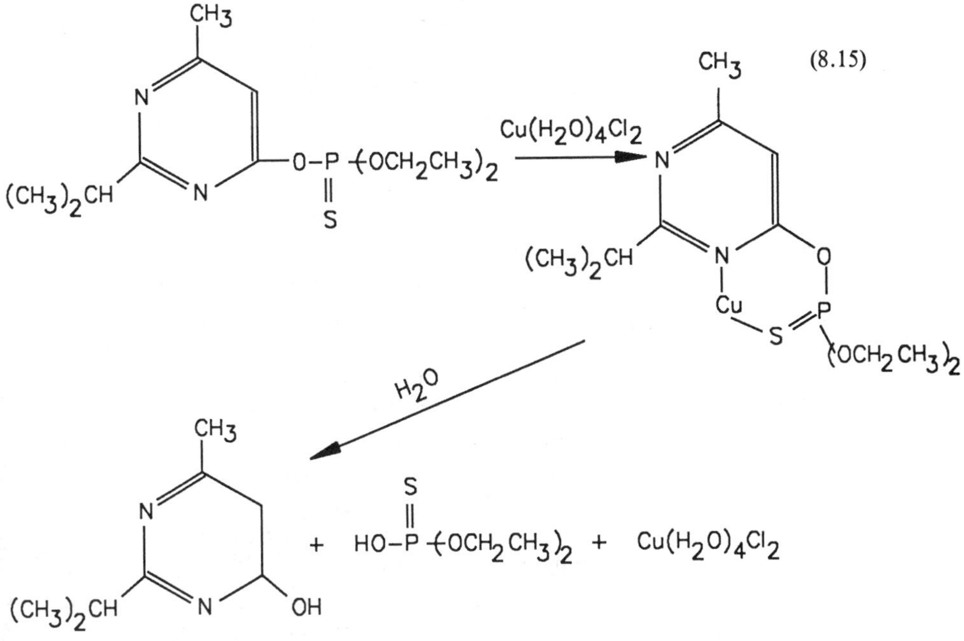

Cu-montmorillonite was nearly as active as a free Cu(II) salt; Cu-beidellite was less active; Cu(II) adsorbed by ion-exchange resins was nearly inactive, and Cu(II) adsorbed by soil organic matter was completely inactive.

Equations 8.8 and 8.10 can be modified to generally describe metal-ion catalysis by including one or more terms of the general form:

$$k_M k_A [M_T] / (k_A + [H^+])$$ (8.16)

where:

k_M = metal-ion catalysis constant

k_A = equilibrium constant for the dissociation of the hydrated ion complex

$[M_T]$ = total metal ion concentration

However, because metals exist as several complexes in soil, the number of terms needed to quantify the complete rate equation for one organic chemical in a soil are too numerous to adequately quantify.

Adsorption can affect the hydrolysis rate by removing the organic chemical from the water phase. For example, the adsorption of the 1-octyl ester of 2,4-dichlorophenoxyacetic acid caused a decrease of the alkaline hydrolysis rate constant by a factor equal to the fraction of the ester associated with humic substances[11]. In a second example, parathion hydrolysis on kaolinite proceeds in two stages because of adsorption[12]. In the first stage, parathion molecules specifically adsorbed at the exchangeable cations are quickly hydrolyzed by contact with the dissociated hydration water molecules. The duration of this stage is short, a matter of days. In the second stage, the parathion molecules which might have initially been bound to the clay surface by different mechanisms are hydrolyzed when they reach the exchangeable cations at the proper orientation. This reaction is slower than that of the first stage.

The water content of soil can affect the hydrolysis rate of organic chemicals, based on the research involving a few chemicals. For example, the addition of water to kaolinites to the limits of adsorbed water, approximately 11 percent, resulted in 95 percent degradation of methyl parathion and parathion in five days[13]. An increase in soil moisture content above 11 percent resulted in a steep decrease in the degradation rate to between 10 and 50 percent in 15 days. The exact mechanisms on how moisture content affects hydrolysis rates are not known; at low soil moisture content, it presumably involves the presence of more organic molecules in the region near the soil surface where pH is lower and where more metal ions are found.

The type of clay mineral present in soil can affect the rate of hydrolysis of an organic chemical in soil. For example, the internal surfaces of an Upton, Wyoming bentonite catalyzed the hydrolysis of acetonitrile to acetamide[14]:

$$CH_3 - CN + H_2O \longrightarrow CH_3C(O)NH_2$$ (8.17)

However, a Libby, Montana vermiculite did not produce appreciable acetamide upon reaction with acetonitrile.

GENERAL SUBSTITUTION AND ELIMINATION REACTIONS

Reaction Mechanisms. The hydrolysis reactions discussed above involve the reaction of an organic chemical with water or OH^- to form an alcohol via the S_N1 or S_N2 nucleophilic substitution reaction mechanisms. Soil systems contain other chemicals that can react with organic chemicals to undergo similar substitution reactions.

An excellent example of how nucleophilic substitution reactions may transform organic chemicals in groundwater involved the release of 1-bromo-3-chloropropane and 13 other halogenated alkanes into a reduced groundwater containing hydrogen sulfide[5]. The proposed reaction scheme for the formation of over 18 reaction products involved hydrolysis and S_N2:

$$R-CH_2BR \xrightarrow[\text{hydrolysis}]{H_2O/OH^-} R-CH_2OH + Br^- \qquad (8.18)$$

$$\Big\downarrow \begin{matrix} H_2S/HS^- \\ \\ S_N2 \end{matrix} \xrightarrow{} R-CH_2SH + Br^- \xrightarrow[\text{Oxidation}]{R^1CH_2SH} R-CH_2-S-S-CH_2-R^1 \quad (8.19)$$

$$R-CH_2SH \rightleftharpoons RCH_2S^- \xrightarrow[S_N2]{R^1CH_2Br} R-CH_2-S-CH_2-R^1 \qquad (8.20)$$

Because HS^- and RS^- are much "softer" nucleophiles than OH^- and H_2O, the activation energies of the reactions of alkyl halides with HS^- and RS^- can be assumed to be significantly smaller than those of hydrolysis[5]; in other words, the reactions of alkyl halides with HS^- and RS^- are preferential over alkyl halide reactions with OH^- and H_2O.

Elimination reactions involve the loss of two leaving groups from adjacent atoms within a molecule. The loss results in the formation of a new double (or triple) bond:

$$\underset{\underset{X_1}{|}\ \underset{X_2}{|}}{R-CH-CH_2} \xrightarrow{} R-CH=CH_2 \qquad (8.21)$$

The above reaction is commonly known as beta elimination because one leaving group (X_2) is lost from the alpha carbon atom and another leaving group (X_1) is lost from the adjacent beta carbon atom.

There are two general types of elimination reaction mechanisms: E1 and E2. The E1 mechanism (elimination, monomolecular) is a two-step process in which the rate determining step is ionization of the organic chemical to produce a carbonium ion which rapidly loses a beta hydrogen proton to a base, which is usually the solvent:

$$R_1-\underset{\underset{H}{|}}{\overset{\overset{R_2}{|}}{C}}-\underset{\underset{R_4}{|}}{\overset{\overset{R_3}{|}}{C}}-X \quad \underset{\text{SLOW}}{\rightleftharpoons} \quad R_1-\underset{\underset{H}{|}}{\overset{\overset{R_2}{|}}{C}}-\underset{\underset{R_4}{|}}{\overset{\overset{R_3}{|}}{C}}{}^+ \; + \; X^- \qquad (8.22)$$

(CARBONIUM ION)

$$R_1-\underset{\underset{H}{|}}{\overset{\overset{R_2}{|}}{C}}-\underset{\underset{R_4}{|}}{\overset{\overset{R_3}{|}}{C}}{}^+ \quad \longrightarrow \quad \underset{R_1}{\overset{R_2}{\diagdown}}C=C\underset{R_4}{\overset{R_3}{\diagup}} \qquad (8.23)$$

The first reaction of the E1 (Eq. 8.22) is the same as that of the S_N1 mechanism. The second reaction is different because the solvent pulls a proton from the beta carbon atom of the carbonium ion rather than attacking it at the positively charged carbon, as in the S_N1 process.

The E2 mechanism (elimination, bimolecular) is a one step process in which two leaving groups depart simultaneously with the proton being pulled off by a base:

(8.24)

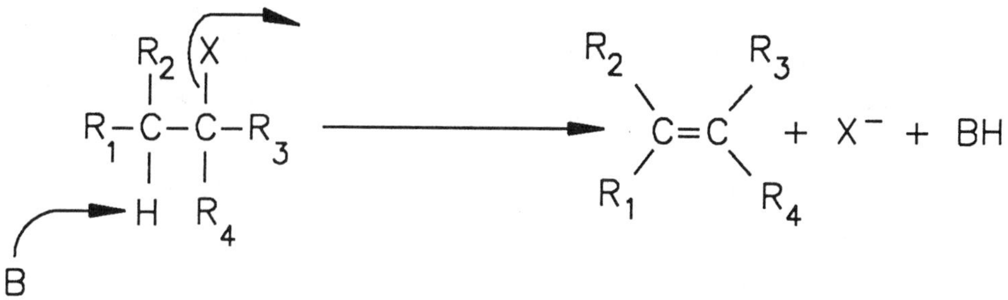

$$R_1-\underset{\underset{H}{|}}{\overset{\overset{R_2}{|}}{C}}-\underset{\underset{R_4}{|}}{\overset{\overset{X}{|}}{C}}-R_3 \longrightarrow \underset{R_1}{\overset{R_2}{>}}C=C\underset{R_4}{\overset{R_3}{<}} + X^- + BH$$

B

This mechanism is analogous to the S_N2 mechanism and often competes with it.

Organic chemicals with branching or with a strong electron withdrawing group on the beta carbon should undergo elimination reactions; the H on the beta carbon should be sufficiently acidic to be removed by H_2O or OH^-, which could act as bases:

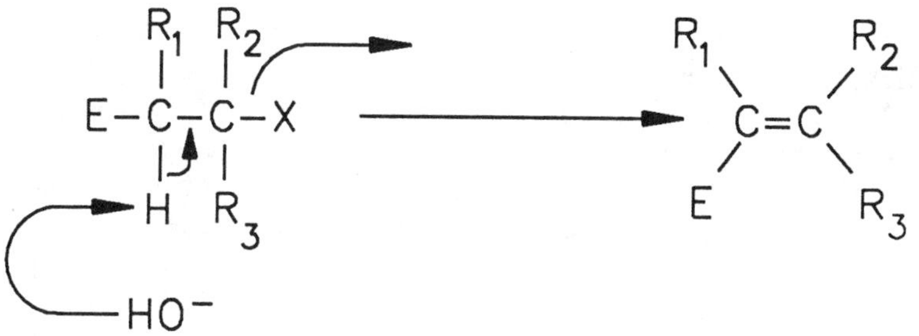

$$E-\underset{\underset{H}{|}}{\overset{\overset{R_1}{|}}{C}}-\underset{\underset{R_3}{|}}{\overset{\overset{R_2}{|}}{C}}-X \longrightarrow \underset{E}{\overset{R_1}{>}}C=C\underset{R_3}{\overset{R_2}{<}} \qquad (8.25)$$

HO^-

For example, two dibromoalkanes with vicinal bromides—1,2-dibromoethane and 1,2-dibromopropane—eliminated HBr in water at pH 7 to yield significant amounts of vinyl bromide and bromopropenes[4]:

(8.26)

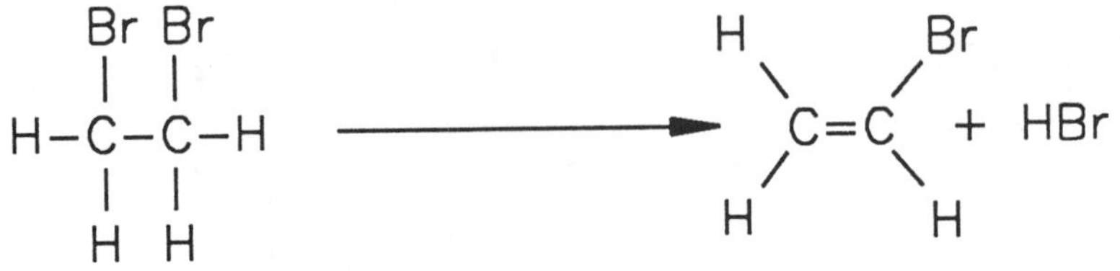

$$H-\underset{\underset{H}{|}}{\overset{\overset{Br}{|}}{C}}-\underset{\underset{H}{|}}{\overset{\overset{Br}{|}}{C}}-H \longrightarrow \underset{H}{\overset{H}{\diagdown}}C=C\underset{\diagdown H}{\overset{\diagup Br}{}} + HBr$$

(8.27)

$$H-\underset{\underset{H}{|}}{\overset{\overset{H}{|}}{C}}-\underset{\underset{H}{|}}{\overset{\overset{Br}{|}}{C}}-\underset{\underset{H}{|}}{\overset{\overset{Br}{|}}{C}}-H \longrightarrow H-\underset{\underset{H}{|}}{\overset{\overset{H}{|}}{C}}-\underset{\underset{H}{|}}{\overset{\overset{H}{|}}{C}}=C\underset{\diagdown H}{\overset{\diagup Br}{}}$$

$$+ HBr$$

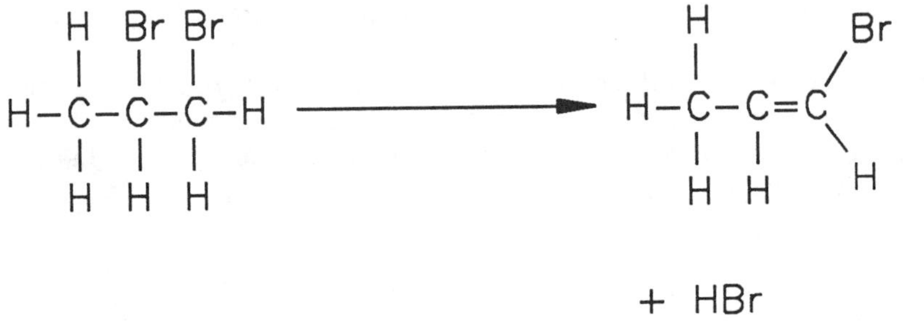

Reaction Rates. In general, the rates of general substitution and elimination reactions in soil systems can be mathematically expressed by Equations 8.6 and 8.7. However, it is most important to recognize that reaction rate constants for these reactions in soil systems generally have not been studied or reported in the published literature. As with hydrolysis reactions, the nature of the soil surface can play a role in determining reaction rates for many organic chemicals. For example, cetyltrimethylammonium bromide underwent no measurable decomposition in the presence of volclay[15]; however, it underwent spontaneous decomposition at ambient temperatures in the presence of montmorillonites from the Mowry formation:

$$CH_3(CH_2)_{13}\text{-}CH_2\text{-}CH_2\text{-}N^+(CH_3)_3 \longrightarrow \qquad (8.28)$$
$$CH_3(CH_2)_{13}\text{-}CH=CH_2 + (CH_3)_3N^+H$$

The nature of the surface catalytic sites responsible for this observed phenomenon is not known.

OXIDATION

Reaction Mechanisms. Oxidation is the removal of electrons from a chemical. In general, it will occur for organic chemicals by two different pathways: in the heterolytic, or polar reaction, pathway, an electrophilic agent attacks an organic molecule and abstracts an electron pair; in the homolytic, or free-radical reaction pathway, an agent abstracts only one electron[16]. In general, free-radical oxidations require much less activation energy than is needed for either cleavage of a covalent bond or a polar oxidation.

Free-radical oxidations frequently consist of two or more steps[16]. The first involves the reversible formation of a free radical through the removal of a single electron from the molecule. This step may happen by (a) thermal (spontaneous) dissociation of molecules that have weak covalent bonds, (b) disruption of molecules by exposure to radiant energy, (c) disruption of molecules by high-energy particles or electrons, such as alpha or beta radiation, or (d) single-electron transfer to ions of transition elements that have incomplete inner electron shells, such as those elements bound in layer lattice silicates in soils and clays.

Addition of the original radical to another molecule is the second step in the free-radical reaction pathway; its occurrence will depend upon the nature of the reactants and reaction conditions[16]. Most free radicals are highly reactive and will react with the first available species they contact. When the total concentration of radicals is low, as is almost invariably true, the free radical will most likely react with a molecule instead of another radical. The resulting product may be another free radical; an example of a neutral radical species adding to an aromatic system would be:

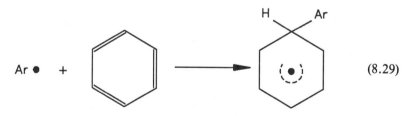

$$(8.29)$$

In this step, the aromatic radical Ar· attacks the ring in the same manner as an electrophile would. This particular intermediate is relatively stable because of resonance. Note that the intermediate could react with another aromatic molecule to form a larger, but chemically very similar, species. Repetition of the process could produce oligomers and, eventually, polymers.

The final step in the free-radical reaction pathway, termination, involves the destruction of the free radicals that appeared in the earlier steps[16]. Termination occurs in three ways: by simple coupling:

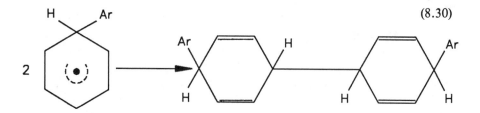

(8.30)

by disproportionation:

(8.31)

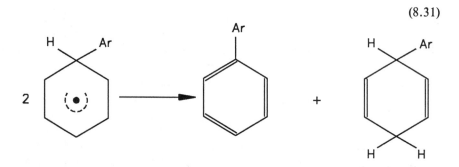

or by abstraction, if a species (R') that can abstract hydrogen is present:

(8.32)

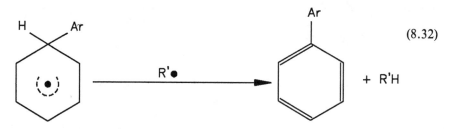

The hydrobiphenyls formed in Equations 8.30 and 8.31 will oxidize readily to the corresponding biphenyls.

The foregoing reaction sequence could also apply in cases where the initiating species is a radical cation rather than an uncharged radical. Because radical cations are electron-deficient species, electron-donating substituents (e.g., OR, NR_2, or alkyl) render an aromatic molecule more susceptible toward attack by radical cations. Aromatic radical cations have been implicated in electrochemical and chemical reaction systems. Radical cations can initiate polymerization of unsaturated organic substrates; however, the kinetics of these processes are complex, and many aspects of the mechanisms are not well understood.

Many variations of and deviations from these simplified descriptions and definitions exist, but their discussion is beyond the scope of this book.

Oxidation and Soil Redox Potentials. The occurrence of oxidation reactions is a function of the electrical potential in the reacting system[16]. This potential is usually stated in volts and is measured in a soil system with a reference electrode in combination with a metallic electrode, such as platinum, which is sensitive and reversible to oxidation-reduction conditions.

The principles of redox reactions are relatively simple when each reaction is considered alone with respect to the reactants and products in true solution; the soil system is very complex, however, and the contribution of each phase to the total potential cannot be deduced. Organic matter, iron, manganese, sulfur, and numerous other substances at equilibrium in the soil—each has its own oxidation state that contributes to the observed redox potential. Furthermore, the contribution of each phase will also vary with such factors as water content, oxygen activity in the soil, and pH.

For oxidation to occur, the potential of the soil system must be greater than that of the organic chemical in question. Very-well-oxidized soils may register a redox potential of 0.8 V while an extremely reduced soil may register −0.5 V. Soil redox potentials can be generally classified as:

Aerated soils:	+800 to +400 mv
Moderately reduced soils:	+400 to +100 mv
Reduced soils:	+100 to −100 mv
Highly reduced soils:	−100 to −500 mv

As long as oxygen is available (reducible at +340 to +320 mv), other oxidized compounds are protected from biological and chemical reduction. Where the oxygen supply is depleted, nitrates are readily reduced (+225 mv) if an energy source is available. Manganese (+200 mv) and iron (+120 mv) tend to buffer redox potentials in the moderately reducing range. Sulfates are reduced at about −150 mv, and their presence can be suspected where highly reducing potentials are observed.

Soil-Catalyzed Oxidation. Soils, clays, and minerals possess the ability to catalyze chemical oxidation of many organic chemicals at ambient pressure and temperature (see Table 8.2). Iron, manganese, aluminum, trace metals within soil minerals and adsorbed oxygen have been identified as the catalysts promoting free-radical oxidation in soil and clay. It is important to recognize, however, that not all soils and clays will catalyze these reactions. Also, quantitative rate constants for chemical oxidation have not been determined,

TABLE 8.2   Organic Chemicals that Undergo Clay-, Mineral-, and Soil-Catalyzed Oxidation[a]

Alizarin Red S
4-aminoacetophenone
4-aminoazobenzene
3-aminobenzenesulfonic acid
2-aminobenzoic acid
4-aminobenzoic acid
4-aminobiphenyl
4-amino-N,N-diethylaniline
4-amino-N,N-dimethylaniline
2-amino-4',5-dimethylazobenzene
4-amino-2',3-dimethylazobenzene
4-aminodiphenylamine
2-amino-3-indolylpropanoic acid
2-aminophenol
4-aminophenol
3-amino-1,2,4-triazole
aniline

benzene
benzidine
bis(4-aminophenyl)methane
4,4'-bis(N,N-dimethylamino)benzophenone
bis(N,N-dimethyl-4-aminophenyl)methane
bis(N,N-dimethyl)-1,4-diaminobenzene
4-bromo-N,N-dimethylaniline
Bromphenol Blue

2-chloro-4-nitroaniline
Congo Red 4B

1,2-diaminobenzene
1,3-diaminobenzene
1,4-diaminobenzene
2,5-diaminofluorene
2,7-diaminofluorene
N,N-diethylaniline
1,2-dihydroxybenzene
1,3-dihydroxybenzene
1,4-dihydroxybenzene
2,5-dihydroxybenzoic acid
3,4-dihydroxybenzoic acid
3,5-dihydroxybenzoic acid
1,2-dihydroxy-3-methoxybenzene
1,2-dihydroxy-4-nitrobenzene
3,4-dihydroxyphenylalanine
3-(3,4-dihydroxyphenyl)propenoic acid
3,5-dihydroxytoluene
3,3'-dimethoxybenzidine
N,N-dimethyl-4-aminoazobenzene

TABLE 8.2   Organic Chemicals that Undergo Clay-, Mineral-, and Soil-Catalyzed Oxidation[a]
(cont.)

N,N-dimethyl-4-aminobenzaldehyde
N,N-dimethylaniline
N,N-dimethyl-4-aminophenyloxalate
3,3'-dimethylbenzidine
diphenylamine
di-B-p-phenylenediamine

2-ethoxyaniline
4-ethoxyaniline
N-ethylaniline
N-ethyl-N-hydroxymethylaniline

Fuchsin Basic

Hexamethyl Violet
hydroquinone
2-hydroxybenzoic acid
4-hydroxybenzoic acid
4-hydroxy-3,5-dimethoxybenzoic acid
4-hydroxy-3-methoxybenzoic acid
N-hydroxymethylaniline
3-(4-hydroxyphenyl)propenoic acid

indole
3-indolylacetic acid
3-indolylpropionic acid
indolylpyruvic acid
isopropoxydiphenylamine
2-methoxyaniline
4-methoxyaniline
methoxybenzene
3-methoxy-4-hydroxybenzoic acid
3-(3-methoxy-4-hydroxyphenyl)propenoic acid
2-methoxyphenol
2-methylaniline
3-methylaniline
4-methylaniline
Methylene Violet

1-naphthylamine
2-naphthylamine
4-nitroaniline
2-nitro-N,N-dimethylaniline
4-nitro-N,N-dimethylaniline
3-nitro-N-methylaniline
4-nitro-N-methylaniline
4-nitroso-N,N-dimethylaniline

phenetole
phenol

TABLE 8.2   Organic Chemicals that Undergo Clay-, Mineral-, and Soil-Catalyzed Oxidation[a]
(cont.)

N-phenyl-2-naphthylamine
Safranine O
Safranine Y

tannin
2-thiobenzoic acid
toluene
1,2,3-trihydroxybenzene
3,4,5-trihydroxybenzoic acid
triphenylmethane
tris(N,N-dimethyl-4-aminophenyl)methane
tyrosine

---

[a]compiled from data presented in references 16 and 17.

but the reactions generally occur very rapidly in soil.

The reaction of benzidine with clays is a classic example of the initiating step of a free-radical oxidation catalyzed by layer lattice silicates in an aqueous environment[17]. An electron from the diamine is abstracted by the clay to form a blue, monovalent, radical cation:

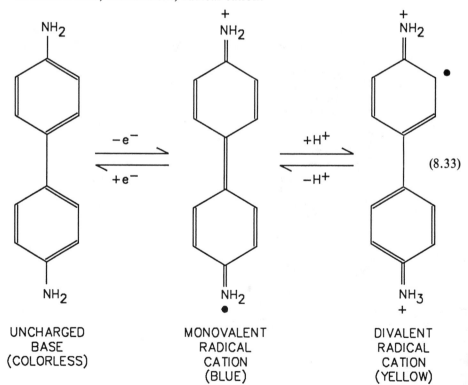

(8.33)

UNCHARGED	MONOVALENT	DIVALENT
BASE	RADICAL	RADICAL
(COLORLESS)	CATION	CATION
	(BLUE)	(YELLOW)

The monovalent species exists only at a pH > 2; at pH < 2, the radical cation accepts a proton on one of its nitrogen atoms to form a yellow, doubly charged radical cation, perhaps with formation of monoprotonated benzidine.

The general guidelines for predicting the occurrence of free-radical oxidation in soils and clays are based on chemical structural characteristics and lower solubility limits[16]. If a chemical has structural characteristics and a water solubility falling within one of these groups, then that chemical can be expected to react as follows:

Group 1: Aromatic chemicals with weak electron-donating ring substituent fragments ($0 > \sigma$-0.50), where σ is the Hammett Sigma constant for the para position) and electron-withdrawing substitutent fragments. Lower water solubility limit is 200 ppm.

Group 2: Aromatic chemicals with strong electron-donating ring substituent fragments ($\sigma < $-0.50) and electron-withdrawing ring substituent fragments. Lower water solubility limit is 112 ppm.

Group 3: Aromatic chemicals with only electron-donating ring substituent fragments. Lower water solubility limit is 29 ppm.

Group 4: Aromatic chemicals containing extensive conjugation. Lower water solubility limit ranges between high parts per billion to low parts per million.

Reaction Rates.   In general, the rates of oxidation reactions in soil systems can be mathematically expressed by Equations 8.6 and 8.7; however, it is most important to recognize that reaction rate constants generally have not been studied or reported in the published literature.

An analysis of the general guidelines for predicting the occurrence of free-radical oxidation in soils[16] does provide insights into the factors affecting reaction rates. This classification of chemicals into groups reveals a compromise between structure and the lower water solubility limit; this trade-off can be explained using simple kinetic theory. The rate of reaction, k, can be expressed as

$$k = CF \times EF \times OF \qquad (8.34)$$

where CF is the collision frequency, EF is the energy factor and OF is the orientation factor[19]. The EF is inversely proportional to the activation energy of a reaction, E_a, or

$$EF \sim 1/E_a \qquad (8.35)$$

The OF is dependent upon the structure of the chemical, the structure of the reaction surface, and the solvent. Assuming that OF was similar for all chemicals and systems referenced in the study, k will remain constant if a decrease in CF is compensated by a decrease in E_a. Because the oxidation reaction site at the clay mineral surface is hydrophilic, the number of collisions (CF) will decrease as the water solubility decreases; therefore, the lower the solubility limit for a group, the lower the CF.

The decrease in CF was apparently compensated for by a decrease in E_a. This trend in E_a among groups can be verified only through an indirect, qualitative comparison of the chemical structures between groups, for quantitative activation energy data are not available for most of these chemicals. Group 2, the group that contained strong electron-donating substituent fragments, should exhibit a lower E_a in transforming to the free-radical intermediate than Group 1, the group having weak electron donating fragments. Because Group 3 chemicals have no electron-withdrawing substituent fragments, their activation energies should be lower than those of either Group 2 or Group 1. Because Group 4 chemicals have extensive conjugation, which should easily stabilize free-radical intermediates, the activation energies needed to transform these chemicals to corresponding intermediates should be the lowest of all the groups.

The guidelines indicate that, in general, many substituted aromatic chemicals may undergo free-radical oxidation. For example, benzene, benzidine, ethylbenzene, naphthalene, and phenol undergo this reaction. However, these guidelines also indicate another large group of organic chemicals that are unlikely to be so oxidized. This second group includes, e.g., tetrachlorodibenzo-p-dioxin (TCDD), hexachlorobenzene, hexachlorocyclopentadiene, polybrominated (PBBs) and polychlorinated (PCBs) biphenyls, and nonaromatics such as kepone and mirex.

REDUCTION

In the organic chemistry literature, the term reduction has several definitions. An organic chemical commonly is said to be reduced if reaction leads to an increase in its hydrogen content or a decrease in its oxygen content. Examples of increases in hydrogen content are:

$$H_2C = CHCl + H_2 \longrightarrow CH_3CH_2Cl \tag{8.36}$$

or

$$Cl_2C = CCl_2 + H^+ \longrightarrow ClCH = CCl_2 + Cl^- \tag{8.37}$$

An example of a decrease in oxygen content is:

$$R - CH \overset{\displaystyle O}{\diagup \ \diagdown} CH_2 + H_2 \longrightarrow R\text{-}CH_2 = CH_2 + H_2O \tag{8.38}$$

Although the above definition is useful, it is deficient because many reduction reactions do not involve changes in a chemical's hydrogen or oxygen content. As a result, reduction is also defined in terms of electron transfer: an organic chemical is said to be reduced if it experiences a net gain of electrons as the result of a chemical reaction.

Reduction reactions involving organic chemicals in clay-water systems can occur. Montmorillonite can initiate the polymerization of hydroxyethyl methacrylate, hydroxypropyl methacrylate, and N-methylol acrylamide by acting as a reducing agent[20]. The accelerating effect of increasing water content suggests the existence of an electron transfer mechanism in which the clay acts as an electron acceptor which is aided by the presence of water.

Of all chemical reactions occurring in soil systems, reduction is probably the least studied and least understood. To further complicate the matter, most published studies on reduction reactions generally focus on biological reduction reactions using experimental methods which cannot exclude totally the contributions of nonbiological processes[21]. For example, autoclaved soils are frequently utilized in laboratory studies to establish the biological origin of observed reactivity. However, the absence of reduction in autoclaved systems does not necessarily establish that such systems require living microorganisms to cause reduction. Autoclaving alters soil surfaces that may mediate the reduction reactions observed in non-autoclaved systems.

Knowledge concerning chemical reduction reactions in soil systems is poor, at best. As a result, the ability to predict organic chemical reduction reactions in soil systems is poor, at best. This situation is especially alarming when one considers the fact that soil-groundwater systems are generally anaerobic systems that should be especially capable of mediating reduction reactions. In general, one should expect an organic chemical to undergo chemical reduction if the potential of the soil system is less than that of the organic chemical in question.

ORGANIC CHEMICALS THAT RAPIDLY REACT WITH WATER

Certain organic chemicals can hydrolyze, oxidize, or reduce very quickly upon contact with water. If these chemicals are present in bulk quantities, the reaction may be violent, because energy is released in quantities or at rates too high to be absorbed by the surrounding environment.

Table 8.3 lists several classes of organic chemicals that react rapidly, and sometimes violently, with water. Because all soils contain water, as discussed in Chapter 2, it is most important to recognize that these chemicals may rapidly react in soil systems. These organic chemicals were grouped together so that general relationships between chemical structure and reactivity would

TABLE 8.3 Organic Chemicals That May Rapidly React With Soil Water and Groundwater.[1]

acetic anhydride
acetyl bromide
acetyl chloride
acrolein[2]
acrylonitrile[2]
3-aminopropiononitrile[2]

bis(difluoroboryl)methane
bis(dimethylaminoborane)aluminum tetrahydroborate
butyldichloroborane

calcium cyanamide
2-chloroethylamine[2]
chlorosulfonyl isocyanate
chlorotrimethylsilane
cyanamide[2]
2-cyanoethanol
cyanoformyl chloride
cyanogen chloride

dichlorodimethylsilane
dichlorophenylborane
dicyanoacetylene
diethylmagnesium
diethylzinc
3,3-difluoro-1,2-dimethoxycyclopropene
diketene[2]
dimethylaluminum chloride
dimethylmagnesium
dimethylzinc
diphenylmagnesium

2,3-epoxypropionaldehyde oxime[2]
N-ethyl-N-propylcarbamoyl chloride

glyoxal[2]

isopropylisocyanide dichloride

maleic anhydride
methacrylic acid[2]
2-methylaziridine[2]
methyl isocyanate
methyl isocyanoacetate

oxopropanedinitrile

perfluorosilanes
peroxyacetic acid
peroxyformic acid

TABLE 8.3 Organic Chemicals That May Rapidly React With Soil Water and Groundwater[1] (cont.)

---

peroxyfuroic acid
peroxypivalic acid
peroxytrifluoroacetic acid
phenylphosphonyl dichloride
phosphorus tricyanide
pivaloyloxydiethylborane
potassium bis(phenylethynyl)palladate
potassium bis(phenylethynyl)platinate
potassium bis(propynyl)palladate
potassium bis(propynyl)platinate
potassium diethynylplatinate
potassium hexaethynylcobaltate
potassium methanediazoate
potassium tert-butoxide
potassium tetracyanotitanate
potassium tetraethynylnickelate
propenoic acid[2]

sulfur trioxide-dimethylformamide
sulfinylcyanamide

2,4,6-trichloro-1,3,5-triazine
trichlorovinylsilane
triethoxydialuminum tribromide
trifluoroacetyltrifluoromethanesulfonate
trifluoromethanesulfenyl chloride

vinyl acetate[2]

---

[1] compiled following an analysis of information presented in reference 22.
[2] indicates potential rapid polymerization.

be more evident.

It is most important to recognize that the absence of a chemical or chemical combination from Table 8.3 in no way implies that no hazard exists for the unlisted chemical. Chemicals possessing similar structures should always be expected to react in a similar fashion. The reader should always consult additional information from the published literature on reactive chemical hazards when dealing with rapidly reacting organic chemicals.

CHEMICAL REACTION PRODUCTS

The hydrolysis, oxidation, or reduction of one organic chemical usually results in the synthesis of one or more new organic chemicals. For example, the soil-catalyzed oxidation of aniline led to the synthesis of seven different chemical reaction products: azobenzene, azoxybenzene, phenazine, formanilide, acetanilide, nitrobenzene, and p-benzoquinone[23].

It would be significantly beyond the scope of any text to attempt to identify all possible reaction products for organic chemicals. The organic chemistry literature abounds with textbooks and journals that discuss the multitude of potential reaction products due to the hydrolysis, oxidation, and reduction of industrial organic chemicals as well as esoteric chemicals in water. These should be consulted if one must identify basic reaction products.

It is most important to recognize that soil can affect the nature of chemical reaction products. This subject is addressed in this section.

Soil Minerals and Reaction Products. Soil minerals may significantly influence the chemical structure of reaction products. Degradation pathways and products can be dependent upon the type of soil mineral present. For example, the degradation of parathion can follow two different pathways:

(8.39)

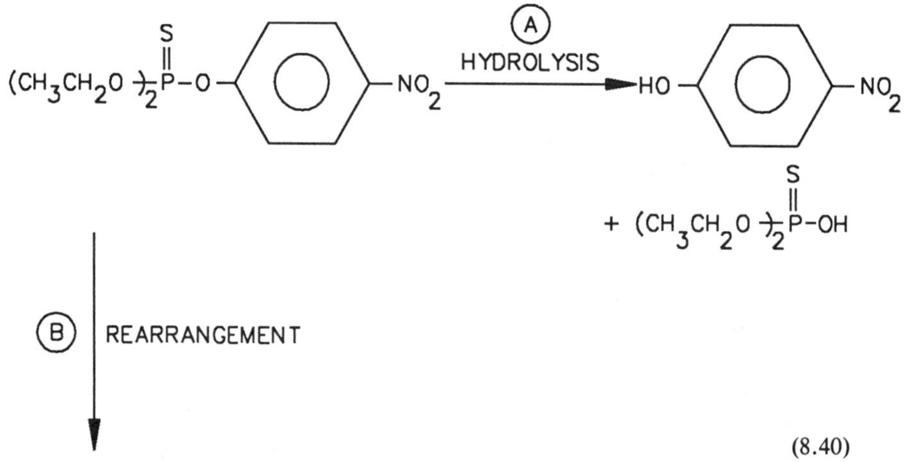

$$(CH_3CH_2O \xrightarrow{}_{\overline{2}} \overset{\overset{\displaystyle S}{\|}}{P}-O- \bigcirc -NO_2 \xrightarrow[\text{HYDROLYSIS}]{\text{Ⓐ}} HO- \bigcirc -NO_2$$

$$+ (CH_3CH_2O \xrightarrow{}_{\overline{2}} \overset{\overset{\displaystyle S}{\|}}{P}-OH$$

Ⓑ REARRANGEMENT

(8.40)

$$(CH_3CH_2O \xrightarrow{}_{\overline{2}} \overset{\overset{\displaystyle O}{\|}}{P}-O- \bigcirc -NO_2 \xrightarrow[\text{HYDROLYSIS}]{} HO- \bigcirc -NO_2$$

$$+ (CH_3CH_2O \xrightarrow{}_{\overline{2}} \overset{\overset{\displaystyle O}{\|}}{P}-OH$$

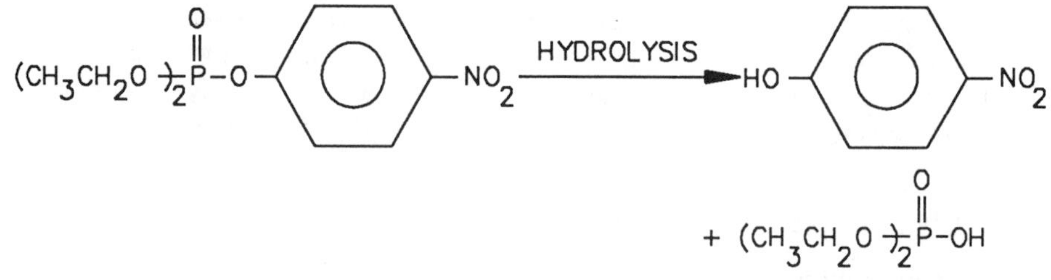

Parathion in the presence of aluminum oxide surfaces will degrade through "A"[24]; in the presence of montmorillonite or silica, it will degrade through "B"; in the presence of kaolinite, it will degrade through both "A" and "B", and the reaction products may be fixed at the soil mineral surface[13].

Another example on how soil mineral surfaces affect reaction pathways and products is the hydrolytic rearrangement of poly(dimethylsiloxane) (i.e. PDMS) in soil[25]. The reaction of PDMS in the presence of montmorillonite resulted in linear oligomerization. The presence of PDMS in the presence of kaolinite resulted predominantly in cyclic oligomerization.

Bound Residues. A bound residue is defined as that unextractable residue of an organic chemical remaining in soil after exhaustive sequential extraction with organic solvents. Bound residues are comprised of the parent organic chemical and/or its first or second generation reaction products which are adsorbed, incorporated, or entrapped in plant, animal, or soil components. Bound residues can comprise from a few percent to as much as 90 percent of an organic chemical applied to soil.

Several organic chemical classes appear to form significant amounts of soil bound residues. These include anilines, phenols, triazines, urea herbicides, carbamates, organophosphates, dodecachloropentacyclodecane insecticides, and cyclodiene insecticides[26].

Ample evidence exists in the published literature showing that many organic chemicals form bound residues in soil by (a) forming stable covalent bonds with soil organic matter, and (b) polymerizing to form soil organic matter. For an example of the latter, twelve aromatic acids and alcohols polymerized after catalytic oxidation by clays, silts, and soils to form organic matter[17]. Also, the organic chemicals 1,2,3-trihydroxybenzene and 1,2-dihydroxybenzene polymerized after catalytic oxidation by aluminum oxide to form organic matter[28].

Present day knowledge about soil bound residues comes almost exclusively from pesticide residue studies. Pesticide bound residues gained scientific attention because of concern that these residues may constitute a potential environmental problem for several reasons[29]. First, the nature and/or identity of pesticide bound residues expected to be present in soil were not known. Second, little was known about the significance of bound residues in terms of their bioavailability, toxicity, and accumulative nature. Third, conventional analytical methods may not detect these types of residues, thus underestimating the soil or plant burden of total pesticide residues. Fourth, the environmental fate of pesticide bound residues were generally not known.

Because the unextractability of a bound residue from soil and plants has little meaning unless it is related to bioavailability, two general categories of bound residues are defined on this basis. Bioavailable bound residues are those which are accumulated by plants and/or animals. Biounavailable bound residues are those which are not accumulated by plants and/or animals. Although the bioavailability of a bound residue is related to the mechanism by which the organic chemical is bound or incorporated into the soil, an analysis of selected published data revealed that plant uptake was usually below one percent of soil bound residue content[26].

Judging from the information published in the scientific literature, bound residues do not appear to pose a significant environmental problem.

REMEDIAL TECHNOLOGIES

The soil system can be manipulated in order to initiate or accelerate the degradation rate of many organic chemicals. For example, soil pH can be altered to favor acid or alkaline hydrolysis. Soil redox potential and oxygen content can be altered to favor oxidation or reduction reactions; this can be accomplished through tillage, surface soil compaction, air injection, irrigation, surface sealing, and through other measures. Inorganic or organic chemicals can be added to the soil system to initiate or accelerate organic chemical transformations.

The in situ initiation and acceleration of chemical reactions for many organic chemicals in soil is technologically feasible. It is unfortunate that this area has received relatively little attention, as evidenced by the very few case studies found in the published literature.

A pesticide aerial application site at the Gila River Indian Community near Phoenix, AZ contained about 3500 yd³ of surface soil contaminated with ethyl parathion and methyl parathion[30]. Soil decontamination was achieved by accelerating the alkaline hydrolysis of the parathions. Soil pH was increased above 9 by adding about 200 grams NaOH/ft² and by thorough mixing with a plow and disk. Soil was disked weekly and watered daily through a sprinkler irrigation system. Treatment was complete in two months.

Approximately 4,200 gallons of acrylate monomer leaked from a corroded underground pipeline at a small industrial plant into a glacial sand and gravel aquifer[31]. The acrylate was slowly polymerizing in the aquifer. A catalyst and activator were injected into the aquifer in order to accelerate reductive polymerization. It was estimated that approximately 85 to 90 percent of the acrylate monomer was converted into a solidified polymer. The remaining monomer was expected to react gradually.

The failure of an underground pipeline resulted in the release of approximately 17,000 gallons of a 50 percent formaldehyde solution into soil consisting of sand underlaid by glacial till[32]. A subsequent subsurface investigation revealed that formaldehyde was present in soil at concentrations up to nine percent at depths of two to ten feet. Formaldehyde had migrated at least two feet into the till. A solution of five to ten percent hydrogen peroxide and a 100 ppm NaBr tracer was injected at a rate less than one gal./min. into the soil system to oxidize formaldehyde. After chemical treatment, formaldehyde concentrations in soil generally were below one ppm.

Bench and field scale research on the dechlorination of chlorinated organics by chemical reagents, which are sprayed on or incorporated into soil, have been performed. 1,2,3,4-tetrachlorodibenzo-p-dioxin in soil was chemically reduced to concentrations below 1 ppb during laboratory experiments by mixing the soil with a reagent[33]; the reagent was comprised of an alkali metal hydroxide, polyethylene glycols, dimethylsulfoxide, and water. The probable mechanisms for this reaction was:

$$ROH + KOH \longrightarrow ROK + H_2O \qquad (8.41)$$

$$Ar\text{-}Cl + ROK \longrightarrow Ar\text{-}OR + KCl \qquad (8.42)$$

PCB concentrations in air-dried soil were reduced from 1000 ppm to below 50 ppm after the addition of potassium or sodium polyethylene glycol during field experiments[34]. However, field tests show that the presence of relatively low amounts of soil moisture can deactivate the polyethylene glycol reagent[35].

REFERENCES

1. Mabey, W. and Mill, T. Critical Review of Hydrolysis of Organic Compounds in Water Under Environmental Conditions. Jour. Phys. Chem. Ref. Data **17(2)**:383-415 (1978).
2. El-Amamy, M. M. and Mill, T. Hydrolysis Kinetics of Organic Chemicals on Montmorillonite and Kaolinite Surfaces as Related to Moisture Content. Clays and Clay Minerals **32**:67-73 (1984).
3. Cowart, R. P., Bonner, F. L., and Epps, E. A. Jr. Rate of Hydrolysis of Seven Organo Phosphate Pesticides. Bull. Environ. Contam. Toxicol. **6**:231-234 (1971).
4. Vogel, T. M. and Reinhard, M. Reaction Products and Rates of Disappearance of Simple Bromoalkanes, 1,2-Dibromopropane, and 1,2-Dibromoethane in Water. Environ. Sci. Technol. **20**:992-997 (1986).
5. Schwarzenbach, R. P., Giger, W., Schaffner, C., and Wanner, O. Groundwater Contamination by Volatile Halogenated Alkanes: Abiotic Formation of Volatile Sulfur Compounds Under Anaerobic Conditions. Environ. Sci. Technol. **19**:322-327 (1985).
6. Armstrong, D. E., Chesters, G., and Harris, R. F. Atrazine Hydrolysis in Soil. Soil Science Society of America Proceedings **31**:61-66 (1967).
7. Dick, W. A. and Tabatabai, M. A. Hydrolysis of Organic and Inorganic

Phosphorus Compounds Added to Soils. Geoderma **21**:175-182 (1978).

8. SRI International. Laboratory Protocols for Evaluating the Fate of Organic Chemicals in Air and Water. Final Report to the U.S. Environmental Protection Agency, Contract No. 68-03-2227. Menlo Park, CA: SRI International (1980).

9. Solomon, D. Minerals, Macromolecules, and Man. Search **10**:363-369 (1977).

10. Mortland, M. M. and Raman, V. V. Catalytic Hydrolysis of Some Phosphate Pesticides by Copper (II). Journal of Agricultural and Food Chemistry **15**:163-167 (1967).

11. Perdue, E. M. and Wolfe, N. L. Modification of Pollutant Hydrolysis Kinetics in the Presence of Humic Substances. Environmental Science and Technology **16**:847-852 (1982).

12. Saltzman, S., Yaron, B., and Mingelgrin, U. The Surface Catalyzed Hydrolysis of Parathion on Kaolinite. Soil Science Society of America Proceedings **38**:231-234 (1974).

13. Saltzman, S., Mingelgrin, U., and Yaron, B. Role of Water in the Hydrolysis of Parathion and Methylparathion on Kaolinite. Journal of Agricultural and Food Chemistry **24(4)**:739-743 (1976).

14. Mortland, M. M. and Berkheiser, V. Triethylene Diamine-Clay Complexes as Matrices for Adsorption and Catalytic Reactions. Clays and Clay Minerals **24**:60-63 (1976).

15. Frenkel, M. and Solomon, D. H. The Decomposition of Organic Amines on Montmorillonites Under Ambient Conditions. Clays and Clay Minerals **25**:463-464 (1977).

16. Dragun, J. and Helling, C. S. Physicochemical and Structural Relationships of Organic Chemicals Undergoing Soil- and Clay- Catalyzed Free-Radical Oxidation. Soil Science **139**:100-111 (1985).

17. Stone, A. T. and Morgan, J. J. Reduction and Dissolution of Manganese (III) and Manganese (IV) Oxides by Organics: 2. Survey of Reactivity of Organics. Environ. Sci. Technol. **18**:617-624 (1984).

18. Solomon, D. H., Loft, B. C., and Swift, J. D. Reactions Catalyzed by Minerals: 4. The Mechanism of the Benzidine Blue Reaction. Clay Miner. **7**:389-397 (1968).

19. Morrison, R. T. and Boyd, R. N. Organic Chemistry. 3rd Edition. New York: Allyn and Bacon (1973).

20. Solomon, D. H. and Loft, B. C. Reactions Catalyzed by Minerals. Part III. The Mechanism of Spontaneous Interlamellar Polymerizations in Aluminosilicates. Journal of Applied Polymer Science **12**:1253-1262 (1968).

21. Macalady, D. L., Tratnyek, P. G., and Grundl, T. J. Abiotic Reduction Reactions of Anthropogenic Organic Chemicals in Anaerobic Sys-

tems: A Critical Review. Journal of Contaminant Hydrology **1**:1-28 (1986).

22. Bretherick, L. Handbook of Reactive Chemical Hazards. Boston: Butterworths (1985).
23. Pillai, P., Helling, C. S., and Dragun, J. Soil-Catalyzed Oxidation of Aniline. Chemosphere **11**:299-317 (1982).
24. Mingelgrin, U. and Saltzman, S. Surface Reactions of Parathion on Clays. Clays and Clay Minerals **27**:72-78 (1979).
25. Buch, R. R. and Ingebrigtson, D. N. Rearrangement of Poly(dimethylsiloxane) fluids on Soil. Environmental Science & Technology **13**:676-679 (1979).
26. Roberts, T. R. Non-Extractable Pesticide Residues in Soils and Plants. Pure and Appl. Chem. **56**:945-956 (1984).
27. Wang, T. S. C., Wang, M. C., and Ferng, Y. L. Catalytic Synthesis of Humic Substances by Natural Clays, Silts, and Soils. Soil Science **135**:350-360 (1983).
28. Wang, T. S. C. and Wang, M. C. Catalytic Synthesis of Humic Substances by Using Aluminas as Catalysts. Soil Science **136**: 226-230 (1983).
29. Khan, S. U. Bound Pesticide Residues in Soil and Plants. Residue Reviews **84**:1-25 (1982).
30. King, J., Tinto, T., and Ridosh, M. In Situ Treatment of Pesticide Contaminated Soils. *In* Proceedings of the 6th National Conference on Management of Uncontrolled Hazardous Waste Sites. November 4-6, 1985, Washington, D.C. Silver Spring, MD: Hazardous Materials Control Research Institute (1985).
31. Williams, E. B. Contamination Containment by In Situ Polymerization. *In* Proceedings of the Second National Symposium on Aquifer Restoration and Groundwater Monitoring, May 26-28, 1982, Columbus, OH. Worthington, OH: National Water Well Association (1982).
32. Cowie, A. M. and Weider, M. F. In Situ Remediation of Formalin Release at Monsanto, Springfield, Massachusetts. *In* 1986 Hazardous Material Spills Conference Proceedings, May 5–8, 1986, St. Louis, MO. Rockville, MD: Government Institutes, Inc. (1986).
33. Peterson, R. L, Milicic, E., and Rogers, C. J. Chemical Destruction/Detoxification of Chlorinated Dioxins in Soils. *In* Incineration and Treatment of Hazardous Waste. Proceedings of the Eleventh Annual Research Symposium, April 29-May 1, 1985, Cincinnati, OH. EPA/600/9-85/028. Cincinnati, OH: U.S. Environmental Protection Agency. (1985).
34. U.S. EPA. Hazardous Waste Engineering Research Laboratory. Destruction of PCBs - Environmental Applications of Alkali Metal Polyethylene Glycolate Complexes. EPA/600/S2-85/108. Cincinnati, OH: U.S. Environmental Protection Agency (1985).

35. U.S. EPA. Hazardous Waste Engineering Research Laboratory. Interim Report on the Feasibility of Using UV Photolysis and APEG Reagent for Treatment of Dioxin Contaminated Soils. EPA/600/S2-85/083. Cincinnati, OH: U.S. Environmental Protection Agency (1985).

9

Organic Chemical Biodegradation in Soil

INTRODUCTION

In addition to the abiotic reactions discussed in Chapter 8, many organic chemicals can be degraded by biotic reactions. In principle, biotic reactions refer to those reactions that involve biota; and in soil systems, these include plants, animals, insects, and microorganisms. Although plants, animals, and insects can degrade many organic chemicals, they do not play a significant role in degrading organic chemicals in soil; as a result, these forms of biota will not be discussed in this chapter. On the other hand, microorganisms play a major role in degrading organic chemicals in soil, and this form of biota will be the focal point of this chapter.

This chapter will first discuss the different kinds of microorganisms found in soil. Then, this chapter will discuss how microorganisms degrade organic chemicals, and how microorganisms can transfer this ability to other microorganisms. Third, this chapter reveals the wide variety of organic chemical structures that can be degraded by soil microorganisms, and discusses the rate at which biodegradation occurs as well as the factors that govern the rate.

Remedial technologies based on biodegradation are utilized today to detoxify organic chemicals present in soil systems. These technologies are under-utilized today in the United States. The final section of this chapter focuses on these remedial technologies.

The biodegradation of organic chemicals in soil is a subject studied not only by soil chemists but also by agronomists, microbiologists, chemists, environmental scientists, and engineers. Numerous journals and countless books have focused on this subject. This chapter could not possibly discuss all aspects of the biodegradation of organic chemicals in soil. However, this chapter (a) addresses the basic principles governing the biodegradation of any organic chemical or waste, (b) reveals the complexity of biodegradation reactions, and (c) relates these to existing remedial technologies for the biodegradation of organic chemicals in soil.

BIOLOGICAL COMPOSITION OF SOIL

Soil serves as the home for numerous microorganisms capable of degrading organic chemicals. Soil microorganisms can be placed in two general categories: animal and plant. Although the animal population plays important roles affecting the biology and chemistry of soil, their impact on the degradation of organic chemicals is small compared to the impact of the plant population. Therefore, this chapter focuses primarily on the plant microorganisms.

Plant microorganisms include bacteria, actinomycetes, fungi, and algae. Bacteria are single-cell organisms that seldom exceed 4 or 5 microns (0.004 to 0.005 mm) in length. Their shape may be nearly round, rodlike, or spiral.

The amount of surface area exposed by soil bacteria is very large. Assuming that a single bacterium is about 1 micron long and 0.5 microns thick, and assuming that one billion bacterial cells are present in one gram of soil, the surface area of soil bacteria would be about 460 acres in the top six inches of soil.

Bacteria outnumber the other microorganisms found in a typical soil (Table 9.1) and possess the ability to rapidly reproduce. If a single bacterium and every subsequent organism were to reproduce every hour, the offspring from the original cell would be about 17,000,000 in 24 hours. In six days, the volume of offspring would surpass the volume of the earth.

Actinomycetes are similar to bacteria; they are unicellular and are of similar size and shape. Some species are similar to molds; they are filamentous, often profusely branched, and produce fruiting bodies. They develop best in moist, well-aerated soil. Actinomycetes are, in general, rather sensitive to acid soil conditions; their growth is seriously inhibited at pH 5.0 or below. Actinomycetes are the second most abundant microorganisms found in soil.

Molds, along with yeasts and mushroom fungi, are the three groups which constitute soil fungi. Molds are filamentous microorganisms comprised of

TABLE 9.1   Size of the Microorganism Populations Found in Surface Soil, Subsoil, and Groundwater.

Organism	Population Size	
	Typical	*Extreme*
Surface Soil (cells / gram soil)		
bacteria	0.1 - 1 billion	> 10 billion
actinomycetes	10 - 100 million	100 million
molds (fungi)	0.1 - 1 million	20 million
algae	10,000 - 100,000	3 million
Subsoil (cells / gram soil)		
bacteris	1000 - 10,000,000	200 million
Groundwater (cells / ml)		
bacteria	100 - 200,000	1 million

TABLE 9.2   Naturally-Occurring Pathogenic Microorganisms Found in Soil.

Genus / Species	*Adverse Human Health Effect(s)*
Acanthamoeba castellani	Fatal meningoencephalitis (inflammation of the brain and meninges).
Actinomadura madurae (Nocardia madurae)	Maduromycosis (severe fungal infections of skin with abscesses that may penetrate bone).
Actinomyces israelii	Inflammatory lesions of lymph nodes draining the mouth.
Alternaria	Allergen in human bronchial asthma. Etiologic agent identified in several diseases of the lung and in skin infections.
Ascaris lumbricoides	Ascariasis, an infection by this eelworm or round-worm type of nematode of the small intestine causing colicky pains and diarrhea.
Ascaris vermiculris (Enterobius vermicularis) (Oxyuris vermicularis)	Infection of the upper part of the large intestine, and occasionally in the female genitals and bladder.
Aspergillus fumigatus	Aspergilloma, tumor-like granulomatous masses formed by colonization of Aspergillus in bronchus or pulmonary cavity; may disseminate through the blood stream to the brain, heart, and kidney. Aspergillosis fungal disease, an infection of ear, lung, nose, and internal abscesses.
Aspergillus nidulans	Aspergilloma (see Aspergillus fumigatus). Onychomycosis, a disease of the nails of fingers and toes. Otomycosis, a severe and very persistent fungal infection of the external auditory meatus and ear canal.
Aspergillus niger	Aspergilloma (see Aspergillus fumigatus). Otomycosis (see Aspergillus nidulans).
Bacillus fusiformis	Trench mouth (Vincent's angina).

TABLE 9.2   Naturally-Occurring Pathogenic Microorganisms Found in Soil. (cont.)

Genus / Species	Adverse Human Health Effect(s)
Bacillus subtilis	Conjunctivitis. Iridochoroiditis, inflammation of the iris and the choroid. Panophthalmitis, inflammation of all the structures or tissues of the eye.
Blastomyces dermatitidis (Mycoderma dermatitidis)	Blastomycosis, a pulmonary infection. Vesicular dermatitis and other skin and mucous membrane infections.
Candida albicans	Bronchocadidias, a pneumonial infection of the respiratory tract. Candidiasis, a chronic infection of the skin, oral mucous membranes, and vagina with lesions. Diarrhea. Onychomycosis (see Aspergillus nidulans).
Clostridium bifermentans	Gas gangrene, a severe and painful condition infecting lacerated wounds that invades muscles and subcutaneous tissues which fill with gas and a serum/blood exudate.
Clostridium botulinum	Botulism, poisoning caused by a neurotoxin (botulin); when absorbed by the stomach and intestine causes vomiting, abdominal pain, difficulty with vision, headache, disturbances of secretion, motor disturbances, dyspepsia, mydriasis, ptosis; fatal at high concentrations.
Clostridium histolyticum	Infection of lacerated wounds producing a toxin responsible for skin lesions and necrosis of muscle tissue.
Clostridium novyi	Gas gangrene (see Clostridium bifermentans).
Clostridium perfringens	Gas gangrene (see Clostridium bifermentans).
Clostridium septicum	Malignant edema.
Clostridium tetani	Tetanus, caused by a potent ecotoxin comprised of tetanospasmin and tetanolysin.

TABLE 9.2 Naturally-Occurring Pathogenic Microorganisms Found in Soil. (cont.)

Genus / Species	Adverse Human Health Effect(s)
Coccidioides immitis	Coccidioidomyeosis, the acute form being a benign respiratory infection; the secondary form is a virulent and severe chronic, progressive disease involving cutaneous and subcutaneous tissue, viscera, central nervous system, and lungs with anemia, phelebitis, and various allergic responses.
Corynebacterium belfantii	Ozena, a nasal infection of varying etiology.
Cryptococcus capsulatus (Histoplasma capsulatum)	Histoplasmosis, an acute pneumonia or disseminated recticuloendothelial hyperplasia with hepatosplenomegaly and anemia, or an influenza-like illness with joint effusion and erythema nodosum.
Cryptococcus neoformans	Acneiform lesions of skin. Cryptococcosis of lungs. Predilection for the brain and meninges.
Diplococcus pneumoniae	Lobar pneumonia.
Entamoeba histolytica	Amoebic dysentery. Skin abscesses. Tropical abscesses of the liver.
Epidermophyton floccosum	Onychomycosis and tinea.
Escherichia coli	Cystitis-urinary tract infection. Gastrointestinal disturbances. Intestinal diarrhea.
Escherichia freundii	Diarrheal infections. Gastrointestinal disturbances.
Fonsecaea compactum	Chromomycosis, a chronic fungal infection of skin producing wart-like nodules or papillomas, which may or may not ulcerate.
Fonsecaea pedrosai	Chromomycosis (see listing under Fonsecaea compactum).
Fusarium oxysporum	Mycotic keratitis, an inflammation of the cornea. Otomycosis externa, an infection of the external auditory meatus and ear canal marked by pruritis and exudative inflammation.

TABLE 9.2 Naturally-Occurring Pathogenic Microorganisms Found in Soil. (cont.)

Genus / Species	*Adverse Human Health Effect(s)*
Fusarium solani	Mycotic keratitis (see Fusarium solani).
Geotrichum candidum	Geotrichosis, an infection of the bronchi, lungs, mouth, or intestinal tract.
Listeria monocytogenes	Upper respiratory disease with angina, lymphadenitis, and conjunctivitis; causes a septicemic disease which may be transmitted transplancentally in pregnant women; may assume an encephalitic form.
Microsporum audouini	Dermatophytosis (i.e. ringworm or tinea).
Microsporum gypseum	Dermatophytosis (see Microsporum audouini).
Mucor corymbifer	Otomycosis (see Fusarium oxysporum).
Mucor racemosus	Otomycosis (see Fusarium oxysporum).
Mucor ramosa (Absidia ramosa)	Mucormycosis, a fungal disease beginning in the upper respiratory tract or lungs in which spores germinate and from which mycelial growths metastasize to other organs. Otomycosis (see Fusarium oxysporum).
Mucor rhizopodiformis	Localized or generalized mycosis.
Mycobacterium tuberculosis	Forms tubercles, caseous necrosis, and tuberculous lesions of lungs, skin, central nervous system, liver, digestive tract, bones, joints, glands, etc.; infection transported throughout body via blood.
Naegleria (Diamastigamoeba)	Sudden and severe meningoencephalitis (see Acanthamoeba castellani).
Necator americanus (Ancylostoma americanum) (Uncinaria americana)	Hookworm disease, a nematode infection that settles in the small intestine; symptoms include abdominal pain, diarrhea, and colic or nausea; anemia occurs in moderate to severe cases along with parasite-induced blood loss.
Nocardia asteroides	Pulmonary infection that simulates tuberculosis.

TABLE 9.2   Naturally-Occurring Pathogenic Microorganisms Found in Soil. (cont.)

Genus / Species	*Adverse Human Health Effect(s)*
Nocardia brasiliensis	Maduromycosis and nocardiosis (see Actinomadura madurae).
Proteus mirabilis	Secondary invader causing wound infection. Summer diarrhea in small children.
Proteus rettgeri	Sporadic and epidemic gastroenteritis (inflammation of the stomach and intestine).
Proteus vulgaris	Cystitis (inflammation of the urinary bladder). Infantile diarrhea. Suppurative lesions.
Pseudomonas aeruginosa	Endocarditis (inflammation of the endothelial lining membrane of the heart and the connective tissue bed on which it lies). Meningitis. Otitis (inflammation of the ear, which may result in pain, fever, hearing abnormalties, deafness, tinnitus, and vertigo). Pneumonia.
Pseudomonas mallei	Glanders, an inflammation of mucous membranes and skin that may result in deep ulcers and necrosis of cartilage and bone.
Salmonella anatum	Intestinal disorders.
Salmonella paratyphi B	Gastroenteritis (inflammation of stomach and intestine). Paratyphoid.
Salmonella schottmulleri	Paratyphoid.
Salmonella typhosa	Typhoid fever.
Sarcocystis lindemanni	Sarcosporidiosis, a parasitic infection manifested by the formation of cysts containing this parasite in muscle.
Shigella alkalescens	Diarrheal disease.
Shigella dysenteriae	Intestinal infections, ulcers, liver abscesses, and severe dysentery.

TABLE 9.2   Naturally-Occurring Pathogenic Microorganisms Found in Soil. (cont.)

Genus / Species	Adverse Human Health Effect(s)
Shigella flexneri	Acute diarrheal disease.
Shigella sonnei	Dysentery.
Streptococcus hemolytica	Bacterial pneumonia.
Streptococcus pyrogenes	Acute glomerulonephritis (inflammation of the capillary loops in kidney). Puerperal sepsis (infections due to presence of the organism or its toxins in tissue and blood following childbirth). Rheumatic fever. Scarlet fever. Septic sore throat.
Streptomyces griseus	Fungal infections.
Streptomyces madurae	Maduromycosis (see Actinomadura madurae).
Streptomyces schenk II	Sporotrichosis, a chronic fungal infection causing nodules, lesions, and ulcers on exposed skin areas which progresses through the body via lymphatic system into muscles, joints, bones, blood, mucous membranes, and central nervous system.
Torula histolytica	Infection of skin or mucous membranes progressing to cerebral and pulmonary granulomata and abscesses.
Trichophyton schoenleinii	Skin lesions with pus and no definitive systemic involvement. Dermatomycosis including various clinical forms of ringworm, favus, tinea cruris, and Dhobie itch.

Compiled from data from references 1,2,3,4, and 5.

simple mycelial threads or profusely branched threads. Over 700 species and over 170 genera of fungi have been identified in soil.

Algae are predominantly chlorophyll-bearing microorganisms that must live at or near the soil surface. Soil algae can be divided into three general groups: blue-green, green, and diatoms.

It may be interesting to note that many naturally-occurring soil microorganisms are responsible for producing adverse health effects in humans (see Table 9.2). The effect may be acute and result in temporary discomfort or it may be chronic and result in severe impairment; sometimes the effect may be fatal.

Most bacteria, actinomycetes, and molds and several species of algae are predominantly heterotrophic: they derive their energy supply from the degradation and utilization of soil organic matter and other organic chemicals present in soil. It is most important to recognize that soil organic matter is comprised of numerous types of organic chemicals. In essence, a top soil containing three percent organic matter contains 30,000 ppm by weight of various organic chemicals.

Table 9.3 lists the different classes of naturally-occurring organic chemicals found in soil. Table 9.4 lists the naturally-occurring organic chemicals that have been specifically identified in soil; Table 9.5 lists the concentration ranges reported in the published literature for some of the naturally-occurring organic chemicals listed in Table 9.4. The soil concentrations of those chemicals listed in Table 9.4, which are not listed in Table 9.5, should be at least a few ppm in soil.

The concentration and types of organic chemicals found in a soil are not constant. Concentrations change as the metabolic activity of the soil microorganism population changes. Also, the concentration and types of organic chemicals change due to input from various sources such as plant roots, plant root exudates, microorganism exudates, and the organic remains of dead plants, animals, and microorganisms.

It is most important to recognize that soils contain many naturally-occurring organic chemicals that can cause adverse health effects to humans. For example, soils possess an inherent level of mutagenic activity; in the *Salmonella typhimurium* and *Aspergillus nidulans* bioassays, organic extracts from three agricultural soils exhibited mutagenic activity ranging from 26 to 716 net revertant colonies[25]. Of the numerous chemicals listed in Tables 9.3, 9.4, and 9.5, 11 are potential carcinogenic agents, 39 are potential mutagenic agents, 8 are potential neoplastigenic agents, and 16 are potential teratogenic agents (see Table 9.6). Louis Pasteur stated that "in the contagion of diseases, the soil is everything."

TABLE 9.3 Classes of Naturally-Occurring Organic Chemicals Found in Soil.

alkaloids	penicillins
n-alkanes, c < 34	pheophytins
n-alkanoic acids, c = 12 − 34	phospholipids
alkanoic diacids	phthalates
n-alkanols, c = 12 − 30	pigments
alkyl alkanoates, c = 12 − 24	polysaccharides (glycans)
alkyl benzenes	polyuronides
n-alkyl methanoates, c = 18 − 34	porphyrins
alkylnaphthalenes	quinones
alkylphenanthrenes	steroids
antibiotics	sterols
auxins	tannins
benzenepolycarboxylic acid	teichoic acids
polymethyl esters	terpenes
branched alkanoic acids	triterpenoids
B vitamins	uronic acids
carotenoids	
catechin-type polymers	
coumarins	
cyclic alkanes	
dialkyl phthalates	
flavonoids	
galloyl derivatives of catechin	
galloyl derivatives of	
leuco-anthocyanins	
hydroxamic acids	
hydroxyalkanoic acids	
hydroxydecanoic acids	
inositols	
leucoanthocyanins	
lichen acids	
lignins	
methyl alkanones, c = 17 − 37	
mucopeptides	
mucopolysaccharides	
nucleic acids	
organic cyanides	

HOW MICROORGANISMS DEGRADE ORGANIC CHEMICALS

The biodegradation of an organic chemical is the modification or decomposition of the chemical by soil microorganisms to produce ultimately microbial cells, carbon dioxide, and water[27]. It is most important to recognize that microorganisms possess numerous enzymes within their cells which are responsible for the biodegradation of organic chemicals.

After an organic chemical has penetrated into the interior of the micro-

TABLE 9.4 Naturally-Occurring Organic Chemicals Found in Soil.

acetic acid
aminoacetic acid
L-2-amino-4-[(2-amino-2-carboxyethyl)-thio]butanoic acid
2-amino-1,4-butanedioic acid
2-aminobutanoic acid
4-aminobutanoic acid
2-amino-4-carbamylbutanoic acid
2-amino-3-carbamoylpropanoic acid
2-amino-3-(3,4-dihydroxyphenyl)propanoic acid
2-aminoethanesulfonic acid
2-aminoethanol
1-amino-4-guanidinopentanoic acid
2-amino-3-hydroxybutanoic acid
2-amino-3-hydroxypropanoic acid
2-amino-4(or 5)-imidazolepropanoic acid
2-amino-3-indoylpropanoic acid
2-amino-3-mercaptopropanoic acid
2-amino-3-methylbutanoic acid
2-amino-4(or 5)-methylimidazolepropanoic acid
2-amino-3-methylpentanoic acid
2-amino-4-methylpentanoic acid
1-amino-2-methylpropane
4-amino-5-methyl-2(1H)-pyrimidinone
2-amino-4-(methylsulfinyl)butanoic acid
2-amino-4-(methylsulfonyl)butanoic acid
2-amino-4-(methylthio)butanoic acid
2-aminooctanoic acid
2-aminopentanedioic acid
3-aminopentanoic acid
2-amino-3-phenylpropanoic acid
1-aminopropane
2-aminopropanoic acid
3-aminopropanoic acid
6-aminopurine
2-aminopurin-6(1H)-one
4-amino-2(1H)-pyrimidinone
2-amino-3-sulfinopropanoic acid
2-amino-3-sulfopropanoic acid
2-amyrenol
anthanthrene
anthracene
9,10-anthraquinone
ascorbic acid

benzo[a]anthracene
benzene
benzenedicarboxylic acid
benzo[b]fluoranthene
benzo[k]fluoranthene

TABLE 9.4   Naturally-Occurring Organic Chemicals Found in Soil. (cont.)

10,11-benzofluoranthene
benzofluorene
benzoic acid
1,2-benzopyrene
3,4-benzopyrene
1,12-benzoperylene
benzylbutyl phthalate
biotin
bis-(2-ethylhexyl)phthalate
1,4-butanediamine
1,4-butanedioic acid
butanoic acid
n-butanol
cis-butenedioic acid
trans-2-butenoic acid

carbazole
carbon disulfide
carbon monoxide
carbonyl sulfide
catechin
chrysene
chrysotalunin
citric acid
coronene
cyanuric acid
cycloheximide

decanoic acid
dehydroabietin
dehydrodivanillin
2,7-diaminoheptanedioic acid
2,6-diaminohexanoic acid
2,5-diaminopentanoic acid
dibutyl phthalate
dicyclohexyl phthalate
diethyl sulfide
1,2-dihydroxybenzene
1,3-dihydroxybenzene
3,4-dihydroxybenzoic acid
2,3-dihydroxy-1,4-butanedioic acid
2,4-dihydroxy-7-methoxy-1,4-benzoxazine-3-one
2,4-dihydroxy-6-methylbenzoic acid

TABLE 9.4   Naturally-Occurring Organic Chemicals Found in Soil. (cont.)

9,10-dihydroxyoctadecanoic acid
dihydroxyperylenquinone
3,4-dihydroxyphenylpropanoic acid
3-(3,4-dihydroxyphenyl)propenoic acid
3,5-dihydroxytoluene
diisoamylphthalate
4,5-dimethoxy-1,2-benzenedicarboxylic acid dimethyl ester
3,5-dimethoxy-4-hydroxybenzoic acid
3-(3,5-dimethoxy-4-hydroxyphenyl)propenoic acid
dimethyldisulfide
dimethyl hexadecane
dimethylsulfide
dimethyl undecane
dioctylphthalate
5-[3-(1,2-dithioanyl)]pentanoic acid
3,3'-dithiobis(2-aminopropanoic acid)
docosane
docosanedioic acid
docosanoic acid
dodecanoic acid
dodecanol
dotriacontane

eicosane
eicosane cyclohexane
eicosanedioic acid
eicosanoic acid
epicatechin
epifriedelanol
ethanedioic acid
ethanol
8-ethoxydecanoic acid
ethylamine
ethylbenzene
ethylene
ethyl methyl pentanol

3,3',4',5,7-flavanpentol
flavonol
fluoranthene
formic acid
N-formyl-N-hydroxyaminoacetic acid
friedelin
friedelin-3B-ol
fumaric acid

gliotoxin
glutathione

TABLE 9.4   Naturally-Occurring Organic Chemicals Found in Soil. (cont.)

heneicosane
n-hentriacontane
heptacosane
heptacosane cyclohexane
heptacosanoic acid
heptadecane
heptadecanedioic acid
heptadecanoic acid
heptanoic acid
hexacosane
hexacosane cyclohexane
hexacosanoic acid
1-hexacosanol
hexadecane
hexadecanedioic acid
hexadecanoic acid
9-hexadecenoic acid
hexanedioic acid
hexanoic acid
hydroxyacetic acid
hydroxyanthraquinone
o-hydroxybenzaldehyde
p-hydroxybenzaldehyde
o-hydroxybenzoic acid
p-hydroxybenzoic acid
2-hydroxy-1,4-butanedioic acid
8-hydroxydecanoic acid
hydroxydocosanoic acid
3-hydroxydodecanoic acid
(B-hydroxyethyl)trimethylammonium hydroxide
3-hydroxyhexadecanoic acid
4-hydroxy-3-methoxybenzoic acid
3-(4-hydroxy-3-methoxyphenyl)-2-oxopropanoic acid
3-(4-hydroxy-3-methoxyphenyl)-2-propen-1-aldehyde
3-(4-hydroxy-3-methoxyphenyl)propenoic acid
3-(4-hydroxy-3-methoxyphenyl)-1,2,3-trihydroxypropane
3-hydroxy-4-methylbenzoic acid
3-hydroxy-5-methylbenzoic acid
8-hydoxynonanoic acid
2-hydroxyoctadecanoic acid
11-hydroxyoctadecanoic acid
3-hydroxyoctanoic acid
8-hydroxyoctanoic acid
3-(4-hydroxyphenyl)-2-aminopropanoic acid
3-(4-hydroxyphenyl)propanoic acid
3-(4-hydroxyphenyl)propenoic acid

TABLE 9.4   Naturally-Occurring Organic Chemicals Found in Soil. (cont.)

2-hydroxypropanoic acid
4-hydroxy-2-pyrrolidinecarboxylic acid
3-hydroxytetradecanoic acid
1-hydroxy-1,2,3-tricarboxypropane
hypoxanthine

2-(4-imidazolyl)ethylamine
2-imino-1-methyl-4-imidazolidinone
indeno(1,2,3-cd)-pyrene
indole-3-acetic acid

1-ketopropionaldehyde

levopimaric acid

methane
methanethiol
methanol
2-methoxy-1,3,4,5-benzenetetracarboxylic acid tetramethyl ester
4-methoxy-1,2,3-benzenetricarboxylic acid trimethyl ester
4-methoxybenzoic acid
3-methoxy-4-hydroxybenzaldehyde
4-methoxy-5-methyl-o-phthalaldehyde-3-carboxylic acid
2-methoxy-1,4-naphthoquinone
methylamine
4-methylbutanoic acid
2-methyl-4-carboxypyridine
20-methyldocosanoic acid
11-methyldodecanoic acid
18-methyleicosanoic acid
20-methylheneicosanoic acid
15-methylheptadecanoic acid
16-methylheptadecanoic acid
methylhexadecane
15-methylhexadecanoic acid
1-methyl-7-isopropylphenanthrene
18-methylnonadecanoic acid
16-methyloctadecanoic acid
17-methyloctadecanoic acid
13-methylpentadecanoic acid
14-methylpentadecanoic acid
3-methylpropanoic acid
5-methyl-2,4(1H,3H)-pyrimidinedione

TABLE 9.4 Naturally-Occurring Organic Chemicals Found in Soil. (cont.)

13-methyltetradecanoic acid
11-methyltridecanoic acid
12-methyltridecanoic acid
22-methyltricosanoic acid

naphthalene
n-nonacosane
nonacosane cyclohexane
nonacosanoic acid
nonadecane
nonadecane cyclohexane
nonadecanedioic acid
nonadecanoic acid
nonanoic acid

octacosane
octacosanoic acid
octadecane
octadecanedioic acid
octadecanoic acid
cis-9-octadecenoic acid
trans-9-octadecenoic acid
octadecyl(4-propionylphenyl)ester
octanoic acid
2-oxobutanedioic acid
2-oxopropanoic acid

pantothenic acid
pentacosane
pentacosane cyclohexane
pentacosanedioic acid
pentacosanoic acid
pentadecane
pentadecanedioic acid
pentadecanoic acid
3,3',4',5,7-pentahydroxyflavone
1,5-pentanediamine
pentanoic acid
pentatriacontane
perylene
phenanthrene
phenylacetic acid
2-phenyl-1,4-benzopyrone
3-phenylpropenoic acid
propanedioic acid

TABLE 9.4   Naturally-Occurring Organic Chemicals Found in Soil. (cont.)

propanoic acid
n-propanol
propanoic acid
2,6(1,3)-purinedione
pyrene
pyridine-3-carboxylic acid
2,4(1H,3H)-pyrimidiendione
2-pyrrolidinecarboxylic acid

quercetin

B-sitosterol
stigmastanol

taraxerol
taraxerone
tetracosane
tetracosanoic acid
tetradecane
tetradecanoic acid
1,3,4,5-tetrahydroxycyclohexanecarboxylic acid
3,4',5,7-tetrahydroxyflavone
1,2,3,4-tetrahydroxyhexanedioic acid
tetramethylhexadecane
tetratriacontane
thiamine
toluene
triacontane
triacontyl octadecanoate
triacosanoic acid
tricosane
tricosanedioic acid
tridecanoic acid
1,2,3-trihydroxybenzene
1,3,5-trihydroxybenzene
3,4,5-trihydroxybenzoic acid
3,4,5-trihydroxy-1-cyclohexene-1-carboxylic acid
2',4',6'-trihydroxy-3-(4-hydroxyphenyl)propiophenone
trihydroxyoctadecanoic acid
1,2,3-trihydroxypropane
trimethylamine
trimethyldecane
trimethyldodecane
trimethylhexadecane
triphenylamine

TABLE 9.4   Naturally-Occurring Organic Chemicals Found in Soil. (cont.)

triphenylene
trithiobenzaldehyde
tritriacontane

uncosane cyclohexane
uncosanedioic acid
uncosanoic acid
undecanoic acid
untriacontane
untriacontane cyclohexane
5-ureidohydantoin

m-xylene
o-xylene
p-xylene

---

bial cell, it will collide with enzymes[27]. Whether or not the chemical transforms as a result of this collision will depend upon (a) the chemical binding
to the enzyme and (b) conformational changes at the enzyme's active site.
In most enzymes, chemical induced conformational changes (i.e. induction)
play a vital role. Degradable chemicals will produce the ideal alignment of
the enzyme and the organic chemical; persistent chemicals will produce a less
favorable alignment. Recalcitrant or non-reacting chemicals are not bound
at all or fail to produce an alignment that leads to reaction.

The reaction of an organic chemical with an enzyme can be expressed like
any other chemical reaction:

$$C + E \underset{\substack{\text{Incorrect} \\ \text{Alignment}}}{\rightleftharpoons} CE \overset{\substack{\text{Correct} \\ \text{Alignment}}}{\longrightarrow} P + E \qquad (9.1)$$

where

C = organic chemical

E = enzyme

CE = enzyme-chemical complex

P = reaction product, a modified or decomposed organic
chemical

TABLE 9.5 Concentrations of Some Naturally-Occurring Organic Chemicals Found in Soils.

Chemical	Concentration (mg/kg soil)	Reference
acetic acid	3.8 - 74.2	6
	265 - 530[a]	2
	995 - 1358[b]	8
alkanes	0 - 286	9, 10
n-alkanoic acids	700 - 2080[c]	11
	< 60	12
n-alkanols	90 - 810[c]	11
alkyl phthalates	0 - 60.6	10
aminoacetic acid	0.21 - 800	6, 13
2-amino-1,4-butanedioic acid	0.2 - 1820	6, 13
4-aminobutanoic acid	0.08 - 128	6, 13
2-amino-3-hydroxybutanoic acid	0.2 - 810	6, 13
2-amino-3-hydroxypropanoic acid	0.16 - 734	6, 13
2-amino-4(or 5)-imidazole-propanoic acid	0.33 - 130	6, 13
2-amino-3-methylbutanoic acid	0.2 - 810	6, 13
2-amino-3-methylpentanoic acid	0.12 - 477	6, 13
2-amino-4-methylpentanoic acid	0.13 - 634	6, 13
2-amino-4-(methylthio)butanoic acid	11 - 33	14
2-aminopentanedioic acid	0.31 - 1370	6, 13
2-amino-3-phenylpropanoic acid	0.12 - 340	6, 13
2-aminopropanoic acid	0.09 - 173	6, 13
3-aminopropanoic acid	0.18 - 1090	6, 13
aromatic acids	< 40	12
benz(a)anthracene	0 - 0.01	15
benzene	1 - 5	16
benzo(b)fluoranthene	0 - 0.03	15
benzo(k)fluoranthene	0 - 0.015	15
benzoic acid	0.14 - 11	12
	7 - 350[a]	7
benzo(g,h,i)perylene	0 - 0.02	15
benz(a)pyrene	0 - 8	15
benzylbutylphthalate	0 - 48[c]	17
bis(2-ethylhexyl)phthalate	150 - 925[c]	17
butanoic acid	0.1 - 0.6	6
n-butanol	0 - 1220	18
carotene	600[c]	11
chrysene	5	11
2,6-diaminohexanoic acid	13 - 700	13
	0.52	6
dibutylphthalate	19 - 56[c]	17
dicyclohexylphthalate	29 - 65[c]	17
diisoamylphthalate	32 - 47[c]	17
4,5-dimethoxy-1,2-benzene dicarboxylic acid dimethyl ester	0 - 19[c]	17

TABLE 9.5   Concentrations of Some Naturally-Occurring Organic Chemicals Found in Soils. (cont.)

Chemical	Concentration (mg/kg soil)	Reference
3,5-dimethoxy-4-hydroxybenxoic acid	0.11 - 5.3	12
dioctylphthalate	0 - 13[c]	17
docosanoic acid	0.16 - 1.7	10
dodecanoic acid	0.14 - 0.78	10
eicosanoic acid	0.05 - 11.6	10
ethanol	0 - 52	18
8-ethoxydecanoic acid	0.28 - 24	12
ethylbenzene	1 - 5	16
fluoranthene	0 - 0.04	15
formic acid	230 - 525[a]	7
	26 - 897[b]	8
heptadecanoate	0.06 - 0.52	10
hexadecanoic acid	0.17 - 6.5	10
9-hexadecnoic acid	0.15 - 1.4	10
hexanoic acid	<0.4	10
hydroxy acids	160 - 3000[c]	11
4-hydroxybenzaldehyde	70 - 550[d]	19
2-hydroxybenzoic acid	0.24 - 11.8	12
4-hydroxybenzoic acid	60 - 11,820[d]	19
	0.2 - 4.3	12
hydroxydecanoic acid	0.05 - 7.34	12
8-hydroxydecanoic acid	0.59 - 29	12
4-hydroxy-3-methoxybenzoic acid	0.33 - 8.7	12
	110 - 230[d]	19
3-(4-hydroxy-3-methoxyphenyl)-propenoic acid	0.02 - 7.9	12
	100 - 7440[d]	19
3-hydroxy-5-methylbenzoic acid	800[a]	7
8-hydroxynonanoic acid	0.03 - 12	12
3-hydroxyoctanoic acid	0.23 - 20	12
8-hydroxyoctanoic acid	0.03 - 13	12
3-(4-hydroxyphenyl)-2-aminopropanoic acid	0.27 - 233	6, 13
3-(4-hydroxyphenyl)propenoic acid	0.01 - 3.3	12
	40 - 7730[d]	19
ideno(1,2,3-cd)pyrene	0 - 0.015	15
inositols	0 - 400[a]	20
methanol	0 - 31	18
2-methoxy-1,3,4,5-benzene-tetracarboxylic acid tetramethyl ester	0 - 12[c]	17

TABLE 9.5 Concentrations of Some Naturally-Occurring Organic Chemicals Found in Soils. (cont.)

Chemical	Concentration (mg/kg soil)	Reference
4-methoxy-1,2,3-benzene-tricarboxylic acid trimethyl ester	0 - 32[c]	17
4-methoxybenzoic acid	0.12 - 5.4	12
3-methoxy-4-hydroxybenzaldehyde	40 - 390[d]	19
2-methoxy-1,4-naphthoquinone	8.6	21
methyl alkanones	100,000 - 400,000[c]	11
4-methylbutanoic acid	0.2 - 1.0	6
20-methylheneicosanoic acid	0.88	10
16-methylheptadecanoic acid	0.08 - 0.1	10
18-methylnonadecanoic acid	0.18 - 0.37	10
3-methylpropanoic acid	0.1 - 0.7	6
22-methyltricosanoic acid	0.7	10
naphthalene	1 - 5	16
nonadecanoate	0.06 - 1.3	10
octadecanoic acid	0.6 - 6.7	10
cis-9-octadecenoic acid	0.15 - 5.7	10
octadecyl(4-propionylphenyl)ester	200[c]	22
pentadecanoic acid	0.1 - 0.7	10
pentanoic acid	0.1 - 2.2	6
phenolic acids (total)	4 - 90[c]	23
polysaccharides	8 - 56,000	14
propanoic acid	0.1 - 9.9	6
	26 - 82[a]	7
n-propanol	0 - 54	18
pyrene	0 - 0.015	15
sterols	0 - 12	11
tetracosanoic acid	0.81	10
tetradecanoic acid	0.14 - 2.2	10
toluene	1 - 5	16
uronic acids	5600	24
	10,000 - 50,000[c]	13
m-xylene	1 - 5	16
o-xylene	1 - 5	16
p-xylene	1 - 5	16

[a] in lbs./acre foot
[b] in kg/hectare
[c] mg/kg organic matter
[d] in mg/kg organic carbon
[e] in mg phosphorus/kg soil

Table 9.6 Selected Toxicological Effects of Naturally-Occurring Organic Chemicals Found in Soil.

Chemical	Effect [a,b]
acetic acid	mut
2-amino-3-mercaptopropanoic acid	mut
2-amino-4-methylpentanoic acid	ter
6-aminopurine	ter
2-aminopurin-6(1H)-one	mut, ter
anthanthrene	carc, eta, mut
anthracene	eta, mut, neo
ascorbic acid	mut
benzo[a]anthracene	carc, eta, mut, neo
benzene	carc, eta, mut, neo, ter
benzo[b]fluoranthene	carc, eta, mut
benzo[k]fluoranthene	eta
1,2-benzopyrene	carc, eta, mut
3,4-benzopyrene	carc, eta, mut, neo, ter
1,12-benzoperylene	mut
biotin	carc
bis-(2-ethylhexyl)phthalate	carc, ter
1,2-butanediamine	mut
butanoic acid	mut
carbon disulfide	mut, ter
catechin	carc, mut
cyanuric acid	eta
cycloheximide	ter
dibutylphthalate	mut, ter
3,4-dihydroxybenzoic acid	mut
dioctylphthalate	ter
3,3'-dithiobis(2-aminopropanoic acid)	mut
dodecanoic acid	neo
ethanol	eta, mut, ter
fluoranthene	eta, mut
N-formyl-N-hydroxyaminoacetic acid	mut, ter
fumaric acid	mut
glutathione	mut
hexadecanoic acid	neo
4-hydroxy-3-methoxybenzoic acid	mut
2-hydroxypropanoic acid	mut
2-(4-imidazolyl)ethylamine	mut
indole-3-acetic acid	eta, mut, ter
methanethiol	mut
3-methoxy-4-hydroxybenzaldehyde	mut
naphthalene	ter
octadecanoic acid	ter
cis-9-octadecenoic acid	eta
3,3',4',5,7-pentahydroxyflavone	mut
perylene	mut
phenanthrene	eta, mut, neo
n-propanol	carc
propenoic acid	ter
2,6(1,3)-purinedione	neo

Table 9.6 Selected Toxicological Effects of Naturally-Occurring Organic Chemicals Found in Soil. (cont.)

Chemical	Effect [a,b]
pyrene	eta, mut
pyridine-3-carboxylic acid	carc
quercetin	mut
toluene	mut
1,2,3-trihydroxybenzene	eta, mut
1,3,5-trihydroxybenzene	mut
3,4,5-trihydroxy-1-cyclohexene-1-carboxylic acid	eta, mut

a - based on a compilation of information presented in reference 26
b - abbreviations for clinical toxicological effects:
   carc = potential carcinogenic agent, based on experimental animal studies.
   eta = potential equivocal tumorigenic agent, based on experimental animal studies.
   mut = potential mutagenic agent, based on experimental studies.
   neo = potential neoplastigenic agent, based on experimental animal studies.
   ter = potential teratogenic agent, based on experimental animal studies.

The enzyme-chemical complex CE forms after the chemical binds to the enzyme. If an incorrect alignment occurs between C and E within the complex, P will not be produced; if a correct alignment occurs, P is produced.

Figure 9.1 illustrates a hypothetical enzyme chemical complex with a correct alignment. There are two important factors illustrated in Figure 9.1 that must be understood in order to grasp a basic understanding about organic chemical biodegradation. First, a proper alignment occurs when the two functional groups, A and B, and one carbon atom bind to the enzyme surface at an active site. Note that the molecular fragment A-C-B must conform to the shape of the active site in order to form a complex having the correct alignment to produce a reaction product.

Second, a proper alignment does not involve the molecular fragments R_1 and R_2. In other words, the molecular fragment A-C-B, but not the rest of the molecule, must conform to the shape of the active site. The practical significance of this factor is that one can usually determine if an organic chemical is susceptible to biodegradation by identifying the molecular fragments within the organic chemical that are susceptible to forming complexes that produce reaction products.

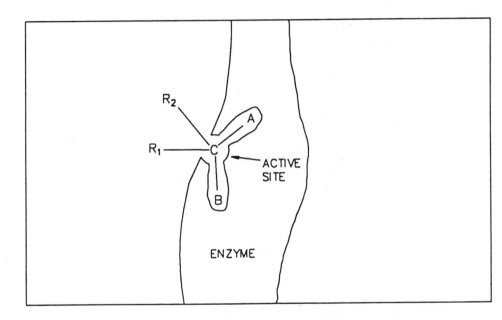

Figure 9.1   A hypothetical complex between an organic chemical
and an enzyme within a soil microorganism.

It is most important to note that not all enzymes that can degrade organic chemicals are found only within cell membranes. Enzymes are found either within the microorganism or outside the microorganism's outer cell membrane. The latter are commonly known as extracellular enzymes. These enzymes generally serve the purpose of generating nutrients for cell growth by catalyzing the degradation of organic chemicals which are too large to pass across the microorganism's outer membrane. These enzymes for the most part degrade polymers such as polysaccharides, protein, lipids, and phos-

TABLE 9.7 Extracellular Enzymes Identified in Soil.

Enzymes Catalyzing Hydrolysis Reactions	*Enzymes Catalyzing Cleavage Reactions and Molecular Rearrangements*
acetylesterases	aminotransferases
acid phosphatase	aromatic-L-amino acid decarboxylase
alkaline phosphatase	aspartic acid decarboxylase
amidase	glutamate decarboxylase
amylase (deaminase)	glutaminase
arylacylamidase	glycosyl transferases
arylesterase	levansucrase
arylsulphatase	lyase
asparaginase	polysaccharases
carboxylesterase	rhodanese
cathepsin	thiosulphate sulfur transferase
cellulase	transaminase
chitinase	transglucosylases
cyanase	tyrosine decarboxylase
dextranase	
beta-fructofuranosidase	
alpha-galactosidase	
beta-galactosidase	*Enzymes Catalyzing Oxidation, Reduction, and Oxidative Coupling Reactions*
alpha-glucosidase	
beta-glucosidase	
inulase	
lichenase	aniline oxidase
lipases	catalase
metaphosphatase	dehydrogenases
nucleases (nucleotidases)	diphenol oxidase
pepsin	glucose oxidase
peptidases	monophenol monooxygenases
phosphodiesterase	(incl. laccases and tyrosinases)
phosphotriesterase	peroxydases
phytase	phenol oxidases
trypsin	urate oxidase
urease	proteinases (proteases)
xylanase	pyrophosphatase

pholipids (see Table 9.7). In addition, they possess a broad ability to degrade other organic chemicals. For example, a phenoloxidase-like enzyme catalyzed the oxidative coupling of 1-naphthol, 4-chloro-1-naphthol, and guaiacol into oligomeric products[28]. An enzyme from the fungus *Trametes versicolor* catalyzed the reaction of 2,6-xylenol and syringic acid to produce dimeric and polymeric hybrid oligomers[29]. An oxidase from the white rot fungus catalyzed the degradation of DDT, several PCB isomers, lindane, 2,3,7,8-TCDD, and benzo[a]pyrene[30]. In general, however, little is known about the ability of extracellular enzymes to degrade organic chemicals of environmental concern.

A microorganism which possesses the enzymes(s) to degrade an organic chemical can transfer this ability to other organisms. Three mechanisms for transferring genetic material are known: conjugation, transduction, and transformation.

Conjugation involves the direct transfer of chromosomes between attached cells. Bacteria generally carry the bulk of their genetic information in a single chromosome. However, many bacterial strains have one or more minor chromosomes, which are one or two percent of the size of the main chromosome. Such small chromosomes are known as plasmids and are generally regarded as nonessential to the normal survival of the bacterium[31]. Most plasmids are transmissable from one bacterial cell to another by the process of conjugation. During the conjugal process, the cell possessing the transmissable plasmid (the donor) produces a conjugal tube which attaches to a second cell which lacks the plasmid (the recipient). The donor cell duplicates its plasmid, transfers one copy to the recipient cell, and retains the second copy. When transfer via the conjugal tube is complete, the conjugating cells separate. Via conjugation, genetic information responsible for degrading a chemical can be transferred to a population of cells that originally did not possess this information.

The breeding of microorganisms containing plasmids that specifically degrade certain organic chemicals is a very promising R&D area. For example, 2,4,5-T is slowly degraded by cometabolism and is not utilized as a sole source of carbon and energy. However, the technique of plasmid-assisted molecular breeding has led to the synthesis of bacterial strains capable of totally degrading 2,4,5-T by using it as a sole carbon source at concentrations greater than 1 mg/ml[32].

Transduction is the process of gene exchange between bacteria which is mediated by a virus[31]. The virus attaches to the bacterial cell, injects its DNA, and the virus multiplies in the host cell; virus multiplication can amount to as many as 1,000 particles/cell. The host cell then lyses and releases virus particles that attach to other bacterial cells and restarts the cycle with DNA injection.

In rare instances, bacterial DNA is packed into a new virus particle instead of viral DNA during the assembly of the virus particles. Upon host cell lysis, the bacterial DNA-bearing particle attaches to another bacterial cell and is injected into the bacterial cell.

There are two limitations which prohibit transduction from being a major mechanism for transferring genetic information to degrade chemicals[31]. First, the amount of DNA transferred by any one viral particle is usually one to two percent of the size of the main chromosome; as a result, the DNA transferred is information usually not responsible for degrading chemicals. Second, successful transduction is usually restricted to very closely related strains of the same species of bacterium.

Transformation is the simplest form of genetic exchange[31]. It involves the release of DNA by the lysis of one bacterium and the uptake of this DNA by a second bacterium. Transformation suffers the same limitations as transduction: it tends to occur between related strains of the same species of bacterium, and only small amounts of DNA are transferred. This mechanism does not appear to be important in nature, but it is employed primarily as a research technique.

Chemicals which possess structures other than simple sugars, amino acids, and fatty acids cannot enter immediately into a microorganism's metabolic pathways; these chemicals require acclimation, a lag period or acclimation period lasting from hours to months, during which little or no degradation occurs. The lag period is usually caused by (a) the transfer of genetic information responsible for degrading a chemical to a population, and (b) the initial phase of exponential growth of the population of organisms capable of degrading the chemical. In addition, the lag period may also be affected by several factors such as the composition and size of the soil microbial population, energy source, acidity and alkalinity, temperature, moisture, presence and concentrations of essential elements, the organic chemical's concentration, oxygen and redox potential, and adsorption; these factors are discussed in detail in a future section of this chapter.

Not all organic chemicals experience a lag period before their biodegradation commences. In addition, the lag period for the initiation of degradation of some organic chemicals can be reduced by the presence of organisms pre-exposed to the chemical of concern. Degradation rates can be as much as 1000X greater in populations that are pre-exposed (i.e. acclimated) to the chemical of concern. For example, acclimated bacteria completely degraded up to 1250 kg parathion per hectare soil within 35 days[33]. Also, the problem of decreased efficacy of several pesticides has been attributed to more rapid biodegradation by soil microorganisms acclimated during the repeated use of the pesticide in successive years on the same soil.

CHEMICAL STRUCTURE AND BIODEGRADATION

A detailed discussion of all molecular fragments that are susceptible to enzymatic degradation is well beyond the scope of this text; there are scores of books and numerous journals which publish hundreds of papers addressing this topic. However, this text will address the biodegradation of several organic chemicals and molecular fragments that are frequently encountered in soil-water systems.

Aliphatic Hydrocarbons. The transformation of aliphatic hydrocarbons involves the use of monooxygenases and the well-known process of beta-oxidation:

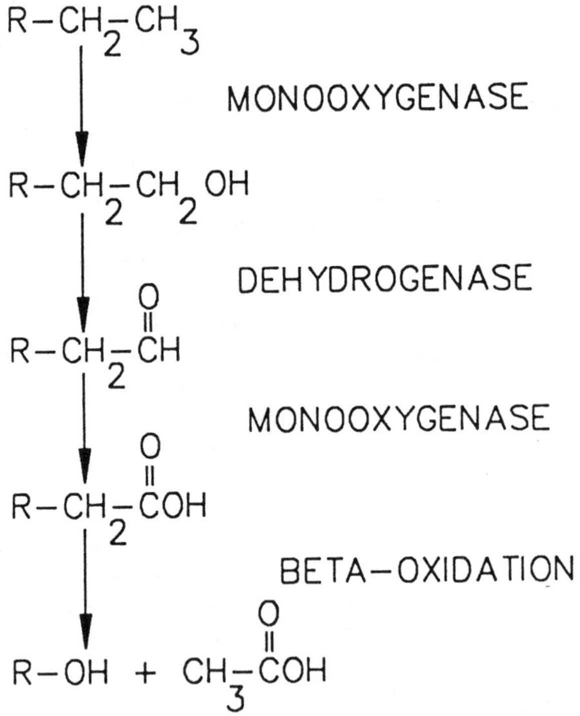

$$
\begin{array}{c}
R-CH_2-CH_3 \\
\downarrow \quad \text{MONOOXYGENASE} \\
R-CH_2-CH_2\,OH \\
\downarrow \quad \text{DEHYDROGENASE} \\
R-CH_2-\overset{\displaystyle O}{\overset{\|}{C}}H \\
\downarrow \quad \text{MONOOXYGENASE} \\
R-CH_2-\overset{\displaystyle O}{\overset{\|}{C}}OH \\
\downarrow \quad \text{BETA-OXIDATION} \\
R-OH \;+\; CH_3-\overset{\displaystyle O}{\overset{\|}{C}}OH
\end{array}
\qquad (9.2)
$$

The aliphatic hydrocarbon is converted in successive reactions into an alcohol, then an aldehyde, and then into an aliphatic acid. During beta-oxidation, two carbon atoms are cleaved from the aliphatic acid to form a new alcohol and acetic acid. The new alcohol can be further degraded by reacting with dehydrogenase, monooxygenase, and beta-oxidation enzymes to form aliphatic hydrocarbons with fewer carbon atoms.

The monooxygenase enzyme may attach the alcohol group at the penultimate carbon instead of the terminal carbon:

$$(9.3)$$

$$R-CH_2-CH_3$$
MONOOXYGENASE

$$\underset{OH}{R-CH-CH_3}$$
DEHYDROGENASE

$$\underset{O}{R-C-CH_3}$$
BETA−OXIDATION

$$R-OH \ + \ HOC-CH_3$$

The secondary alcohol is also degraded by beta-oxidation enzymes to produce a new aliphatic alcohol with fewer carbon atoms. This new alcohol can be further degraded by reacting with dehydrogenase, monooxygenase, and beta-oxidation enzymes to form aliphatic hydrocarbons with fewer carbon atoms.

Branched aliphatic hydrocarbons are generally more resistant to biodegradation than unsubstituted hydrocarbons. In general, as the number of substituents increases, the greater the resistance to biodegradation.

Cyclic Hydrocarbons. The transformation of cyclic hydrocarbons involves a sequence of reactions similar to those discussed under aliphatic hydrocarbons. For example, cyclohexane is oxidized and then converted to an aliphatic hydrocarbon:

(9.4)

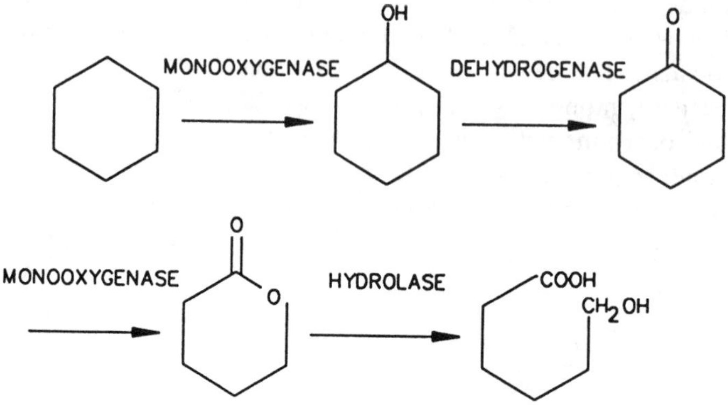

The aliphatic hydrocarbon degradation product is readily degraded to other products by beta-oxidation. The transformation of cyclic hydrocarbons with functional groups becomes more complicated because more than one reaction pathway is available. For example, cyclohexane carboxylic acid can be transformed into an aliphatic acid or into a derivative of benzoic acid:

(9.5)

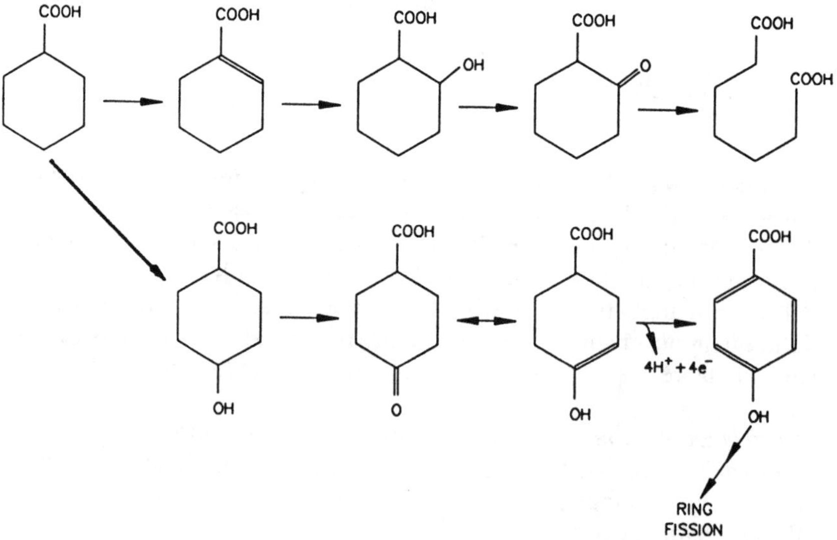

Benzene and Benzenoid Chemicals.   The transformation of benzene and benzenoid chemicals involves the class of enzymes known as dioxygenases. Molecular oxygen is incorporated into the aromatic ring structure to form a dihydroxybenzene degradation product:

(9.6)

The aromatic ring is destroyed by cleavage, and new aliphatic degradation products are formed. Several aromatic chemicals possessing one or more six-carbon rings—benzoic acid, ethyl benzene, phthalic acid, phenanthrene, naphthalene, anthracene, toluene, phenol, and naphthol—follow very similar degradation reaction pathways[34].

Many benzenoid chemicals possess aliphatic hydrocarbon functional groups. If the functional group is relatively long, it usually is the molecular fragment which is first degraded. If the functional group is relatively short, then biodegradation can occur by transformation of the aliphatic hydrocarbon functional group or by oxidation of the aromatic ring. For example, toluene can be transformed at the methyl group or at the aromatic ring:

(9.7)

The biodegradation of aromatic chemicals can occur in anaerobic systems. The pathway generally involves the reduction of the aromatic ring to form a cyclic hydrocarbon that can be degraded to other products[35]. For example, benzoic acid can be degraded anaerobically:

(9.8)

COOH $\quad$ COOH $\qquad\qquad$ O

$+6H \longrightarrow \qquad \longrightarrow \cdots \longrightarrow \xrightarrow{-CO_2}$

$+H_2O \longrightarrow \quad$ H—, —OH, —OH, H $\quad \longrightarrow \cdots \longrightarrow \quad$ COOH, COOH

The degradation of organic chemicals in anaerobic systems is an area that has not been studied extensively in the past but presently is receiving considerable attention.

Molecular Fragments Containing Carbon and Heteroatoms. Organic chemicals contain a wide array of molecular fragments. These fragments can contain carbon, hydrogen, oxygen, nitrogen, sulfur, halogens, and other atoms that are connected by single, double, and triple bonds. Soil microorganisms possess enzymes that can degrade a wide array of molecular fragments under aerobic and anaerobic conditions (see Table 9.8).

It is most important to recognize that the existence of such a diversity of potential transformations leads to the generation of a wide variety of degradation products. This will be illustrated by tracking the degradation of trichloroethylene (TCE) by using only three of the reactions listed in Table 9.8: dehydrohalogenation, reductive dehalogenation, and carbon-carbon double bond reduction.

TABLE 9.8 Potential Transformations of Organic Chemical Molecular Fragments by Soil Microorganisms[a]

MOLECULAR FRAGMENT[b]	TRANSFORMATION
Ar	ArH hydroxylation to ArOH ArH nitration to $ArNO_2$
$-C \equiv C-$	$RC \equiv CH$ reduction to $RCH = CH_2$
$= CH-$	$R(R')CHR''$ hydroxylation to $R(R')C(OH)R'$
$-CH_2-$	RCH_2R' hydroxylation to $RCH(OH)R''$ and/or $RC(O)R'$ RCH_2R' metabolism to $R(R')CH-O-CHR(R')$
$-CH_3$	RCH_3 hydroxylation to RCH_2OH and/or oxidation to $RCH(O)$ and/or $RCOOH$
$-HC = CH-$	formation of an epoxide $AlkCH = CH_2$ oxidation to $AlkCHO$ and/or $AlkCH_2CHO$ and/or $\quad AlkC(O)CH_3$ and/or $AlkCH(OH)CH_2OH$ $AlkCH = CH_2$ reduction to $AlkCH_2CH_3$ $Ar_2C = CH_2$ hydration to Ar_2CHCH_2OH $Ar_2C = CH_2$ reduction to Ar_2CHCH_3
$-CH_2-CH_2-$	ether formation to $-CHOCH-$
$-CHO$	$RCHO$ oxidation to $RCOOH$ $RCHO$ reduction to RCH_2OH
$-COOH$	$AlkCOOH$ reduction to $AlkCH_2OH$ $ArCOOH$ decarboxylation to ArH $ArCOOH$ reduction to $ArOH$ $ArN(R)COOH$ decarboxylation to $ArN(R)H$
$-C(O)-$	$RC(O)R'$ oxidation to $RCOOR'$ $RC(O)R'$ reduction to $RCH(OH)R'$
$-COO-$	$RC(O)OR'$ ester cleavage to $RCOOH$ and $R'OH$
$-C(O)NH-$	$RC(O)NHCH_2R'$ amide hydrolysis to $RC(O)NH_2$ and R' CHO $RC(O)N(CH_3)_2$ demethylation to $RC(O)NH_2$ lactam (cyclic amide) hydrolysis to $NH_2(CH_2)_nCOOH$
$-C(O)NHOH$	$ArC(O)NHOH$ hydroxamic acid hydrolysis to $ArCOOH$ and NH_2OH (which is converted to NO_2-)

TABLE 9.8 Potential Transformations of Organic Chemical Molecular Fragments by Soil Microorganisms[a] (cont.)

MOLECULAR FRAGMENT[b]	TRANSFORMATION
$-C(O)NH_2$	amide hydrolysis to $-COOH$ and NH_3
$-CH=NOH$	$RCH=NOH$ oxime hydrolysis to $RCHO$ and NH_2OH (which is converted to NO_2-) $RCH=NOH$ oxime metabolism to $RC\equiv N$
$-C\equiv N$	$RC\equiv N$ nitrile metabolism to $RC(O)NH_2$ and/or $RCOOH$
$-C(S)-$	$RC(S)R'$ transformation to $RC(O)R'$
$-C(S)NH_2$	thioamide hydrolysis to $-CN$
$-Cl$	$AlkCH_2CCl_3$ dehydrohalogenation to $AlkCH=CCl_2$ $AlkCH_2CHCl_2$ dehydrohalogenation to $AlkCH=CHCl$ $AlkCH_2CH_2Cl$ dehydrohalogenation to $AlkCH=CH_2$ $AlkCCl_3$ reductive dehalogenation to $AlkCHCl_2$ $AlkCHCl_2$ reductive dehalogenation to $AlkCH_2Cl$ $AlkCH_2Cl$ reduction to $AlkCH_2OH$ $ArCl$ dehalogenation to ArH $ArCl$ dehalogenation to $ArOH$ $HetCl$ dehalogenation to $HetOH$
Halides (Br, Cl, F, I)	$Ar-$halide dehalogenation to $ArOH$
$-Hg-$	$R-Hg-R'$ cleavage to $R-H$ and/or Hg
$=N-$	$R(R')N-R''$ cleavage to $R-NH-R'$ and/or $R-NH_2$ $R(R')NR''$ N oxidation to $R(R')N(O)R'$ $RN(Alk)_2$ cleavage to $RNH(Alk)$ and/or RNH_2
$-NH-$	$(Alk)_2NH$ N-nitrosation with NO_2- to $(Alk)_2NNO$ $RNH-CHR'(R'')$ cleavage to RNH_2 and $R'(R'')CHO$ $RNH-CH_2R'$ cleavage to RNH_2 and $R'CHO$
$-NH_2$	$ArNH_2$ deamination to $ArOH$ and NH_3 $ArNH_2$ dimerization to $ArN=NAr$

TABLE 9.8 Potential Transformations of Organic Chemical Molecular Fragments by Soil Microorganisms[a] (cont.)

MOLECULAR FRAGMENT[b]	TRANSFORMATION
$-NH_2$(cont.)	$ArNH_2$ N-acylation to $ArNHC(O)H$ and/or $ArNHC(O)CH_3$ and/or $ArNHC(O)CH_2CH_3$ RNH_2 metabolism to $RCHO$ and NH_3 RNH_2 oxidation to RNO_2 RNH_2 oxidation to $RNHC(O)R'$
$-NHOH$	RCH_2NHOH hydroxylamine hydrolysis to $RCHO$ and NH_2OH (which is converted to NO_2^-)
$-NHC(O)-$	$RNHC(O)R'$ cleavage to RNH_2 and/or $HOOCR'$ $R(R')NC(O)R''$ cleavage to $R(R')NH$ and $HOOCR''$
$-NC(O)O-$	$R_2NC(O)OR'$ hydrolysis to $R'OH$ and R_2NCOOH (which spontaneously decomposes to R_2NH and CO_2) $CH_3NHC(O)OAr$ N-demethylation to $HCHO$ and $H_2NC(O)OAr$, then to NH_3, CO_2, and $ArOH$
$-NHC(O)N=$	$ArNHC(O)NH_2$ hydrolysis to $ArNH_2$ $ArNHC(O)N(CH_3)_2$ N-demethylation to $ArNHC(O)NHCH_3$, then N-demethylation to $ArNHC(O)NH_2$
$-NO_2$	$AlkNO_2$ metabolism to $AlkCHO$ RNO_2 nitro reduction to RNH_2 RNO_2 nitro metabolism to ROH
$=NOC(O)-$	$RCH=NOC(O)R$ cleavage to $RCH=NOH$
$-O-$	$ArOR$ ether cleavage to $ArOH$ $ROCH_2R'$ ether cleavage to ROH
$-NHSO_3H$	$RNHSO_3H$ sulfamic acid cleavage to RNH_2 and SO_4^{2-}
$-OCH_3$	$ArOCH_3$ oxidative O-dealkylation to $ArOCH_2OH$, then to $ArOH$ and $HCHO$
$-OH$	$AlkCH_2OH$ oxidation to $AlkCHO$ $AlkCH(OH)CH_3$ oxidation to $AlkC(O)CH_3$ $ArOH$ methylation to $ArOCH_3$ $AROH$ oxidative coupling to form organic matter and bound residues

TABLE 9.8 Potential Transformations of Organic Chemical Molecular Fragments by Soil Microorganisms[a] (cont.)

MOLECULAR FRAGMENT[b]	*TRANSFORMATION*
$- OS(O_2)O -$	sulfate ester cleavage to $- OH$ and/or $HOS(O_2)O -$
$- P \, (S) -$	reduction to $- P(O) -$
$- S -$	RSR' oxidation to RS(O)R' and/or $RS(O)_2R'$ RSR' cleavage to $R - OH$ and/or HSR' and/or R'CHO RSR' reductive cleavage to RSH and R'H
$- SS -$	RSSR cleavage to RSH
$- SC(O)N =$	thiolcarbamate hydrolysis to $- SH$, CO_2, and $HN =$
$- SC(S)N =$	dithiocarbamate hydrolysis to $- OH$, CS_2, and $HN =$
$- SH$	RSH aerobic dimerization to RSSR RSH anaerobic metabolism to ROH and/or RH
$- S(O) -$	RS(O)R' sulfoxide reduction to RSR'
$- S(O_2) \, NH_2$	$ArS(O)_2NH_2$ cleavage to $ArS(O_2)OH$ and NH_3
$- SnOH$	R (R') (R'') SnOH cleavage to R (R') SnO, then to RSnOOH

a – Compiled from data from References 36 & 37.

b – Abbreviations: Alk – branched or unbranched aliphatic fragment with or without heteroatom functional groups.
                 Ar – substituted or unsubstituted aromatic fragment.
                 Het – heterocyclic fragment
                 R – Alk or Ar fragment

First, TCE can undergo reductive dehalogenation, the removal of one Cl atom and the addition of one H atom. Three reaction products can be formed: 1,1-dichloroethylene (1,1-DCE), cis-1,2-dichloroethylene (cis-1,2-DCE), and/or trans-1,2-dichloroethylene (trans-1,2-DCE).

Second, 1,1-DCE can undergo reductive dehalogenation to form vinyl chloride. Also, it's carbon-carbon double bond can be reduced to form 1,1-dichloroethane (1,1-DCA).

Third, cis-1,2-DCE and trans-1,2-DCE can undergo reductive dehalogenation to form vinyl chloride. Also, these two organic chemicals have carbon-carbon double bonds that can be reduced to form 1,2-dichloroethane (1,2-DCA).

Fourth, 1,1-DCA and 1,2-DCA can undergo dehydrohalogenation to form vinyl chloride. Also, these two chemicals can undergo reductive dehalogenation to form chloroethane.

The degradation pathway of TCE to DCEs, DCAs, vinyl chloride, and chloroethane leads to the production of six chlorinated volatile hydrocarbons. The degradation pathway of tetrachloroethylene (TCE) leads to the production of seven chlorinated volatile hydrocarbons. The degradation pathway of 1,1,1-trichloroethane (1,1,1-TCA) leads to the production of four chlorinated volatile hydrocarbons. These pathways are illustrated in Figure 9.2. It is important to recognize that these degradation pathways are responsible for the presence of several chlorinated volatile hydrocarbons in groundwater in which only one chlorinated hydrocarbon was present initially. It is also important to note that all the chlorinated volatile hydrocarbons are degraded via other reactions listed in Table 9.8 to products such as alkanes, alcohols, acetates, aldehydes, carbon dioxide, and chloride, which are not listed in Figure 9.2.

Not all microbial degradation reactions create organic chemicals with lower molecular weights. Some reactions result in products having higher molecular weights; these reactions fall into two general classes. The first reaction class involves conjugation reactions: an enzyme masks an existing functional group on a molecule by adding a new functional group onto it. The conjugation reaction synthesizes a new chemical with a somewhat larger molecular weight. There are nine examples of this reaction class listed in Table 9.8: ArOH methylation, $ArNH_2$ N-acylation (three examples), ArH nitration, $(Alk)_2NH$ N-nitrosation, $ArNH_2$ dimerization, RSH dimerization, and $-CH_2-CH_2-$ ether formation.

The second reaction class involves bound residues. Bound residues were discussed in Chapter 8; they are comprised of the parent organic chemical and/or its first or second generation reaction products which are adsorbed, incorporated, or entrapped in soil components. Oxidative coupling enzymes, both intracellular and extracellular (see Table 9.7), can create stable covalent bonds between parent organic chemicals and between a parent organic

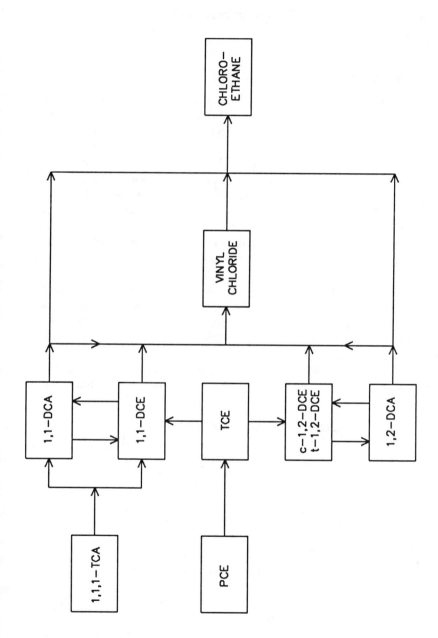

FIGURE 9.2   Transformation pathways for various chlorinated volatile hydrocarbons in soil systems. Based on information from Table 9.8.

chemical and soil organic matter. The latter reaction results in the increase in total soil organic matter.

BIODEGRADATION RATE CONSTANTS

Earlier chapters in this text have established the basic principle that organic chemicals possessing different chemical structures will behave differently. Organic chemicals possessing different chemical structures will biodegrade at different rates.

There are two basic definitions for biodegradation that must be understood in order to properly interpret biodegradation rate data. Primary degradation is any biologically induced structural change in an organic chemical. For example, the primary biodegradation of tetrachloromethane results in the replacement of one Cl atom by one H atom to form trichloromethane. Those organic chemical biodegradation rates reported in terms of the removal, disappearance, or loss of the initial chemical over time refer to primary degradation.

Ultimate degradation is the biologically mediated degradation of an organic chemical into carbon dioxide, water, oxygen, and other inorganic products of metabolism. Biodegradation rates reported in terms of BOD, COD, oxygen uptake, methane or carbon dioxide evolution, or loss of dissolved organic carbon refer to ultimate degradation.

Scientists have generally expressed the rate at which organic chemicals degrade in terms of two rate laws: the power rate law and the hyperbolic rate law.

The power rate law states that the biodegradation rate of an organic chemical is proportional to the organic chemical's concentration:

$$-dC/dt = kC^n \tag{9.9}$$

where:

C = concentration of the organic chemical.

k = biodegradation rate constant.

n = the order of the reaction (1,2, etc).

A review of the scientific literature reveals that the biodegradation of many organic chemicals is generally first-order with respect to the organic chemical's concentration. As a result, the biodegradation rate constant k can be calculated using the following equation:

$$k = (2.303/t) \log(a/[a - x]) \tag{9.10}$$

where:

 k = rate constant, 1/time.

 t = time.

 a = initial concentration.

 $a - x$ = concentration at time t.

The half life, the time needed for half of the concentration to react, can be estimated if k is known and by assigning a and x the values of 1 and 0.5, respectively, into Equation 9.10:

$$t_{1/2} = 0.693/k \qquad\qquad (9.11)$$

where:

 $t_{1/2}$ = half life.

The hyperbolic rate law is used commonly to express the growth of microorganism populations and is based upon Monod kinetics. The Monod kinetics rate equation reveals that the growth rate of a single-species or mixed-species population of microorganisms, utilizing one organic chemical as an energy source, is dependent upon chemical concentration and the growth rate of the microorganism:

$$V = V_{max} \, C/(K + C) \qquad\qquad (9.12)$$

where:

 V = specific growth rate of microorganisms.

 V_{max} = maximum growth rate of microorganisms.

 C = concentration of organic chemical.

 K = organic chemical concentration supporting a growth rate equal to one-half of the maximum ($V_{max}/2$).

Equation 9.12 can be simplified, by making various assumptions, to yield a simplified form of the second-order rate equation:

$$-dc/dt = KBC \qquad\qquad (9.13)$$

where:

 t = time.

 B = concentration of the microorganism population.

Second order biodegradation rate constants are usually reported in these units: ml/gram sludge/day, mg/gram bacteria/day, ml/cell/day, and mgCOD/gram biomass/hr. The half-life of the chemical can be calculated:

$$t_{1/2} = 0.693/(KB) \tag{9.14}$$

Table 9.9 lists biodegradation rate constants for many organic chemicals.

FACTORS AFFECTING BIODEGRADATION RATES

A careful analysis of the biodegradation and disappearance rates listed in Table 9.9 reveals that several rates for one chemical have been reported in the published literature. For example, the time to attain 100 percent disappearance for toluene in soil incubation studies ranged from seven days to about 120 weeks. It is most important to recognize that, in addition to chemical structure, a number of soil factors affect the biodegradation and disappearance rates of organic chemicals. The manipulation and optimization of these factors is needed in order to neutralize the effects of Leibig's Law of the Minimum. This law states that the rate of a biological process such as growth or metabolism is limited by the factor present at its minimum level. In other words, if all the soil factors discussed in this section are adjusted to their optimum level, then the biodegradation or disappearance rate of an organic chemical will also be at its optimum. On the other hand, if all factors except one were at their optimum levels, the rate of biodegradation or disappearance would be significantly reduced due to the one factor.

The Composition and Size of the Soil Microbial Population. The biodegradation rate of an organic chemical is generally dependent upon (a) the presence of soil microorganisms capable of degrading the chemical, and (b) the number of these organisms present in the soil system. The relationship between degradation and population size should be obvious: the greater the number of microorganisms capable of degrading the chemical, the faster the degradation of the chemical.

The size of the soil microorganism population is greatest generally in the surface horizons of soil. In this region the soil temperature, moisture, aeration, and energy supply are at relatively more favorable levels for supporting microorganisms.

The size of the soil microorganism population is not constant. It may change as the soil environment changes. For example, large population changes have been observed due to the addition of oil to soil. The naturally-occurring soil microorganism population includes several genera of bacteria and fungi capable of degrading petroleum products. In decreasing order *Pseu-*

TABLE 9.9   Biodegradation and Disappearance Rates for Several Organic Chemicals[a]

Organic Chemical	Biodegradation or Disappearance Rate	Medium and Inoculum	Reference
acenaphthene	98% in 7d	scf, sdw	38
	< 60d half life	si, nmf	39
	6.6%/w − 100% in 1w	si, naf	40
acenaphthylene	96% in 7d	scf, sdw	38
	100% in 4 mo	si, nmf	41
acetanilide	14.7 mgCOD/g/h	bss, as	42
acetophenone	4d half life	sgw, fo	43
acrolein	100% in 7d	scf, sdw	38
acrylonitrile	100% in 7d	scf, sdw	38
aldrin	0% in 7d	scf, sdw	38
alkanes (C6 to C10)	< 4d half life	sgw, fo	43
4-aminoacetanilide	11.3 mgCOD/g/h	bss, as	42
2-aminobenzoic acid	27.1 mgCOD/g/h	bss, as	42
3-aminobenzoic acid	7.0 mgCOD/g/h	bss, as	42
4-aminobenzoic acid	12.5 mgCOD/g/h	bss, as	42
2-aminopentanedioic acid	2.5-18.1h aerobic half life	si, nmf	44
	1.7-16.7h anaerobic half life	si, nmf	44
	11-14d half life	gwi, nmf	45
2-aminophenol	21.1 mgCOD/g/h	bss, as	42
3-aminophenol	10.6 mgCOD/g/h	bss, as	42
4-aminophenol	16.7 mgCOD/g/h	bss, as	42
aminophenol-sulphonic acid	7.1 mgCOD/g/h	bss, as	42
2-aminopyridine	55% in 64d	si, nmf	46
3-aminopyridine	64% in 64d	si, nmf	46
4-aminopyridine	6% in 64d	si, nmf	46
2-aminotoluene	15.1 mgCOD/g/h	bss, as	42
3-aminotoluene	30 mgCOD/g/h	bss, as	42
4-aminotoluene	20 mgCOD/g/h	bss, as	42
ammonium oxalate	9.3 mgCOD/g/h	bss, as	42
aniline	19 mgCOD/g/h	bss, as	42
anthracene	35% in 7d	scf, sdw	38
	93% in 16 mo	si, nmf	41
	200-460d half life	si, nmf	39
aroclor 1016	33% in 7d	scf, sdw	38
aroclor 1221	100% in 7d	scf, sdw	38
aroclor 1232	100% in 7d	scf, sdw	38
aroclor 1242	36% in 7d	scf, sdw	38
aroclor 1248	0% in 7d	scf, sdw	38
aroclor 1254	11% in 7d	scf, sdw	38
aroclor 1260	0% in 7d	scf, sdw	38
benzaldehyde	119 mgCOD/g/h	bss, as	42
benz(a)anthracene	8% in 7d	scf, sdw	38
	36% in 16 mo	si, nmf	41

TABLE 9.9   Biodegradation and Disappearance Rates for Several Organic Chemicals[a] (cont.)

Organic Chemical	Biodegradation or Disappearance Rate	Medium and Inoculum	Reference
benz(a)anthracene (contd)	240-680d half life	si, nmf	41
benzene	43% in 7d	scf, sdw	38
	110d half life	sgw, fo	43
	68d half life	sgw, fo	47
	48d half life	gwi, nmf	47
	20-90% in 80d	si, nmf	48
	100% in 434d	sgw, fo	48
	>99% in 120w	si, nmf	49
m-benzene-disulphonic acid	3.4 mgCOD/g/h	bss, as	42
benzenesulphonic acid	10.6 mgCOD/g/h	bss, as	42
benzo(b)fluoranthene	360-610d half life	si, nmf	39
benzo(k)fluoranthene	910-1400d half life	si, nmf	39
benzoic acid	88.5 mgCOD/g/h	bss, as	42
	7.3h (ring) aerobic half life	si, nmf	44
	3.9h (carboxyl) aerobic half life	si, nmf	44
	18.2h (ring) anaerobic half life	si, nmf	44
	26d (ring) half life	gwi, nmf	45
	41d (carboxyl) half life	gwi, nmf	45
benzo(g,h,i)perylene	590-650d half life	si, nmf	39
benzo(a)pyrene	28% in 16 mo	si, nmf	41
	220-530d	si, nmf	39
alpha-BHC	0% in 7d	scf, sdw	38
beta-BHC	0% in 7d	scf, sdw	38
delta-BHC	0% in 7d	scf, sdw	38
gamma-BHC	0% in 7d	scf, sdw	38
biphenyl	37d half life	sgw, fo	43
bis-(2)chloroethoxy)-methane	0% in 7d	scf, sdw	38
bis-(2-chloroethyl)ether	100% in 7d	scf, sdw	38
bis-(2-chloroisopropyl) ether	74% in 7d	scf, sdw	38
bis-(2-ethylhexyl)-phthalate	0% in 7d	scf, sdw	38
borneol	8.9 mgCOD/g/h	bss, as	42
bromochloromethane	100% in 7d	scf, sdw	38
bromodichlorobenzene	<4.5%/w	si, nmf	50
bromodichloromethane	>99% in 2d	cfc, bm	51
bromoform	8% in 7d	scf, sdw	38
	>99% in 2d	cfc, bm	51
4-bromodiphenyl ether	0% in 7d	scf, sdw	38
1,4-butanediol	40 mgCOD/g/h	bss, as	42

TABLE 9.9 Biodegradation and Disappearance Rates for Several Organic Chemicals[a] (cont.)

Organic Chemical	Biodegradation or Disappearance Rate	Medium and Inoculum	Reference
n-butanol	84 mgCOD/g/h	bss, as	42
sec-butanol	55 mgCOD/g/h	bss, as	42
tert-butanol	30 mgCOD/g/h	bss, as	42
sec-butylbenzene	100% in 7d	bgw, nmf	52
	100% in 192h	sp, nmf	53
butylbenzoate	4d half life	sgw, fo	43
butylbenzylphthalate	100% in 7d	scf, sdw	38
camphor	37d half life	sgw, fo	43
caprolactam	16 mgCOD/g/h	bss, as	42
3-carboxy-4-hydroxy-benzenesulfonic acid	11.3 mgCOD/g/h	bss, as	42
2-carboxypyridine	100% in 8d	si, nmf	46
3-carboxypyridine	100% in 4d	si, nmf	46
4-carboxypyridine	100% in 16d	si, nmf	46
chloramphenicol	3.3 mgCOD/g/h	bss, as	42
chlordane	0% in 7d	scf, sdw	38
2-chloroaniline	16.7 mgCOD/g/h	bss, as	42
3-chloroaniline	6.2 mgCOD/g/h	bss, as	42
4-chloroaniline	5.7 mgCOD/g/h	bss, as	42
chlorobenzene	60% in 7d	scf, sdw	38
	37d half life	sgw, fo	43
	<3.8%/w	si, nmf	50
	0.2-1.9%/w	si, nmf	54
chlorodibromomethane	18% in 7d	scf, sdw	38
4-chlorodiphenyl ether	0% in 7d	scf, sdw	38
2-chloroethyl vinyl ether	64% in 7d	scf, sdw	38
2-chloronaphthalene	100% in 7d	scf, sdw	38
2-chloro-4-nitrophenol	5.3 mgCOD/g/h	bss, as	42
2-chlorophenol	85% in 7d	scf, sdw	38
	25 mgCOD/g/h	bss, as	42
4-chlorophenol	11 mgCOD/g/h	bss, as	42
2-chloropyridine	100% in 8d	si, nmf	46
3-chloropyridine	100% in 4d	si, nmf	46
4-chloropyridine	100% in 16d	si, nmf	46
chrysene	3% in 7d	scf, sdw	38
	16% in 16 mo	si, nmf	41
m-cresol	55 mgCOD/g/h	bss, as	42
o-cresol	54 mgCOD/g/h	bss, as	42
p-cresol	55 mgCOD/g/h	bss, as	42
cresols	4d half life	sgw, fo	43
1,2-cyclohexanediol	66 mgCOD/g/h	bss, as	42
cyclohexanol	28 mgCOD/g/h	bss, as	42
cyclohexanolone	51.5 mgCOD/g/h	bss, as	42
cyclohexanone	30 mgCOD/g/h	bss, as	42
	1.1d half life	sgw, fo	43

TABLE 9.9 Biodegradation and Disappearance Rates for Several Organic Chemicals[a] (cont.)

Organic Chemical	Biodegradation or Disappearance Rate	Medium and Inoculum	Reference
cyclopentanol	55 mgCOD/g/h	bss, as	42
cyclopentanone	57 mgCOD/g/h	bss, as	42
2,4-D	5000h half life	gwi, nmf	45
p,p '-DDD	0% in 7d	scf, sdw	38
p,p '-DDE	0% in 7d	scf, sdw	38
p,p '-DDT	0% in 7d	scf, sdw	38
2,4-diaminophenol	12 mgCOD/g/h	bss, as	42
2,3-diaminopyridine	27% in 64d	si, nmf	46
2,6-diaminopyridine	40% in 64d	si, nmf	46
1,2,3,4-dibenzanthracene	17% in 16 mo	si, nmf	41
dibenz(a,h)anthracene	750-940d half life	si, nmf	39
dibenzofuran	100% in 1w	si, naf	40
dibromochloromethane	>99% in 2d	cfc, bm	51
1,2-dibromoethane	99% in <1m at 6-8 ppb	swi, nmf	55
	32-70% in 110d at 15-18 ppm	swi, nmf	55
	100% in 16w	si, nmf	49
di-n-butylphthalate	100% in 7d	scf, sdw	38
2,3-dicarboxypyridine	100% in 8d	si, nmf	46
2,4-dicarboxypyridine	100% in 8d	si, nmf	46
1,2-dichlorobenzene	33% in 7d	scf, sdw	38
1,3-dichlorobenzene	59% in 7d	scf, sdw	38
1,4-dichlorobenzene	46% in 7d	scf, sdw	38
dichlorobenzenes	110d half life	sgw, fo	43
dichlorobromomethane	35% in 7d	scf, sdw	38
1,1-dichloroethane	40% in 7d	scf, sdw	38
	<1.2-<2.6%/w	si, nmf	54
1,2-dichloroethane	23% in 7d	scf, sdw	38
	>99% in 2d	cfc, bm	51
1,1-dichloroethylene	62% in 7d	scf, sdw	38
	68% in 4d	swi, nmm	56
	92% in 40w	si, nmf	49
	110d half life	swi, nmf	57
cis-1,2-dichloroethylene	49% in 7d	scf, sdw	38
	100% in 50h	swi, nmm	56
	100% in 16w	si, nmf	49
	140d half life	swi, nmf	57
trans-1,2-di- chloroethylene	54% in 7d	scf, sdw	38
	100% in 50h	swi, nmm	56
	92% in 40w	si, nmf	49
	139d half life	swi, nmf	57
2,4-dichlorophenol	100% in 7d	scf, sdw	38
	1.05 mgCOD/g/h	bss, as	42
1,2-dichloropropane	39% in 7d	scf, sdw	38

TABLE 9.9 Biodegradation and Disappearance Rates for Several Organic Chemicals[a] (cont.)

Organic Chemical	Biodegradation or Disappearance Rate	Medium and Inoculum	Reference
1,3-dichloropropylene	55% in 7d	scf, sdw	38
2,3-dichloropyridine	100% in 16d	si, nmf	46
2,6-dichloropyridine	100% in 8d	si, nmf	46
dieldrin	0% in 7d	scf, sdw	38
diethanolamine	19.5 mgCOD/g/h	bss, as	42
1,3-diethylbenzene	100% in 9d	bgw, nmf	52
	100% in 192h	sp, nmf	53
diethylene glycol	13.7 mgCOD/g/h	bss, as	42
diethylphthalate	100% in 7d	scf, sdw	38
1,2-dihydroxybenzene	55.5 mgCOD/g/h	bss, as	42
1,3-dihydroxybenzene	57.5 mgCOD/g/h	bss, as	42
2,5-dihydroxy-benzoic acid	80 mgCOD/g/h	bss, as	42
2,3-dihydroxypyridine	100% in 64d	si, nmf	46
2,4-dihydroxypyridine	100% in 64d	si, nmf	46
2,3-dimethylaniline	12.7 mgCOD/g/h	bss, as	42
2,5-dimethylaniline	3.6 mgCOD/g/h	bss, as	42
3,4-dimethylaniline	30 mgCOD/g/h	bss, as	42
1,3-dimethyl-5-tert-butylbenzene	100% in 10d	bgw, nmf	52
	100% in 192h	sp, nmf	53
dimethylcyclohexanol	21.6 mgCOD/g/h	bss, as	42
1,2-dimethyl-3-ethylbenzene	100% in 12d	bgw, nmf	52
	100% in 192h	sp, nmf	53
1,2-dimethyl-4-ethylbenzene	100% in 11d	bgw, nmf	52
	100% in 192h	sp, nmf	53
1,3-dimethyl-2-ethylbenzene	100% in 9d	bgw, nmf	52
	100% in 192h	sp, nmf	53
1,3-dimethyl-4-ethylbenzene	100% in 11d	bgw, nmf	52
1,4-dimethyl-2-ethylbenzene	100% in 7d	bgw, nmf	52
	100% in 192h	sp, nmf	53
1,4-dimethylnaphthalene	100% in 9d	bgw, nmf	52
	100% in 192h	sp, nmf	53
2,3-dimethylnaphthalene	100% in 9d	bgw, nmf	52
	100% in 192h	sp, nmf	53
2,3-dimethylphenol	35 mgCOD/g/h	bss, as	42
2,4-dimethylphenol	100% in 7d	scf, sdw	38
	28.2 mgCOD/g/h	bss, as	42
2,5-dimethylphenol	10.6 mgCOD/g/h	bss, as	42
2,6-dimethylphenol	9.0 mgCOD/g/h	bss, as	42

TABLE 9.9 Biodegradation and Disappearance Rates for Several Organic Chemicals[a] (cont.)

Organic Chemical	Biodegradation or Disappearance Rate	Medium and Inoculum	Reference
3,4-dimethylphenol	13.4 mgCOD/g/h	bss, as	42
3,5-dimethylphenol	11.1 mgCOD/g/h	bss, as	42
dimethylphthalate	100% in 7d	scf, sdw	38
2,4-dimethylpyridine	100% in 32d	si, nmf	46
2,6-dimethylpyridine	100% in 32d	si, nmf	46
2,4-dinitrophenol	64% in 7d	scf, sdw	38
	6.0 mgCOD/g/h	bss, as	42
2,4-dinitrotoluene	64% in 7d	scf, sdw	38
2,6-dinitrotoluene	70% in 7d	scf, sdw	38
di-n-octylphthalate	0% in 7d	scf, sdw	38
diphenylether	11d half life	sgw, fo	43
1,2-diphenylhydrazine	76% in 7d	scf, sdw	38
docosane	4.5-50.6% in 4w	si, nmf	58
dotriacontane	0.6-43.3% in 4w	si, nmf	58
alpha-endosulfan	0% in 7d	scf, sdw	38
beta-endosulfan	0% in 7d	scf, sdw	38
endosulfan sulfate	0% in 7d	scf, sdw	38
endrin	0% in 7d	scf, sdw	38
ethylbenzene	85% in 7d	scf, sdw	38
	100% in 12d	bgw, nmf	52
	37d half life	sgw, fo	43
	>99% in 120w	si, nmf	49
	100% in 192h	sp, nmf	53
ethylene diamine	9.8 mgCOD/g/h	bss, as	42
ethylene glycol	41.7 mgCOD/g/h	bss, as	42
2-ethyltoluene	100% in 12d	bgw, nmf	62
3-ethyltoluene	100% in 10d	bgw, nmf	52
4-ethyltoluene	100% in 7d	bgw, nmf	52
fluoranthene	0% in 7d	scf, sdw	38
	140-440d half life	si, nmf	39
fluorene	74% in 7d	scf, sdw	38
	32-60d half life	si, nmf	39
	92%/w	si, naf	40
furfuryl alcohol	41 mgCOD/g/h	bss, as	42
furfurylaldehyde	37 mgCOD/g/h	bss, as	42
glucose	180 mgCOD/g/h	bss, as	42
	4.6-25.6h aerobic half life	si, nmf	44
	1.2-19h anaerobic half life	si, nmf	44
glycerol	85 mgCOD/g/h	bss, as	42
heptachlor	0% in 7d	scf, sdw	38
heptachlor epoxide	0% in 7d	scf, sdw	38
hexachlorobenzene	39% in 7d	scf, sdw	38

TABLE 9.9 Biodegradation and Disappearance Rates for Several Organic Chemicals[a] (cont.)

Organic Chemical	Biodegradation or Disappearance Rate	Medium and Inoculum	Reference
hexachloro-1,3-butadiene	100% in 7d	scf, sdw	38
hexachloro-cyclopentadiene	100% in 7d	scf, sdw	38
hexachloroethane	100% in 7d	scf, sdw	38
hydroquinone	54.2 mgCOD/g/h	bss, as	42
2-hydroxybenzoic acid	95 mgCOD/g/h	bss, as	42
4-hydroxybenzoic acid	100 mgCOD/g/h	bss, as	42
2-hydroxypyridine	100% in 64d	si, nmf	46
3-hydroxypyridine	100% in 32d	si, nmf	46
4-hydroxypyridine	100% in 32d	si, nmf	46
indan	100% in 11d	bgw, nmf	52
	1y half life	sgw, fo	43
	100% in 192h	sp, nmf	53
ideno(1,2,3-c,d)pyrene	600-730d half life	si, nmf	39
isophorone	100% in 7d	scf, sdw	38
isophthalic acid	76 mgCOD/g/h	bss, as	42
isopropanol	52 mgCOD/g/h	bss, as	42
isopropylbenzene	100% in 11d	bgw, nmf	42
	100% in 192h	sp, nmf	53
menthol	17.7 mgCOD/g/h	bss, as	42
4-(methylamino)-phenol sulfate	0.8 mgCOD/g/h	bss, as	42
3-methyl-4-chlorophenol	77% in 7d	scf, sdw	38
methylcresols	110d half life	sgw, fo	43
4-methylcyclohexanol	40 mgCOD/g/h	bss, as	42
4-methylcyclohexanone	61.5 mgCOD/g/h	bss, as	42
methylene chloride	100% in 7d	scf, sdw	38
1-methyl-2-ethylbenzene	100% in 192h	sp, nmf	53
1-methyl-3-ethylbenzene	100% in 192h	sp, nmf	53
1-methyl-4-ethylbenzene	100% in 192h	sp, nmf	53
5-methyl-2-isopropyl-1-phenol	15.6 mgCOD/g/h	bss, as	42
1-methylnaphthalene	100% in 9d	bgw, nmf	52
	100% in 1w	si, naf	40
2-methylnaphthalene	100% in 9d	bgw, nmf	52
	100% in 1w	si, naf	40
	100% in 192h	sp, nmf	53
methylparathion	410.1h aerobic half life	si, nmf	44
2-methylpyridine	100% in 16d	si, nmf	46
3-methylpyridine	100% in 32d	si, nmf	46
4-methylpyridine	100% in 32d	si, nmf	46
naphthalene	100% in 7d	scf, sdw	38
	100% in 9d	bgw, nmf	52

TABLE 9.9   Biodegradation and Disappearance Rates for Several Organic Chemicals[a] (cont.)

Organic Chemical	Biodegradation or Disappearance Rate	Medium and Inoculum	Reference
naphthalene (contd)	110d half life	sgw, fo	43
	100% in 1w	si, naf	40
	100% in 192h	sp, nmf	53
1-naphthalene-sulfonic acid	18 mgCOD/g/h	bss, as	42
naphthoic acid	15.5 mgCOD/g/h	bss, as	42
1-naphthol	38.4 mgCOD/g/h	bss, as	42
2-naphthol	39.2 mgCOD/g/h	bss, as	42
1-naphthol-2-sulfonic acid	18 mgCOD/g/h	bss, as	42
1-naphthylamine	0 mgCOD/g/h	bss, as	42
1-naphthylamine-6-sulphonic acid	0 mgCOD/g/h	bss, as	42
nitrilotriacetate	86.6-161.2 aerobic half life	si, nmf	44
	49.4-125.8h anaerobic half life	si, nmf	44
	31h half life	gwi, nmf	45
4-nitroacetophenone	5.3 mgCOD/g/h	bss, as	42
2-nitrobenzaldehyde	13.8 mgCOD/g/h	bss, as	42
3-nitrobenzaldehyde	10.0 mgCOD/g/h	bss, as	42
4-nitrobenzaldehyde	13.8 mgCOD/g/h	bss, as	42
nitrobenzene	94% in 7d	scf, sdw	38
	14.0 mgCOD/g/h	bss, as	42
2-nitrobenzoic acid	20 mgCOD/g/h	bss, as	42
3-nitrobenzoic acid	7.0 mgCOD/g/h	bss, as	42
4-nitrobenzoic acid	19.7 mgCOD/g/h	bss, as	42
2-nitrophenol	100% in 7d	scf, sdw	38
	14.0 mgCOD/g/h	bss, as	42
3-nitrophenol	17.5 mgCOD/g/h	bss, as	42
4-nitrophenol	100% in 7d	scf, sdw	38
	17.5 mgCOD/g/h	bss, as	42
N-nitroso-di-N-propylamine	14% in 7d	scf, sdw	38
N-nitrosodi-phenylamine	67% in 7d	scf, sdw	38
2-nitrotoluene	32.5 mgCOD/g/h	bss, as	42
3-nitrotoluene	21.0 mgCOD/g/h	bss, as	42
4-nitrotoluene	32.5 mgCOD/g/h	bss, as	42
nonadecane	7.5-54% in 4w	si, nmf	58
octacosane	1.3-39.1% in 4w	si, nmf	58
octadecane	19.5-31.9% in 4w	si, nmf	58
1-octadecene	16.4-32.3% in 4w	si, nmf	58
octadecenoic acid	82-312.2h aerobic half life	si, nmf	44

TABLE 9.9   Biodegradation and Disappearance Rates for Several Organic Chemicals[a] (cont.)

Organic Chemical	Biodegradation or Disappearance Rate	Medium and Inoculum	Reference
pentachlorophenol	18% in 7d	scf, sdw	38
perylene	0% in 16 mo	si, nmf	41
phenanthrene	100% in 7d	scf, sdw	38
	100% in 4 mo	si, nmf	41
	<60-200d half life	si, nmf	39
phenol	97% in 7d	scf, sdw	38
	98.5 mgCOD/g/h	bss, as	42
phenylisocyanate	37d half life	sgw, fo	43
phthalic acid	78.4 mgCOD/g/h	bss, as	42
phthalimide	20.8 mgCOD/g/h	bss, as	42
n-propanol	71 mgCOD/g/h	bss, as	42
propylbenzene	100% in 11d	bgw, nmf	52
	100% in 192h	sp, nmf	53
pyrene	41% in 7d	scf, sdw	38
	97% in 16 mo	si, nmf	41
	210-1900d half life	si, nmf	39
pyridine	100% in 8d	si, nmf	46
sodium acetate	8.6h aerobic half life	si, nmf	44
	1.4-15.6h anaerobic half life	si, nmf	44
styrene	2.3-12.0%/w	si, nmf	54
sulphanilic acid	4.0 mgCOD/g/h	bss, as	42
1,1,2,2-tetra-chloroethane	0% in 7d	scf, sdw	38
	97% in 2d	cfc, bm	51
tetrachloroethylene	38% in 7d	scf, sdw	38
	0% in 190h	swi, nmm	56
	300d half life	sgw, fo	59
	87-99.98% in 2-4d	cfc, nmm	60
	86% in 2d	cfc, bm	51
	68% in 21d	swi, nmf	61
	0.9-1.8%/w	si, nmf	54
tetrachloromethane	84% in 7d	scf, sdw	38
	>99% in 2d	cfc, bm	51
tetrahydrofurfuryl alcohol	40 mgCOD/g/h	bss, as	42
1,2,3,4-tetra-hydronaphthalene	100% in 9d	bgw, nmf	52
	100% in 192h	sp, nmf	53
1,2,3,4-tetra-methylbenzene	100% in 11d	bgw, nmf	52
	100% in 192h	sp, nmf	53
1,2,3,5-tetra-methylbenzene	100% in 9d	bgw, nmf	52

TABLE 9.9   Biodegradation and Disappearance Rates for Several Organic Chemicals[a] (cont.)

Organic Chemical	Biodegradation or Disappearance Rate	Medium and Inoculum	Reference
1,2,4,5-tetra-methylbenzene	100% in 10d	bgw, nmf	52
	100% in 192h	sp, nmf	53
toluene	100% in 7d	scf, sdw	38
	100% in 10d	bgw, nmf	52
	37d half life	sgw, fo	43
	39d half life	sgw, fo	47
	37d half life	gwi, nmf	47
	100% in 30-80d	si, nmf	48
	100% in 80d	sgw, fo	48
	>99% in 120w	si, nmf	49
	>93%/w	si, nmf	50
	0.9-3.2%/w	si, nmf	54
	100% in 192h	sp, nmf	53
p-toluenesulphonic acid	8.4 mgCOD/g/h	bss, as	42
1,2,4-trichlorobenzene	48% in 7d	scf, sdw	38
trichlorobenzenes	11d half life	sgw, fo	43
1,1,1-trichloroethane	26% in 7d	scf, sdw	38
	300d half life	sgw, fo	59
	98% in 2d	cfc, bm	51
	<1.1-<3.2%/w	si, nmf	54
1,1,2-trichloroethane	3% in 7d	scf, sdw	38
trichloroethylene	51% in 7d	scf, sdw	38
	69% in 4d	swi, nmm	56
	300d half life	sgw, fo	59
	<3.5%/w	si, nmf	50
	89% in 40w	si, nmf	49
	<1.2-<2.3%/w	si, nmf	54
trichlorofluoromethane	49% in 7d	scf, sdw	38
trichloromethane	48% in 7d	scf, sdw	38
	68% in 27d	swi, nmf	62
	96% in 2d	cfc, bm	51
	3% in 5d	si, nmf	63
	<1.0-<2.8%/w	si, nmf	54
2,4,6-trichlorophenol	100% in 7d	scf, sdw	38
triethylene glycol	27.5 mgCOD/g/h	bss, as	42
1,3,5-trihydroxybenzene	22.1 mgCOD/g/h	bss, as	42
3,4,5-trihydroxy-benzoic acid	20 mgCOD/g/h	bss, as	42
tri-isobutylphosphate	37d half life	sgw, fo	43
1,2,3-trimethylbenzene	100% in 12d	bgw, nmf	52
	100% in 192h	sp, nmf	53
1,2,4-trimethylbenzene	100% in 7d	bgw, nmf	52
	100% in 192h	sp, nmf	53
1,3,5-trimethylbenzene	100% in 12d	bgw, nmf	52
	100% in 192h	sp, nmf	53

TABLE 9.9   Biodegradation and Disappearance Rates for Several Organic Chemicals[a] (cont.)

Organic Chemical	Biodegradation or Disappearance Rate	Medium and Inoculum	Reference
1,3,3-trimethyl-2-norcamphanone	110d half life	sgw, fo	43
vinyl chloride	100% in 23d	swi, nmm	56
m-xylene	100% in 7d	bgw, nmf	52
	37d half life	sgw, fo	43
	15d half life	sgw, fo	47
	29d half life	gwi, nmf	47
	100% in 65d	si, nmf	48
	100% in <300d	sgw, fo	48
o-xylene	100% in 12d	bgw, nmf	52
	11d half life	sgw, fo	43
	32d half life	sgw, fo	47
	31d half life	gwi, nmf	47
	100% in 25-60d	si, nmf	48
	100% in <300d	sgw, fo	48
	>99% in 120w	si, nmf	49
	100% in 192h	sp, nmf	53
p-xylene	100% in 7d	bgw, nmf	52
	37d half life	sgw, fo	43
	17d half life	sgw, fo	47
	100% in <300d	sgw, fo	48

[a] abbreviations:

as	=	activated sludge as microbial inoculum.
bgw	=	batch test using groundwater.
bm	=	bacterial inoculum produced in a methanogenic environment.
bss	=	batch test using distilled water, dissolved salts, and the organic chemical as the sole carbon source.
cfc	=	continuous-flow, fixed film laboratory study using glass bead columns.
COD	=	chemical oxygen demand.
d	=	day(s).
fo	=	estimation based on field observation.
gwi	=	groundwater incubation study.
h	=	hour(s).
mo	=	month(s).
naf	=	natural acclimated microbial flora.
nmf	=	natural microbial flora used as inoculum.
nmm	=	natural microbial flora under methanogenic conditions.
scf	=	static-culture flask biodegradation test, original culture.
sdw	=	settled domestic wastewater utilized as microbial inoculum.
sgw	=	naturally-occurring soil-groundwater system.
si	=	soil incubation study.
sm	=	soil microcosm study.
sp	=	soil percolation study.
swi	=	soil-water or sediment-water incubation study.
w	=	week(s).
y	=	year(s).

domonas, Arthrobacter, Alcaligenes, Corynebacterium, Flavobacterium, Achromobacter, Micrococcus, Nocardia, and *Mycobacterium* appear to be the most consistently isolated hydrocarbon-degrading bacteria in soil[27]. In decreasing order *Trichoderma, Penicillium, Aspergillus,* and *Mortierella* appear to be the most consistently isolated hydrocarbon-degrading fungi in soil[27]. In one oil in soil study, eight months after the addition of oil to soil, the number of oil-degrading bacteria in soil increased tenfold and comprised almost 50 percent of the total soil bacterial population[64]. In another oil in soil study, soil receiving an application of 39.2 percent crude oil possessed the highest number of microorganisms relative to soil receiving less amounts of oil[65].

If the naturally-occurring soil microorganisms are not capable of degrading an organic chemical or waste at a sufficiently rapid rate, mutant microorganisms may work. Seeding a mutant population into a soil-groundwater system is a promising area for the biodegradation of organic chemicals other than bulk hydrocarbons. Although a number of successful case histories have been reported in the published literature, these cases lack the experimental designs needed to differentiate the effect of the mutant microorganisms from those of the naturally-occurring microorganisms. Also, no information is available regarding the relative risk to human health and the environment resulting from the presence of a mutant microorganism for an unknown, possibly indefinite, period of time in soil and groundwater.

Energy. One major factor limiting microorganism growth and metabolism in soil is the presence of a suitable and available source of energy. Soil microbiologists have long observed that wherever an available energy source is abundant in soil, microorganisms capable of utilizing that source are usually present in abundant numbers[66]. A substantial fraction of the soil microorganism population is probably in a dormant state most of the time because of the inadequacy of the average soil's energy supply[66]. Many industrial organic chemicals, when added to soil, stimulate soil microorganisms because they serve as energy sources.

There are many organic chemicals that are transformed by soil microorganisms which do not utilize the chemical as a carbon or energy source. The process in which an organic chemical is transformed but not utilized by an organism that derives its energy from other organic chemicals is known as cometabolism.

When cometabolism is affecting the transformation of an organic chemical, several distinct soil microorganisms are usually needed in order to substantially degrade the chemical. One organism causes an initial modification via the cometabolic process such that the second and subsequent microorganisms can use the modified chemical as an energy source to cause successive modifications. It is important to note that cometabolism does not mineralize the organic chemical to CO_2 and H_2O; it causes an alteration in the chemical structure to form a modified chemical.

Cometabolism has been identified as a process that influences the degradation of several organic chemicals (see Table 9.10). Although our understanding of the process is far from complete, research interest in this area is high at the present time and should lead to interesting findings in the future.

The presence of some organic chemicals can have a particularly significant stimulating effect on the microbiological degradation of some organic chemicals either through cometabolism or by another mechanism. For example, the presence of fulvic acid enhanced the biodegradation of 2(methylthio)benzothiazole in a fermentor broth containing activated sludge bacteria[70]. The presence of sodium ligninsulfonate enhanced the biodegradation of various mixtures of commercial PCBs (Aroclors) in a growth medium containing PCB degrading bacteria[71]. The adaptation to increasing concentrations of amino acids, carbohydrates, or fatty acids enhanced the ability of the microbial community of a mesotrophic reservoir to degrade mcresol, m-aminophenol, and p-chlorophenol[72].

The presence of an organic chemical, which possesses a chemical structure similar to the structure of the chemical of concern, can have a stimulating effect on the microbiological degradation of the chemical of concern. For example, the addition of aniline to soil containing 0.2 to 100 ppm 3,4-dichloroaniline increased the mineralization rate of 3,4-dichloroaniline severalfold[73]. The addition of small amounts of Aroclor 1221 to a growth medium containing *Pseudomonas sp.* 7509 enhanced the degradation of Aroclor 1254[71]. The addition of biphenyl to an Altamont soil enhanced the degradation of Aroclor 1242[74]. The process in which the addition of one chemical stimulates the degradation of another chemical with a similar chemical structure is known as analog enrichment.

The presence of some organic chemicals in soil can have an inhibitory effect on the microbiological degradation of some organic chemicals. For example, the adaptation to increasing concentrations of humic acids reduced the ability of the microbial community of a mesotrophic reservoir to degrade m-cresol, m-aminophenol, and p-chlorophenol[75]. The degradation of benzene and naphthalene by a mixed microbial community from an oil refinery settling pond was inhibited until phenol was degraded[76]. The mineralization of 2 ppb phenol by *Pseudomonas acidovorans* was delayed 16 hours by the presence of 70 ppb acetate, and the delay was lengthened by increasing acetate concentrations[77]. When *Pseudomonas sp.* strain ANL was grown in a salts solution supplemented with 300 ppb each of glucose and aniline, glucose was mineralized first, and aniline was mineralized only after much of the glucose was converted to carbon dioxide[77].

TABLE 9.10 Organic Chemicals Modified by Cometabolism.

acenaphthalene
alkyl benzene sulfonate
anthracene
benzene
bis(4-chlorophenyl) acetic acid
butane
1-butene
cis-2-butene
trans-2-butene
n-butylbenzene
n-butylcyclohexane
carbon monoxide
3-chlorobenzoate
4-chlorotoluene
cumene
cyclohexane
cycloparaffins
p-cymene
DDT
n-decane
1,2-diethylbenzene
diethyl ether
9,10-dimethylanthracene
1,3-dimethylnaphthalene
2,3-dimethylnaphthalene
1,6-dimethylnaphthalene
2,6-dimethylnaphthalene
2,7-dimethylnaphthalene
dodecane
ethane
ethene
ethylbenzene
heptadecane
hexadecane
4-isopropyltoluene
limonene
2-methylanthracene
2-methylnaphthalene
3-methylphenanthrene
naphthalene
octadecane
pentadecane
phenylcyclohexane
propane
propene
n-propylbenzene
retene

TABLE 9.10   Organic Chemicals Modified by Cometabolism. (cont.)

tetradecane
thianaphthene
toluene
2,4,5-trichlorophenoxyacetate
tridecane
1,2,4-trimethylbenzene
undecane
m-xylene
p-xylene

Compiled from data in references 67, 68 & 69.

All the mechanisms by which organic chemicals retard the biodegradation of other organic chemicals are not known. However, one mechanism that has received little attention is diauxie or sparing[76]; diauxie is an antagonistic interaction of organic chemicals in which microorganisms preferentially degrade one chemical in a mixture before synthesizing the enzymes needed to degrade other chemicals in the mixture. Microbiologists have extensively studied the preferential metabolism of sugars but not of environmentally significant organic chemicals.

Acidity and Alkalinity.   The majority of soil microorganisms will thrive best in the pH range of 6 to 8. Most will tolerate well a pH range of about 4 to 9. Strong acid or alkaline conditions will inhibit the growth and metabolism of most soil microorganisms.

Not all soil microorganisms or metabolic processes should be expected to respond equally to acidity or alkalinity. For example, ammonification was relatively insensitive to acidity in a perfusion study in which soil was exposed to pH 2.0 simulated acid rain[78]. In the same study, nitrification was more sensitive, being retarded in NH_4-N supplemented soils exposed to pH 3.0 simulated acid rain and inhibited at pH 2.5[78]. Acid rain at pH 3.7 and 3.0 did not significantly alter soil respiration but did significantly reduce nitrification[79]; in another study, however, acidification had little effect on soil respiration (COD_2 evolution) until the pH was lowered below three[80]. In this same study, glucose was not degraded at approximately pH 2, but was degraded after soil pH was raised to about pH 4.1 - 4.3. Nitrogen fixation in soil cores was not significantly altered by 690 days of exposure to acid rain[81].

The effect of acid rain on enzymatic activities depends on the enzyme type. Protease activity was not significantly altered in any of five soils in 97 or

690 day experiments in which pH 3.7 and 3.0 acid rain was percolated through soil columns[81]. In the same study, dehydrogenase and phosphatase activities decreased in soils exposed to acid rain for 690 days.

Temperature.   Microbiological reactions follow the general rule that the rate of a chemical's reaction increases as the temperature increases[34]. As a result, warmer temperatures favor relatively faster biodegradation rates.

Temperature limits to microorganism activity do exist. Because microorganisms require liquid water, the lower temperature limit to microorganism activity is the freezing point of water[34]. A number of researchers have reported the degradation of hydrocarbons at or slightly above the freezing point of water[82]. Because most microorganisms contain essential enzymes that are denatured at or above 50° Celsius, the higher temperature limit to microorganism activity is about 50° Celsius[34]. Groundwater temperatures in the U.S.A. fall within these limits (see Figure 9.3).

Moisture.   Soil microorganisms need water to support their metabolic processes. As a result, microorganisms are expected to respond to changes in soil moisture content through a complex series of interactions involving nutrient fluxes, soil temperature, pore size changes, and soil atmosphere changes.

An interesting series of published experiments gives good information on how soil microorganisms respond to changes in soil moisture content. In field plots receiving rainfall, the number of bacteria doubled within three days[84]; however, during a period of drought immediately following the rainfall period, the number of bacteria decreased by about 30 percent, then increased again as rainfall commenced. In field plots receiving irrigation, the number of bacteria increased by 50 percent and then remained constant[84]. The change in microbial activity due to a change in soil moisture content may be substantial under some circumstances; for example, rewetting a dry soil caused as much as a 40X increase in soil respiration[85].

In general, extreme moisture conditions should be unfavorable for microorganism growth and metabolism in unsaturated zone soil. Because individual species are seldom eliminated entirely in extremely wet or dry soil moisture conditions, the drying of a wet soil or the rewetting of a very dry soil should reestablish microorganism activity. Between these extreme conditions, soil moisture content should have an undramatic effect on the microbiological degradation of organic chemicals, as evidenced by experiments on the land treatment of a refinery and a petrochemical sludge[86].

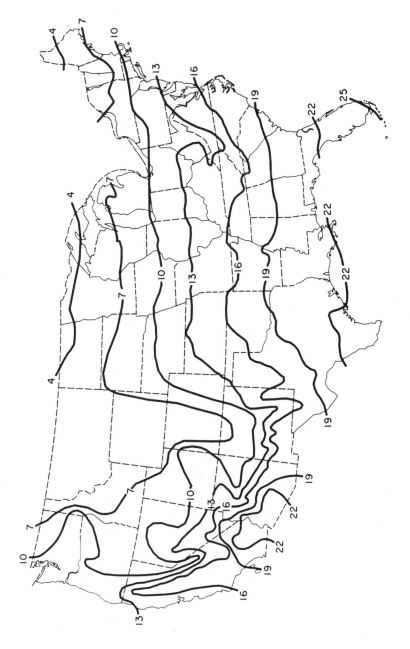

FIGURE 9.3   Approximate temperature of groundwater, in degrees Celsius, in the continental United States, at depths of 10 to 25 meters.[83]

Essential Elements.   Research has shown that certain elements are necessary for the normal growth and nutrition of biota, including microorganisms. These essential elements include: nitrogen, phosphorus, potassium, calcium, magnesium, sulfur, iron, manganese, copper, zinc, boron, molybdenum, chlorine, and cobalt.

It is most important to recognize that microorganism growth and metabolism is dependent upon all of these elements, and that any one of them, if out of balance, can reduce or entirely prevent growth or metabolism. These elements must be present and available to the microorganism in (a) a usable form, (b) appropriate concentrations, and (c) proper ratios.

Under normal circumstances, soil is the provider of essential elements to microorganisms. However, when an organic chemical enters a soil system in bulk quantities, the soil's supply of elements is almost always inadequate to support desirable biodegradation rates. The addition of these elements usually results in accelerated biodegradation rates.

When a soil receives a relatively large amount of an essential or nonessential element, several events may occur. First, an initial reduction in the number of soil microorganisms may occur. Second, the number of species of soil microorganisms may decrease. Third, soil processes performed by soil microorganisms may be adversely affected. Fourth, element-resistant microorganisms may adapt to the soil and its relatively high elemental concentration.

Table 9.11 lists information on microorganism processes affected by metals in soil. An analysis of the information in Table 9.11 revealed several important facts regarding the effect of metals on soil microorganism processes. First, a concentration of a metal may adversely affect one process, yet have no effect on another. For example, 1000 ppm Cd in a pH 4.8 sandy loam soil retards nitrification but has no effect on ammonification.

Second, the soil type can have a very significant effect on how metals affect microorganism processes. For example, 100 ppm Pb in a pH 5 loamy sand caused a 25 percent decrease in respiration. However 1000 ppm Pb in a pH 5 sandy loam soil had no effect on respiration. Because sandy loam soil should have a larger surface area relative to loamy sand soil, the sandy loam soil probably fixed more Pb than the loamy sand soil. Fixed Pb is not available to microorganisms. In summary, it is not sufficient to know just the total metal concentration when assessing the effect of metals on soil microorganisms; the fixation reactions discussed in Chapter 3 will significantly influence the effect of metals on soil microorganisms; the fixation reactions discussed in Chapter 3 will significantly influence the effect of metals on soil microorganisms.

Third, soil pH can have a very significant effect on how metals affect microorganism processes. For example, 1000 ppm Zn had no effect on nitrification and N mineralization at pH 6.0, a slight effect at pH 7.0, and significantly retarded these two processes at pH 7.7. Because pH affects the solubility of Zn, changes in pH should change the amount of fixed Zn, which changes the amount of Zn available to the microorganism.

Fourth, the presence of other bulk organic materials in soil can have a very significant effect on how metals affect microorganism processes. For example, 10,000 ppm Pb in a pH 5 sandy loam soil retards respiration. However, 15,000 ppm Pb plus two percent humic acid in a pH 5 sandy loam soil had no effect on soil respiration. In addition, 20,000 ppm Pb plus four percent compost in a pH 5 sandy loam soil had no effect on soil respiration after 20 days of incubation.

Fifth, in neutral pH soils, relatively large amounts of metals must be present in soils to have an adverse impact on microorganism processes. On the other hand, in acidic soils, relatively small amounts of metals have an adverse impact on microorganism processes.

Some microorganisms have adapted mechanisms to maintain low intracellular concentrations of metals while surviving in soils with relatively high metal concentrations[92]. An understanding of the biochemical basis for microorganism resistance to metal toxicity is still emerging. Several mechanisms have been identified. Some microorganisms have energy-driven efflux pumps that keep intracellular concentrations of metals low by pumping the metal out of its cell. Some microorganisms can convert enzymatically and intracellularly a more toxic form of an element or metal into a less toxic form. Some microorganisms can synthesize intracellular polymers that trap and remove metals from the intracellular solution. Some microorganisms can bind large amounts of metal ions to their cell surfaces via precipitation or by covalent or ionic bonding. Also, some microorganisms can biomethylate metals; the methylated species can then be transported out of the microorganism by diffusion-controlled processes.

Organic Chemical Concentration. The concentration of an organic chemical in a soil system affects its biodegradation rate. For some chemicals, the biodegradation rate is limited by low concentrations; for others, the rate is limited by high concentrations. At the present time, published scientific studies can only be utilized to derive generalizations on the effect of organic chemical concentration on biodegradability.

Low concentrations of an organic chemical can affect its degradation rate in several ways. First, the lower concentration may become a limiting factor because it may not induce the enzymes responsible for degradation. Second, the lower concentration may result in a prolonged acclimation period. Third, the low concentration may prohibit the chemical from serving as an energy source for microorganism metabolism.

Many biodegradation studies have been performed while utilizing chemi-

TABLE 9.11 Effect of Various Metals on Microorganisms.

Metal	Soil Type	Soil Concentration	Microbial Process	Effect	Ref.
Ag	loamy sand, pH 5	10 ppm	Respiration	43% Decr	87
	loamy sand, pH 5	100 ppm	Respiration	72% Decr	87
Bi	loamy sand, pH 5	10 ppm	Respiration	11% Decr	87
	loamy sand, pH 5	100 ppm	Respiration	4% Decr	87
Cd	loamy sand, pH 5	10 ppm	Respiration	17% Decr	87
	loamy sand, pH 5	100 ppm	Respiration	11% Decr	87
	silt loam, pH 6.75	100 ppm	Denitrification	Signif. Retard.	88
	sandy loam, pH 4.8	500 ppm	Nitrification	Retard.	89
	sandy loam, pH 4.8	1000 ppm	Ammonification	None	89
	sandy loam, pH 4.8	1000 ppm	Nitrification	Retard.	89
Co	loamy sand, pH 5	10 ppm	Respiration	4% Decr	87
	loamy sand, pH 5	100 ppm	Respiration	23% Decr	87
Cu	loamy sand, pH 5	10 ppm	Respiration	3% Decr	87
	loamy sand, pH 5	100 ppm	Respiration	25% Decr	87
	silt loam, pH 6.75	250 ppm	Denitrification	Retard.	88
Hg	loamy sand, pH 5	10 ppm	Respiration	33% Decr	87
	loamy sand, pH 5	100 ppm	Respiration	55% Decr	87
Ni	loamy sand, pH 5	10 ppm	Respiration	6% Decr	87
	loamy sand, pH 5	100 ppm	Respiration	28% Decr	87

TABLE 9.11 Effect of Various Metals on Microorganisms. (cont.)

Metal	Soil Type	Soil Concentration	Microbial Process	Effect	Ref.
Pb	loamy sand, pH 5	10 ppm	Respiration	6% Decr	87
	loamy sand, pH 5	100 ppm	Respiration	25% Decr	87
	sandy loam, pH 5	1000 ppm	Respiration	None	90
	silt loam, pH 6.75	1000 ppm	Denitrification	Retard.	88
	sandy loam, pH 5	10,000 ppm	Respiration	Retard.	90
	sandy loam, pH 5	15,000 ppm + 2% humic acid	Respiration	None	90
	sandy loam, pH 5	20,000 ppm + 4% compost	Respiration	Initial retard; none after 20 days	90
Sb	loamy sand, pH 5	10 ppm	Respiration	18% Decr	87
	loamy sand, pH 5	100 ppm	Respiration	31% Decr	87
Sn	loamy sand, pH 5	10 ppm	Respiration	16% Decr	87
	loamy sand, pH 5	100 ppm	Respiration	35% Decr	87
Ti	loamy sand, pH 5	10 ppm	Respiration	4% Decr	87
	loamy sand, pH 5	100 ppm	Respiration	28% Decr	87
Zn	loamy sand, pH 5	10 ppm	Respiration	21% Decr	87
	loamy sand, pH 5	100 ppm	Respiration	45% Decr	87
	silt loam, pH 6.75	250 ppm	Denitrification	Retard.	88
	—	1000 ppm	Nitrification	None at pH 6.0 Slight at pH 7.0 Retard. at pH 7.7	91
	—	1000 ppm	N mineralization	None at pH 6.0 Slight at pH 7.0 Retard. at pH 7.7	91

[a] Abbreviations:

    Decr.   = decrease
    Retard. = retardation
    Signif. = significant

cal concentrations that are higher than those encountered in the field. Many researchers have assumed that if a chemical is readily biodegradable at a moderate or high concentration, then ppb or ppt concentrations of the same chemical should also be readily biodegradable. Because this assumption does not hold for many chemicals, one should always check published studies to determine the concentration range studied during a particular biodegradation test, especially if one is interested in biodegradation at relatively low concentrations. Also, it is important to remember that studies utilizing very low concentrations may result in a reaction rate so slow that the chemical was reported as nondegrading, when in fact it was degrading.

All microorganisms are not affected to the same extent by a chemical or its metabolite at a certain concentration. The data on the effects of DDT on selected species of soil microorganisms listed in Tables 9.12 and 9.13 exemplify this effect. An organic chemical at a given concentration (a) may be lethal to one specie, (b) may serve as an energy source for another specie with metabolic stimulation being the end result, (c) may be degraded by another specie as a cometabolite, or (d) may have no significant metabolic effect in yet another specie.

In general, relatively large concentrations of an organic chemical are usually needed in order to significantly affect all microorganisms in a soil. For example, an analysis of the data presented in Table 9.13 will reveal that very large concentrations of DDT, greater than 20,000 ppm in soil, may be needed in order to adversely affect all four of the most important soil microbial processes. It is important to note that rarely is DDT present as a sole pesticide in soil on viable agricultural and horticultural farms; therefore, the soil loading rates listed in Tables 9.12 and 9.13 only reflect DDT concentrations and not the concentrations of DDT metabolites or other pesticides which were present.

Oxygen and the Redox Potential.   The degradation of organic chemicals can occur under aerobic or anaerobic conditions, i.e., with or without oxygen. Under aerobic oxidation, molecular oxygen serves as an electron acceptor; one atom of an oxygen molecule is incorporated into the structure of the organic chemical, while the second combines with hydrogen to form water. The general process can be described by the following equation:

$$\text{Bacteria} + O_2 + \text{Organics} + \text{Nutrients} \longrightarrow CO_2 + H_2O + \text{Byproducts} + \text{Cell Biomass} \quad (9.15)$$

Approximately 5 to 50 percent of the organic material metabolized will be transformed into cell biomass. The more refractive a compound, the less carbon there is available for cell growth. Therefore, an increase in cell number

TABLE 9.12   The Effects of DDT on Selected Species of Microorganisms.

Organism	DDT Concentration (ppm)	Effect
Bacteroides fragilis	0.01	Inhibition
Fusarium oxysporum	0.1	Inhibition
Heliscus submersus	0.1-60	Stimulation
Nitrifying bacteria	0.5-10.0	No Effect
Mycorrhiza	<1.0	Stimulation
Mycorrhiza	1.0-10.0	Inhibition
Aquatic hyphomycetes	>2.0	Stimulation
Phycomycetes	2.0-60	Stimulation
Hyphomycetes	2.0-60	Stimulation
Nitrogen fixing bacteria	5.0-500	No Effect
Spore forming bacteria	5.0-500	No Effect
Azotobacter	5.0-500	No Effect
Actinomycetes	5.0-500	No Effect
Nitrifying bacteria	1,000	Inhibition
Ammonifying bacteria	1,000	Inhibition
Sulfur oxidizing bacteria	1,000	Inhibition

Compiled from data presented in Ref. 93.

TABLE 9.13   A Summary of the Effects of DDT on Microbial Processes in Soil.

Process	DDT Concentration (ppm)	Effect	Reference
CO_2 evolution	1.0-2,500	No Effect	93
	100	Stimulation	94
Nitrate production	25-100	Stimulation	93,94
	500-20,000	No Effect	93
Ammonification	0.5-500	No Effect	93
	1000	Inhibition	95
Nitrification	200	No Effect	96

is directly related to the biodegradability of the compound.

Some organic chemicals can degrade in anaerobic environments at substantially greater rates than in aerobic environments. For example, no toxaphene degradation was observed in a Crowley silt loam that was incubated aerobically in the laboratory for six weeks[97]; however, extensive degradation occurred during anaerobic degradation. During anaerobic conditions, molecules other than oxygen are used as the final electron acceptor (see Table 9.14).

For many organic chemicals, anaerobic biodegradation generally proceeds

TABLE 9.14   Relationship Between Respiration, Redox Potential, and Typical Electron Acceptors and Products[98]

Form of Respiration	Typical Redox Potential	Electron Acceptors	Products
Aerobic respiration	$+400$ mV	O_2	H_2O
Nitrate respiration & Denitrification	-100 mV	NO_3^-	NO_2^-, N_2
Sulfate reduction	-160 to -200 mV	SO_4^{2-}	HS^-
Methanogenesis	-300 mV	CO_2	CH_4

at a much lower rate than aerobic biodegradation. However, the introduction of oxygen into an anaerobic soil system can stimulate biodegradation. It is important to remember, however, that the amount of oxygen needed will depend upon the concentration of the organic chemical(s). The theoretical amount of oxygen required to degrade 1 mg/l of a hydrocarbon substrate can be calculated by performing a stoichiometric analysis for the given substance, as shown by the following equation:

$$C_xH_y + \left[x + (y/4) \right] O_2 \longrightarrow xCO_2 + (y/2)H_2O \qquad (9.16)$$

Usually, about 3 to 4 mg/liter of oxygen is required to degrade 1 mg/liter of a medium-length hydrocarbon compound. If 50 percent of the organic material is converted to bacterial cell matter and the other half oxidized to carbon dioxide and water, only 4 to 6 mg/liter of organic material can be converted and oxidized under oxygen saturation conditions. Thus, for contaminated groundwaters having organic concentrations significantly higher than the above values, in-line aeration prior to injection is insufficient, because only about 10 mg/liter dissolved oxygen can be attained on a single pass, and the reinjected groundwater will use up all available oxygen in a very short period of time.

Organic chemicals present in high concentrations in groundwater are not degraded aerobically until (a) dispersion during transport decreases the chemical's concentration in groundwater, or (b) oxygen is added to the soil-groundwater system. In soils containing hydrocarbon at residual saturation, the estimated volumes of water containing sufficient dissolved oxygen to completely renovate the hydrocarbon saturated soil are enormous: 5000 volumes for stony to coarse gravelly soils, 8000 volumes for gravelly to coarse sandy soils, 15,000 volumes for coarse to medium sandy soils, 25,000 volumes for medium to fine sandy soils, and 32,000 volumes for fine sandy to silt soils[99].

Oxygen can be added to a soil-groundwater system by air sparging. The solubility of air in water is about 40 to 50 ppm; the amount of oxygen, therefore, that could be added to the system would be at most 8-10 ppm. This amount can be rapidly depleted by an active microorganism population[100], as discussed above.

Oxygen can be added to a soil-groundwater system by injecting pure oxygen into the system. Use of pure oxygen limits the available oxygen to a 40-50 ppm level[100]. However, because the hydrostatic pressure of shallow aquifers is essentially atmospheric pressure, degassing usually occurs immediately[100].

Oxygen can be added to soil-groundwater systems in the form of colloidal gas aphrons[101]. Colloidal gas aphrons (CGAs) are a microdispersion of air or gas encapsulated in a thin film of water considerably thicker than a monolayer. CGAs are similar to soap bubbles in structure but are colloidal in size. Almost any water soluble surfactant and any gas of limited solubility can be used to produce a typical CGA dispersion of 60-70 percent air in the form of 25 to 50 micron bubbles. CGAs were first produced by passing a dilute surfactant solution through a venturi throat into which a very small gas entry port had been placed. If the velocity of the solution flowing through the venturi exceeds a critical velocity, air will be sucked into the venturi at the throat and shear off by solution passage. This ingestion of air will cause very small, very uniform bubbles to be introduced into solution; these bubbles do not coalesce when they collide, unlike air bubbles created by sparging or electrolysis that are 2 to 1000 times larger and tend to coalesce and rise rapidly to the surface.

Laboratory studies on the in situ biodegradation of hexadecane utilizing CGAs gave results revealing that CGAs were effective carriers of oxygen needed for biodegradation[101]. CGAs made with sodium dodecyl benzene sulfonate were injected into an unconsolidated saturated sand containing 200 ppm hexadecane. One series of experimental units were injected with air CGAs; another, with pure oxygen CGAs. *Pseudomonas putida* and other hexadecane degrading organisms isolated from primary sludge were inoculated into the units. Approximately 90 percent of the hexadecane was degraded in units containing oxygen CGAs and 70 percent was degraded in units containing air CGAs in 96 hrs with reaerations at 48 and 72 hrs.

The addition of hydrogen peroxide to groundwater can substantially increase oxygen levels. Because hydrogen peroxide is miscible with water, the amount of oxygen added to the system is limited only by the reactivity of hydrogen peroxide. One molecule of hydrogen peroxide can generate one-half part of oxygen:

$$H_2O_2 \longrightarrow H_2O + 1/2\ O_2 \qquad\qquad (9.17)$$

Although hydrogen peroxide can be toxic to microorganisms, it can be added to soil-groundwater systems at concentrations up to 100 or 200 ppm without being toxic[102]; concentrations as high as 1000 ppm can be attained without toxic effects if a proper acclimation period is provided[102].

It is most important to recognize that the hydrogen peroxide added to a soil system to enhance microorganisms can react with the organic chemical of concern and with naturally-occurring soil organic matter. Hydrogen peroxide is an oxidizing agent. The addition of significant amounts of hydrogen peroxide to a Canadian podzol subsurface soil and to two tropical volcanic surface soils produced many water-soluble organic compounds such as alkanes, aliphatic acids, phenols, phenolic acids, benzenecarboxylic acids, and organonitrogen and organosulfur chemicals (see Table 9.15)[103].

In addition, iron catalyzes the decomposition of hydrogen peroxide in groundwater. A standard practice to avoid decomposition by iron is to add phosphate into treated, injected water in sufficient amounts to precipitate iron. Organic inhibitors can be added to stabilize the degradation rate of hydrogen peroxide so that the oxygen demand of soil microorganisms is balanced by the oxygen from decomposing hydrogen peroxide.

Adsorption. Adsorption can either increase or decrease a microorganism's ability to degrade an organic chemical. The increase in the degradation of some adsorbed organic chemicals may be related to the distribution of the microorganism population in a soil system. There is a greater population density of microorganisms on or near soil particle surfaces than in the water phase; as a result, the adsorption of an organic chemical increases the concentration of the chemical in areas where microorganisms abound, and the potential for the microorganism to attack the chemical is enhanced.

The increase in the degradation of some adsorbed organic chemicals may be due to the influence of microbially produced surface active agents or biosurfactants. Some species of *Clostridium, Corynebacterium, Bacillus, Norcardia,* and *Pseudomonas* produce biosurfactants, which are broadly grouped as carbohydrate-containing, amino acid-containing, phospholipids, fatty acids, and neutral acids. These biosurfactants may aid the transport of the adsorbed organic chemical to the active enzyme site where degradation is catalyzed.

The decrease in the degradation rate of some adsorbed organic chemicals may be due to the ability of microorganisms to attack only those chemicals dissolved in the water phase. The adsorbed chemical is protected from degradation even though microorganisms are present in both solid and water phases. A general review of the published literature revealed that adsorbed organic chemicals generally tend to be less subject to degradation by soil microorganisms.

TABLE 9.15   Organic Chemicals Produced by the Addition of Hydrogen Peroxide to Three Soils[a].

1,2-benzenedicarboxylic acid dimethyl ester
1,3-benzenedicarboxylic acid dimethyl ester
1,4-benzenedicarboxylic acid dimethyl ester
benzenepentacarboxylic acid pentamethyl ester
1,2,3,4-benzenetetracarboxylic acid tetramethyl ester
1,2,3,5-benzenetetracarboxylic acid tetramethyl ester
1,2,4,5-benzenetetracarboxylic acid tetramethyl ester
1,2,3-benzenetricarboxylic acid trimethyl ester
1,2,4-benzenetricarboxylic acid trimethyl ester
bis-(2-ethylhexyl)phthalate
6-carbomethoxy-4-pyridinecarboxaldehyde
decyl methyl ester
di-isobutyl phthalate
3,4-dimethoxyacetophenone
3,4-dimethoxy-1,5-benzenedicarboxylic acid dimethyl ester
3,5-dimethoxybenzoic acid methyl ester
dioctyl adipate
docosane
docosyl methyl ester
dodecyl methyl ester
eicosyl methyl ester
ethylbenzylsulfonate
hexadecyl methyl ester
1,6-hexanedicarboxylic acid dimethyl ester
3-methoxy-1,2-benzenedicarboxylic acid dimethyl ester
2-methoxy-1,3,4,5-benzenetetracarboxylic acid tetramethyl ester
2-methoxy-1,3,5-benzenetricarboxylic acid trimethyl ester
3-methoxy-1,2,4-benzenetricarboxylic acid trimethyl ester
3-methoxybenzoic acid methyl ester
6-methylacetate-4-pyridinecarboxaldehyde
octadecyl methyl ester
1,2,3-propanetricarboxylic acid trimethyl ester
tetracosyl methyl ester
tetradecyl methyl ester

a – compiled from data presented in reference 103.

PREDICTING BIODEGRADATION RATES

Unlike other physical-chemical reactions and properties of organic chemicals that may be estimated using mathematical techniques (e.g. water solubility, solvent solubility, adsorption, bioconcentration, dissociation constants, hydrolysis rate, photolysis rate), no quantitative, accurate procedure for estimating the rate of biodegradation of organic chemicals exists at this time. There are several reasons why no procedure exists at the present time.

First, the many factors that control biodegradation rates have not been examined methodically for different classes of chemicals, for different microorganism species, and for different microorganism populations. The interaction between some of these soil factors and microorganism species and populations are not well understood.

Second, most data found in the published literature cannot be compiled and analyzed. Experimental methods utilized to measure biodegradation are numerous and varied in their procedures and materials. As a result, these data are difficult to compare because they frequently apply only to a particular set of experimental conditions.

Third, some experimental procedures and materials are known to affect measured biodegradation rates; however, these have not been adequately studied or remedied by either developing quantitative correction factors or by improved experimental designs. A classic example of this phenomenon is the bottle effect. As early as 1943, it was realized that placing a natural water sample in a container may greatly enhance microbial activity, possibly due to adsorption of organic matter and chemicals on the container's surfaces. Yet few changes in experimental designs have been implemented to address this phenomenon.

Fourth, the vast majority of published biodegradation studies do not focus on biodegradation rates. These studies focus primarily on either (a) the identification of the organisms responsible for the degradation of a specific substance, (b) the classification of metabolic pathways, and (c) the metabolic products of degradation. These studies become most useful when used in conjunction with rate data to determine if a change in mechanism occurs as a result of a change in microorganism species, factors, and/or environmental conditions.

Generalizations or "rules of thumb" regarding the effects of chemical structure on biodegradation rates can be found in the scientific literature. The typical rule of thumb focuses on whether a functional group or molecular fragment either increases or decreases the biodegradation rate of an organic chemical. For example, research on detergents in the 1950s and 1960s revealed that (a) beta oxidation is an important metabolic pathway for degrading alkyl chains in surfactants, and (b) branching from the main chain decreases the biodegradation rate. The latter is a typical rule of thumb: alkyl chain branching decreases the rate of biodegradation relative to an organic chemical identical in structure except for the chain branching. Other rules of thumb are:

- hydroxyl and carboxyl functional groups on benzene rings usually increase biodegradation rates.

- halogen, nitro, and sulfonate functional groups on benzene rings usually decrease biodegradation rates.
- as the number of chlorine atoms within the molecule increases, the biodegradation rate decreases.
- the presence of hydroxyl, aldehyde, carboxyl, ester, and amide functional groups on organic chemicals usually causes faster biodegradation rates.
- water soluble chemicals are usually degraded faster than less soluble chemicals.
- n-alkanes, n-alkylaromatics, and aromatic compounds in the C_{10} to C_{22} range are usually readily biodegradable.
- n-alkanes, n-alkylaromatics, and aromatic hydrocarbons in the C_5 to C_9 range are biodegradable, but in most environments volatilization competes very effectively with biodegradation as a fate process.
- gaseous n-alkanes (C_1 - C_4) are biodegradable but are usually utilized by a narrow range of specialized hydrocarbon degraders.
- the n-alkanes, alkylaromatics, and aromatic compounds above C_{22} have very low water solubilities which result in slow rates of microbial degradation.
- condensed or fused aromatic and cycloparaffinic molecules with four or more rings have very low biodegradation rates.
- rate of oxidation of straight chain aliphatic hydrocarbons is correlated to chain length; in general, short chains are not as quickly degraded as long chains.
- unsaturated aliphatic organics have faster biodegradation rates than corresponding saturated aliphatic organics.

It is most important to recognize one major limitation of these rules of thumb: they are applicable primarily within a class of chemicals possessing very similar chemical structures which are biodegrading in the same environment. For example, one should not deduce from these rules of thumb that pentachlorophenol (an aromatic chemical with five chlorine atoms) will degrade at a slower rate than 1,1,1-TCA (an aliphatic chemical with three chlorine atoms) because pentachlorophenol possesses more chlorine atoms than 1,1,1-TCA. However, one may conclude that 1,1,1-TCA should degrade at a slower rate than 1,1-DCA because it contains one more chlorine atom.

Although the present day state-of-the-science does not allow one to make accurate quantitative predictions, gross estimates of biodegradation rates can be made. These gross estimates are usually reported in terms of half lives or ultimate biodegradation occurring in a few days, a few weeks, a few months, or a year or more; these gross estimates are usually made under the assumption that all factors affecting biodegradation are at favorable levels. These estimates are based primarily on the expertise of the scientist in (a) discerning the biodegradability of the molecular fragments that comprise the organic chemical, (b) discerning the physical/chemical properties of the organic chemical that affect its movement to the enzyme, and (c) discerning how actual soil conditions affect the biodegradation process. The third area of expertise is acquired primarily by conducting field, greenhouse, and laboratory studies on organic chemical biodegradation. It should be most evident that estimating biodegradation rates is difficult due to the complexity of the biodegradation process, and few scientists are capable of making gross estimates.

In spite of the complexity of the biodegradation process, the development of accurate, quantitative methods to predict the biodegradation rates of organic chemicals will prove to be a formidable challenge to the soil chemist and microbiologist for many, many years.

TREATMENT TECHNOLOGIES

Introduction

Several treatment technologies for the removal of organic chemicals and wastes in soil and groundwater are based upon degradation by soil microorganisms. The basic principles discussed previously in this chapter will govern the success or failure of each of the technologies discussed below.

In-situ Groundwater Bioreclamation.   Biological techniques for degrading organic chemicals in groundwater are classified into two general categories. The first category involves the withdrawal and treatment of groundwater by conventional biological wastewater treatment processes such as activated sludge, aerobic lagoons, anaerobic lagoons, facultative lagoons, rotating biological filters, and trickling filters. Simply stated, this first category involves the removal of contaminated water to a treatment process. Because general information regarding these techniques can be found in a standard wastewater engineering text, these techniques will not be discussed in this chapter.

The second category is in-situ groundwater bioreclamation. This treatment technology is also commonly known as biorestoration, enhanced biodegradation, or in-situ treatment. In-situ bioreclamation transforms the volume of contaminated soil and groundwater into a treatment sytem by enhancing soil microorganism's ability to degrade the organic chemical or waste. Dissolved nutrients and oxygen are introduced through water injection wells into the soil-groundwater system; they are circulated through the zone of contamination as groundwater travels from the injection well, through the zone of contamination, and into a groundwater recovery well. Groundwater from the recovery well can be (a) recirculated into the injection well after being replenished with oxygen and nutrients through either a batch or continuous feed process, or (b) discharged to a sewer or receiving water body. Because the rate in which oxygen and nutrients reach microorganisms is directly dependent upon the rate and direction of groundwater flow, the proper control of groundwater flow is a critical element in the successful operation of a groundwater bioreclamation system.

In principle, any industrially-derived organic product or waste containing metabolizable organic chemicals is amenable to groundwater bioreclamation. However, this technology has been applied almost exclusively to the restoration of soil-groundwater systems containing bulk hydrocarbons. Because petroleum hydrocarbons are comprised of a large number of organic chemicals possessing many different structures (see Table 9.16), laboratory studies are usually run to determine if the native soil microorganisms can degrade the hydrocarbons.

Laboratory studies are run for another important reason: to determine the optimum combination of nutrients that will result in maximum metabolism and chemical degradation under aerobic conditions. Soil systems vary substantially regarding their concentrations of macronutrients and micronutrients needed to sustain microorganism growth. Although many reports on the application of in-situ groundwater bioreclamation are found in the literature, very few contain sufficient information on the amounts of nutrients added and the procedures and materials utilized to optimize microorganism growth, metabolism, and degradation rates. Table 9.17 lists the nutrients utilized to sustain microorganisms and optimize degradation rates during three field and pilot studies. In general, nutrient solutions usually contain inorganic salts at concentrations between 0.001 and 0.02 percent by weight for each nutrient. While the size of the microorganism population increases substantially during nutrient addition, the population size decreases to its normal level after nutrient addition terminates.

Not all contaminated soil-groundwater systems will need nutrients to sustain in-situ bioreclamation. One example is a soil-groundwater system underneath a railyard in Europe, which contained several petroleum products; in-situ bioreclamation only required the addition of one gram of ozone per gram of dissolved organic carbon in injected water to restore this aquifer as a drinking water source[115]. Naturally-occurring soil microorganisms are used in the majority of cases to degrade organic chemicals during in-situ groundwater bioreclamation. In principle, acclimated and mutant microorganisms can be utilized during in-situ groundwater bioreclamation. It is most unfortunate that studies found in the published literature generally lack appropriate experimental designs to determine the true effect of mutant microorganisms on the degradation of chemicals.

In-situ groundwater bioreclamation offers several distinct advantages over other technologies. It is a technology that degrades chemicals and does not result in temporary containment or storage. Because it is primarily a subsur-

TABLE 9.16   **Organic Chemicals Identified in Various Petroleum Hydrocarbons**[a]

Chemical	Source[b]
acenaphthalenes	fo
acenaphthenes	fo
acetic acid	co
adamantane	co
alkanes	fo
aniline	fo
anthracene	fo
benzene	co, fo, g, hog, ug
1,2-benzofluorene	co
benzoic acid	fo
benzothiophenes	fo
bicyclic sesquiterpanes	co
bicyclo-[3.2.1] octane	co
cis-bicyclo-[3.3.0]-octane	co
biphenyl	co, fo, k
1,3-butadiene	g
n-butane	co, g, hog, ug
1-butanethiol	co
2-butanethiol	co
butanoic acid	co
trans-2-butene	g, hog
1-butene	g
cis-2-butene	g
trans-2-butene	g
t-butyl alcohol	g
n-butylbenzene	co, g
sec-butylbenzene	co, g
t-butylbenzene	co, g
sec-butylcyclohexane	g
t-butylcyclohexane	g
d1-2-sec-butyl-4,5-dimethylpyridine	co
butylthiophane	co
carbazole	co
m-cresol	co
o-cresol	co
p-cresol	co
cycloalkanes	fo
cyclobutane	co
cycloheptane	co, g
cyclohexane	co, g, hog
cyclohexanecarboxylic acid	co
cyclohexanethiol	co
cyclohexene	g
cyclopentane	co, g

TABLE 9.16 Organic Chemicals Identified in Various Petroleum Hydrocarbons[a] (cont.)

cyclopentaneacetic acid	co
cyclopentanecarboxylic acid	co
cyclopentanethiol	co
cyclopentene	g
3-cyclopentylpyridine	co
4-cyclopentylpyridine	co
m-cymene	fo, g
o-cymene	fo, g
p-cymene	fo, g
trans-decahydronaphthalene	co
n-decane	co, g, k
decanoic acid	fo
1-decene	g
cis-2-decene	g
trans-2-decene	g
decylthiophane	co
diasteranes	co
1,2-dibromoethane	g
1,2-dichloroethane	g
1,2-diethylbenzene	co, fo, g
1,3-diethylbenzene	g
1,4-diethylbenzene	co, g
diethylphenol	co
2,6-dimethylanthracene	co
2,7-dimethylanthracene	co
2,3-dimethylbenzo-[h]-quinoline	co
2,4-dimethylbenzo-[h]-quinoline	co
2,2-dimethylbutane	co, g, hog
2,3-dimethylbutane	co, g, hog, ug
2,3-dimethyl-1-butene	g
2,3-dimethyl-2-butene	g
3,3-dimethyl-1-butene	g
2,3-dimethyl-2-butenedioic acid	co
1,3-dimethyl-5-t-butylbenzene	g
2,4-dimethyl-8-sec-butylquinoline	co
1,1-dimethylcyclohexane	co, g
cis-1,2-dimethylcyclohexane	co, g, hog
cis-1,3-dimethylcyclohexane	co, hog
cis-1,4-dimethylcyclohexane	co, hog
trans-1,2-dimethylcyclohexane	co, g, hog
trans-1,3-dimethylcyclohexane	co, hog
trans-1,4-dimethylcyclohexane	co, hog
1,1-dimethylcyclopentane	co, g, hog
cis-1,2-dimethylcyclopentane	co, g
trans-1,2-dimethylcyclopentane	co, g
cis-1,3-dimethylcyclopentane	co, g, hog, ug
trans-1,3-dimethylcyclopentane	co, g, hog, ug

TABLE 9.16   Organic Chemicals Identified in Various Petroleum Hydrocarbons[a] (cont.)

2,2-dimethylcyclopentanecarboxylic acid	co
2,3-dimethylcyclopentylacetic acid	co
dimethylcyclopropane	g
1,8-dimethyldibenzothiophene	co
2,3-dimethyl-4-8-diethylquinoline	co
1,2-dimethyl-3-ethylbenzene	co, g, k
1,2-dimethyl-4-ethylbenzene	co, fo, g, k
1,3-dimethyl-2-ethylbenzene	co, g, k
1,3-dimethyl-4-ethylbenzene	co, g, k
1,3-dimethyl-5-ethylbenzene	co, fo, g, k
1,4-dimethyl-2-ethylbenzene	co, g
2,2-dimethyl-3-ethylpentane	g
2,4-dimethyl-3-ethylpentane	g
2,3-dimethyl-4-ethyl-8-n-propylquinoline	co
2,3-dimethyl-8-ethylquinoline	co
2,4-dimethyl-8-ethylquinoline	co
2,3-dimethyl-4-ethylthiophene	co
2,4-dimethyl-3-ethylthiophene	co
3,4-dimethyl-2-ethylthiophene	co
2,2-dimethylheptane	g, hog
2,3-dimethylheptane	co, g
2,4-dimethylheptane	g
2,5-dimethylheptane	g
2,6-dimethylheptane	co, g
3,3-dimethylheptane	g
3,4-dimethylheptane	g
3,5-dimethylheptane	g
2,2-dimethylhexane	co, g, hog
2,3-dimethylhexane	co, g, hog
2,4-dimethylhexane	co, g, hog
2,5-dimethylhexane	co, g, hog
3,3-dimethylhexane	co, g
3,4-dimethylhexane	co, g, hog
2,2-dimethyl-trans-3-hexene	g
2,3-dimethyl-1-hexene	g
2,3-dimethyl-2-hexene	g
2,3-dimethyl-trans-3-hexene	g
2,5-dimethyl-trans-3-hexene	g
1,2-dimethyl-4-hydroxybenzene	fo
1,3-dimethyl-5-hydroxybenzene	fo
2,4-dimethyl-1-hydroxybenzene	fo
1,1-dimethylindan	g
1,6-dimethylindan	g
1,2-dimethyl-3-isopropylbenzene	g
2,3-dimethyl-6-isopropylpyridine	co
dimethylmaleic anhydride	co
1,2-dimethylnaphthalene	co, fo, k
1,3-dimethylnaphthalene	fo

TABLE 9.16   Organic Chemicals Identified in Various Petroleum Hydrocarbons[a] (cont.)

1,4-dimethylnaphthalene	fo
1,6-dimethylnaphthalene	co, k
1,7-dimethylnaphthalene	fo
2,6-dimethylnaphthalene	co, fo, k
2,6-dimethyloctane	g
2,2-dimethylpentane	co, g, hog
2,3-dimethylpentane	co, g, hog, ug
2,4-dimethylpentane	co, g, hog, ug
3,3-dimethylpentane	co, g, hog
2,3-dimethyl-1-pentene	g
2,3-dimethyl-2-pentene	g
2,4-dimethyl-1-pentene	g
2,4-dimethyl-2-pentene	g
3,3-dimethyl-1-pentene	g
cis-3,4-dimethyl-2-pentene	g
trans-3,4-dimethyl-2-pentene	g
4,4-dimethyl-l-pentene	g
cis-4,4-dimethyl-2-pentene	g
trans-4,4-dimethyl-2-pentene	g
1,8-dimethylphenanthrene	co
2,6-dimethylphenol	fo
2,2-dimethylpropane	co, g
2,2-dimethyl-1-propanethiol	co
1,2-dimethyl-3-propylbenzene	g
1,2-dimethyl-4-propylbenzene	co, g, k
1,3-dimethyl-4-propylbenzene	co, g, k
1,3-dimethyl-5-n-propylbenzene	g
1,4-dimethyl-2-n-propylbenzene	g
2,3-dimethyl-8-n-propylquinoline	co
2,4-dimethyl-8-n-propylquinoline	co
2,3-dimethylpyridine	co
2,3-dimethylquinoline	co
2,4-dimethylquinoline	co
2,8-dimethylquinoline	co
2,6-dimethylstyrene	g
3,3-dimethyl-2-thiabutane	co
2,3-dimethylthiacyclopentane	co
2,4-dimethylthiacyclopentane	co
cis-2,5-dimethylthiacyclopentane	co
trans-2,5-dimethylthiacyclopentane	co
2,6-dimethyl-4-thiaheptane	co
2,2-dimethyl-3-thiapentane	co
2,4-dimethyl-3-thiapentane	co
2,6-dimethylundecane	g
dinaphthenebenzenes	fo
diterpanes	co
n-docosane	co, fo
n-dodecane	co, fo, g, k

TABLE 9.16 Organic Chemicals Identified in Various Petroleum Hydrocarbons[a] (cont.)

n-dotriacontane	co
n-eicosane	co, fo
eicosanoic acid	co
ethane	co
ethanethiol	co
ethanol	fo
ethylbenzene	co, fo, g, hog, ug
2-ethyl-1-butene	g
ethylcyclohexane	co, g
ethylcyclopentane	co, g, hog
3-ethylcyclopentene	g
3-ethylheptane	g
4-ethylheptane	g
3-ethylhexane	co, g
2-ethyl-1-hexene	g
3-ethyl-3-hexene	g
1-ethyl-3-isopropylbenzene	g
1-ethyl-4-isopropylbenzene	g
1-ethyl-4-methylbenzene	fo
1-ethylnaphthalene	fo
2-ethylnaphthalene	fo
4-ethyloctane	g
3-ethylpentane	co, g, hog
3-ethylpentanoic acid	co
2-ethyl-1-pentene	g
3-ethyl-l-pentene	g
3-ethyl-2-pentene	g
1-ethyl-2-n-propylbenzene	g
1-ethyl-3-n-propylbenzene	g
ethylstyrene	g
2-ethylthiacyclopentane	co
formic acid	co
n-heneicosane	co, fo
n-hentriacontane	co
heptacosane	co, fo
n-heptadecane	co, fo
n-heptane	co, g, hog
heptanoic acid	co,
l-heptene	g
cis-2-heptene	g
cis-3-heptene	g
trans-2-heptene	g
trans-3-heptene	g
heptylthiophane	co
hexacosane	co, fo

TABLE 9.16   Organic Chemicals Identified in Various Petroleum Hydrocarbons[a] (cont.)

n-hexadecane	co, fo, k
hexadecanoic acid	co
hexadecylthiophane	co
n-hexane	co, g, hog, ug
2-hexanethiol	co
3-hexanethiol	co
n-hexanoic acid	co
l-hexene	g, hog
cis-2-hexene	g
cis-3-hexene	g
trans-2-hexene	g
trans-3-hexene	g
hexylthiophane	co
hopanes	co
indan	co, g
indans	fo
indanol	fo
indene	g
isobutane	co, g
isobutylbenzene	co, g
isobutylcyclohexane	g
isobutylcyclopentane	g
isooctylthiophane	co
isopentylbenzene	g
isoprenoids	co
isopropyl alcohol	g
isopropylbenzene	co, fo, g
isopropylcyclohexane	g
isopropylcyclopentane	co, g
isoquinoline	fo
2-isopropylphenol	fo
methane	co
methanethiol	co
methanol	fo
8-methyl-1,2-benzofluorene	co
methylbicyclo-[2.2.1]-heptane	co
3-methylbiphenyl	co
2-methyl-1,3-butadiene	g
2-methylbutane	co, g
3-methylbutanoic acid	co
2-methyl-1-butanethiol	co
2-methyl-2-butanethiol	co
3-methyl-1-butanethiol	co
3-methyl-2-butanethiol	co
2-methyl-1-butene	g
2-methyl-2-butene	g

TABLE 9.16 Organic Chemicals Identified in Various Petroleum Hydrocarbons[a] (cont.)

3-methyl-1-butene	g
1-methyl-3-n-butylbenzene	co
1-methyl-3-t-butylbenzene	g
l-methyl-4-t-butylbenzene	g
methyl-t-butyl ether	g
1-methylchrysene	co
methylcyclohexane	co, g, hog, ug
4-methylcyclohexanecarboxylic acid	co
1-methylcyclohexene	g
4-methylcyclohexene	g
methylcyclopentane	co, g, hog, ug
2-methylcyclopentanecarboxylic acid	co
3-methylcyclopentanecarboxylic acid	co
cis-2-methylcyclopentanethiol	co
1-methylcyclopentene	g
3-methylcyclopentene	g
3-methylcyclopentylacetic acid	co
2-methylcyclopentylpropanoic acid	co
4-methyldecane	g
1-methyl-2-ethylbenzene	co, g, k
1-methyl-3-ethylbenzene	co, g, k, ug
1-methyl-4-ethylbenzene	co, g, k, ug
1-methyl-1-ethylcyclopentane	co, g
1-methyl-cis-2-ethylcyclopentane	co, g
1-methyl-trans-2-ethylcyclopentane	co, g
1-methyl-cis-3-ethylcyclopentane	co, g
1-methyl-trans-3-ethylcyclopentane	co, g
2-methyl-3-ethylhexane	g
2-methyl-4-ethylhexane	g
2-methyl-3-ethylpentane	co, g
3-methyl-3-ethylpentane	co, g
3-methyl-5-ethylpyridine	co
2-methyl-8-ethylquinoline	co
2-methylheptane	co, g, hog
3-methylheptane	co, g, hog
4-methylheptane	co, g, hog
2-methyl-1-heptene	g
2-methyl-2-heptene	g
6-methyl-1-heptene	g
2-methylhexane	co, g, hog
3-methylhexane	co, g, hog, ug
2-methylhexanoic acid	co
3-methylhexanoic acid	co
4-methylhexanoic acid	co
5-methylhexanoic acid	co
2-methyl-1-hexene	g
2-methyl-2-hexene	g
2-methyl-cis-3-hexene	g

TABLE 9.16 Organic Chemicals Identified in Various Petroleum Hydrocarbons[a] (cont.)

2-methyl-trans-3-hexene	g
3-methyl-1-hexene	g
3-methyl-cis-2-hexene	g
3-methyl-trans-2-hexene	g
3-methyl-cis-3-hexene	g
3-methyl-trans-3-hexene	g
4-methyl-1-hexene	g
4-methyl-cis-2-hexene	g
4-methyl-trans-2-hexene	g
5-methyl-1-hexene	g
5-methyl-cis-2-hexene	g
1-methyl-2-isopropylbenzene	co, g
1-methyl-3-isopropylbenzene	co, g
1-methyl-4-isopropylbenzene	co, g
1-methyl-cis-4-isopropylcyclohexane	g
1-methyl-trans-4-isopropylcyclohexane	g
1-methylindan	co, g, k
2-methylindan	co, g, k
4-methylindan	co, g, k
5-methylindan	g
1-methylnaphthalene	co, fo, k
2-methylnaphthalene	co, fo, g, k
methylnaphthalenes	fo
2-methylnonane	co, g
3-methylnonane	co, g
4-methylnonane	co, g
5-methylnonane	co, g
2-methyl-1-nonene	g
2-methyloctane	co, g, ug
3-methyloctane	co, g, ug
4-methyloctane	co, g, ug
2-methyl-1-octene	g
2-methyl-2-octene	g
2-methylpentane	co, g, hog, ug
3-methylpentane	co, g, hog, ug
2-methylpentanoic acid	co
3-methylpentanoic acid	co
4-methylpentanoic acid	co
2-methyl-2-pentanethiol	co
2-methyl-3-pentanethiol	co
3-methyl-3-pentanethiol	co
4-methyl-2-pentanethiol	co
2-methyl-1-pentene	g, ug
2-methyl-2-pentene	g, ug
3-methyl-1-pentene	g
3-methyl-2-pentene	g
cis-3-methyl-2-pentene	g
trans-3-methyl-2-pentene	g

TABLE 9.16 Organic Chemicals Identified in Various Petroleum Hydrocarbons[a] (cont.)

4-methyl-1-pentene	g
4-methyl-cis-2-pentene	g
4-methyl-trans-2-pentene	g
2-methylpropane	g, hog
2-methyl-1-propanethiol	co
2-methyl-2-propanethiol	co
2-methylpropanoic acid	co
2-methylpropene	g
1-methyl-2-n-propylbenzene	co, g, k
1-methyl-3-n-propylbenzene	g
1-methyl-4-n-propylbenzene	co, g, k
1-methylpyrene	co
3-methylpyridine	co
2-methylquinoline	co
2-methylstyrene	g
2-methyl-1,2,3,4-tetrahydronaphthalene	co
5-methyl-1,2,3,4-tetrahydronaphthalene	co
6-methyl-1,2,3,4-tetrahydronaphthalene	co
1-methyltetralin	k
2-methyltetralin	k
4-methyl-2-thia[0.3.3]-bicyclononane	co
3-methyl-2-thia[0.3.3]-bicyclooctane	co
methyl-8-thia[3.2.1]-bicyclooctane	co
3-methyl-2-thiabutane	co
2-methylthiacyclopentane	co
3-methylthiacyclopentane	co
4-methylthiacyclopentane	co
2-methyl-3-thiahexane	co
4-methyl-3-thiahexane	co
2-methyl-3-thiapentane	co
2-methylthiophene	co
3-methylthiophene	co
naphthalene	co, fo, g, k
naphthalenes	fo
2-naphthol	co
n-nonacosane	co
n-nonadecane	co, fo
n-nonane	co, fo, g, k
nonanoic acid	co, fo
nonanol	g
1-nonene	g
nonhopanes	co
nonylthiophane	co
n-octacosane	co
n-octadecane	co, fo
octadecanoic acid	co

TABLE 9.16 Organic Chemicals Identified in Various Petroleum Hydrocarbons[a] (cont.)

octadecylthiophane	co
n-octane	co, fo, g, hog, k
n-octanoic acid	co
2-octanethiol	co
octanoic acid	fo
1-octene	g
cis-2-octene	g
trans-2-octene	g
trans-4-octene	g
octylthiophene	co
pentacosane	co, fo
n-pentadecane	co, fo, k
1,4-pentadiene	g
pentamethylbenzene	g
n-pentane	co, fo, g, hog, ug
1-pentanethiol	co
2-pentanethiol	co
3-pentanethiol	co
pentanoic acid	co
1-pentene	g, hog, ug
cis-2-pentene	g, hog, ug
trans-2-pentene	g, hog, ug
n-pentylbenzene	g
n-pentylcyclohexane	g
n-pentylcyclopentane	g
pentylthiophane	co
perylene	co
phenanthrene	fo
phenol	co, fo
polynaphthenes	fo
porphyrins	co
propane	co, g, hog
1-propanethiol	co
2-propanethiol	co
propanoic acid	co
n-propylbenzene	co, g
propylcyclohexane	g
n-propylcyclopentane	co, g
4-n-propylheptane	g
quinoline	fo
steranes	co
n-tetracosane	co, fo
n-tetradecane	co, fo, k
tetradecanoic acid	co
tetradecylthiophane	co
1,2,3,4-tetrahydronaphthalene	co, g
5,6,7,8-tetrahydroquinoline	co
tetralin	g, k

TABLE 9.16 Organic Chemicals Identified in Various Petroleum Hydrocarbons[a] (cont.)

tetralins	fo
1,3,5,7-tetramethylanthracene	co
1,3,6,7-tetramethylanthracene	co
2,3,6,7-tetramethylanthracene	co
1,2,3,4-tetramethylbenzene	co, g, k
1,2,3,5-tetramethylbenzene	co, fo, g, k
1,2,4,5-tetramethylbenzene	co, fo, g, k
2,2,3,3-tetramethylbutane	co
1,1,3,3-tetramethylcyclopentane	co
1,1-cis-2-cis-3-tetramethylcyclopentane	co
1,1-cis-2-trans-3-tetramethylcyclopentane	co
1,1-cis-2-trans-4-tetramethylcyclopentane	co, g
1,1-cis-3-trans-4-tetramethylcyclopentane	co
1-cis-2-trans-3-cis-4-tetramethylcyclopentane	co
1-trans-2-cis-3-trans-4-tetramethylcyclopentane	co
1-trans-2-trans-3-cis-4-tetramethylcyclopentane	co
2,2,3,3-tetramethylhexane	g
2,6,10,14-tetramethylhexadecane	fo
1,4,5,7-tetramethylnaphthalene	co
2,3,6,7-tetramethylnaphthalene	co
2,6,10,14-tetramethylpentadecane	fo
2,2,3,4-tetramethylpentane	g
2,2,4,4-tetramethylpentane	co
2,3,4,8-tetramethylquinoline	co
2,3,4,5-tetramethylthiophene	co
thiaadamantane	co
6-thia[0.3.4]bicyclononane	co
9-thia[3.3.1]bicyclononane	co
2-thia[2.2.2]bicyclooctane	co
2-thia[3.3.0]bicyclooctane	co
cis-3-thia[3.3.0]bicyclooctane	co
2-thia[3.2.1]bicyclooctane	co
3-thia[3.2.1]bicyclooctane	co
6-thia[3.2.1]bicyclooctane	co
8-thia[3.2.1]bicyclooctane	co
2-thiabutane	co
thiacyclohexane	co
thiacyclopentane	co
5-thiadecane	co
3-thiaheptane	co
2-thiahexane	co
3-thiahexane	co
cis-1-thiahydrindane	co
trans-1-thiahydrindane	co
thianaphthene	co
5-thianonane	co
3-thiaoctane	co
2-thiapentane	co

TABLE 9.16 Organic Chemicals Identified in Various Petroleum Hydrocarbons[a] (cont.)

3-thiapentane	co
2-thiapropane	co
7-thiatridecane	co
6-thiaundecane	co
thiophene	co
toluene	co, fo, g, hog, ug
n-triacontane	co
n-tricosane	co, fo
tricyclic terpanes	co
n-tridecane	co, fo, k
triethylphenol	co
2,3,6-trimethylanthracene	co
1,2,3-trimethylbenzene	co, g, k
1,2,4-trimethylbenzene	co, g, k, ug
1,3,5-trimethylbenzene	co, g, k
2,2,3-trimethylbutane	co, g, hog
2,3,3-trimethyl-1-butene	g
1,1,2-trimethylcyclohexane	g
1,1,3-trimethylcyclohexane	co, g
1,1,4-trimethylcyclohexane	g
1-cis-2-trans-3-trimethylcyclohexane	g
1-trans-2-cis-3-trimethylcyclohexane	g
1-trans-2-cis-4-trimethylcyclohexane	co, g
1-trans-2-trans-4-trimethylcyclohexane	co, g
1-cis-3-cis-5-trimethylcyclohexane	g
1-cis-3-trans-5-trimethylcyclohexane	g
cis-2,2,6-trimethylcyclohexanecarboxylic acid	co
trans-2,2,6-trimethylcyclohexanecarboxylic acid	co
2-(2,2,6-trimethylcyclohexyl)-4,6-dimethylpyridine	co
1,1,2-trimethylcyclopentane	co, g
1,1,3-trimethylcyclopentane	co, g
1-cis-2-cis-3-trimethylcyclopentane	co, g
1-cis-2-cis-4-trimethylcyclopentane	co, g
1-cis-2-trans-3-trimethylcyclopentane	co, g
1-cis-2-trans-4-trimethylcyclopentane	co, g
1-trans-2-cis-3-trimethylcyclopentane	co, g
1-trans-2-cis-4-trimethylcyclopentane	co, g
1,1,2-trimethylcyclopentanecarboxylic acid	co
1,2,2-trimethylcyclopentane-1,3-dicarboxylic acid	co
3,3,4-trimethylcyclopentylacetic acid	co
2,6,10-trimethyldodecane	fo
1,2,3-trimethyl-4-ethylbenzene	g
1,2,3-trimethyl-5-ethylbenzene	g
1,2,4-trimethyl-3-ethylbenzene	g
1,2,4-trimethyl-5-ethylbenzene	g
1,2,5-trimethyl-3-ethylbenzene	g
1,3,5-trimethyl-2-ethylbenzene	g
2,3,8-trimethyl-4-ethylquinoline	co

TABLE 9.16 Organic Chemicals Identified in Various Petroleum Hydrocarbons[a] (cont.)

2,3,4-trimethyl-8-ethylquinoline	co
2,3,4-trimethyl-5-ethylthiophene	co
2,2,4-trimethylheptane	g
2,2,5-trimethylheptane	g
2,2,5-trimethylheptane	g
2,2,6-trimethylheptane	g
2,4,4-trimethylheptane	g
2,4,5-trimethylheptane	g
2,5,5-trimethylheptane	g
3,3,4-trimethylheptane	g
3,3,5-trimethylheptane	g
3,4,4-trimethylheptane	g
3,4,5-trimethylheptane	g
2,2,3-trimethylhexane	g
2,2,4-trimethylhexane	co, g
2,2,5-trimethylhexane	co, g
2,3,3-trimethylhexane	g
2,3,5-trimethylhexane	co, g
2,4,4-trimethylhexane	g
3,4,4-trimethyl-2-hexene	g
2,3,4-trimethyl-8-isopropylquinoline	co
1,2,5-trimethylnaphthalene	co
1,2,6-trimethylnaphthalene	co
1,2,7-trimethylnaphthalene	co
1,2,8-trimethylnaphthalene	co
1,3,5-trimethylnaphthalene	co
1,3,7-trimethylnaphthalene	co
1,3,8-trimethylnaphthalene	co
1,6,7-trimethylnaphthalene	co
2,3,6-trimethylnaphthalene	co
2,2,3-trimethylpentane	g, hog, ug
2,2,4-trimethylpentane	co, g, hog, ug
2,2,5-trimethylpentane	g, hog, ug
2,3,3-trimethylpentane	co, g, hog, ug
2,3,4-trimethylpentane	co, g, hog, ug
2,3,6-trimethylphenol	fo
2,4,6-trimethylphenol	fo
2,4,4-trimethyl-1-pentene	g
2,4,4-trimethyl-2-pentene	g
1,2,8-trimethylphenanthrene	co

TABLE 9.16   Organic Chemicals Identified in Various Petroleum Hydrocarbons[a] (cont.)

2,3,4-trimethyl-8-n-propylquinoline	co
2,3,4-trimethylpyridine	co
2,3,5-trimethylpyridine	co
2,3,8-trimethylquinoline	co
2,4,8-trimethylquinoline	co
2,3,5-trimethylthiacyclopentane	co
2,3,4-trimethylthiophene	co
triterpanes	co
n-undecane	co, fo, g, k
undecylthiophane	co
vinyl-2-ethylhexylether	g
m-xylene	co, fo, g, hog, ug
o-xylene	co, fo, g, hog, ug
p-xylene	co, fo, g, hog, ug
1,2,3-xylenol	co
1,2,4-xylenol	co
1,3,4-xylenol	co
1,3,5-xylenol	co

a - compiled from information in references 104, 105, 106, 107, & 108.
b - abbreviations: co = crude oil
           fo = fuel oil
           g   = gasoline
           hog = high octane gasoline
           k   = kerosene
           ug = unleaded gasoline

TABLE 9.17 Composition of Nutrient Solutions Utilized in In-situ Groundwater Bioreclamation Field and Pilot Scale Studies Utilizing Air Sparging and Naturally-Occurring Soil Microorganisms.

Chemical(s)	Nutrients	Results	Reference
acetone (11,000 lbs) n-butyl alcohol (67,000 lbs) dimethylaniline (26,000 lbs) methylene chloride (182,000 lbs)	0.9 mg/l $CaCl_2$ 0.9 mg/l $CaCO_3$ 0.45 mg/l $FeSO_4$ 410[a] & 15 mg/l K_2HPO_4 270[a] & 44 mg/l KH_2PO_4 18 mg/l $MgSO_4$ 18[a] & 1.8 mg/l $MnSO_4$ 9 mg/l Na_2CO_3 500[a] mg/l NH_4Cl	95% removal of organics in 3 yrs; methylene chloride reduced from 70-90 mg/l to 0.02-1.0 mg/l in one yr; acetone reduced from 54 mg/l to <1 mg/l in one yr.	109 110
high octane gasoline	10 mg/l $CaCl_2 \cdot 2H_2O$ 5 mg/l $FeSO_4 \cdot 7H_2O$ 400 mg/l KH_2PO_4 200 mg/l $MgSO_4 \cdot 7H_2O$ 20 mg/l $MnSO_4 \cdot H_2O$ 100 mg/l Na_2CO_3 600 mg/l Na_2HPO_4 10 mg/l NH_4NO_3	no gasoline in producing well water 10 months after biostimulation.	111 112 113
1,1,1-TCA	0.03 mg/l $CaCl_2 \cdot 2H_2O$ 20 µg/l $CoCl_2 \cdot 6H_2O$ 0.8 mg/l $FeCl_3 \cdot 6H_2O$ 0.002 mg/l H_3BO_3 11 mg/l KH_2PO_4 14 mg/l K_2HPO_4 4 mg/l $MgCl_2 \cdot 6H_2O$ 0.02 mg/l $MnCl_2 \cdot 4H_2O$ 480 mg/l $NaHCO_3$ 0.4 µg/l $NaMoO_2 \cdot 2H_2O$ 21 mg/l NH_4Cl 212 mg/l $(NH_4)_2SO_4$ 2 µg/l $NiCl_2 \cdot 6H_2O$ 0.001 mg/l $ZnCl_2$ 1.2 g/l glucose and/or 500-1000 ppm ethanol to maintain C/N ratio at 10/1.	1-4 mg/l TCA removed to below detection limit (20µg/l).	114

[a] excess concentration utilized at onset only.

face technology, in-situ bioreclamation results in minimal air pollution problems and minimal worker safety problems. Energy requirements usually are not high because it depends upon naturally-occurring reactions to remove the chemical in question.

In-situ groundwater bioreclamation, like any treatment technology, has several potential disadvantages. First, it cannot be utilized for all organic chemicals because (a) some organic chemicals may degrade slowly under site-specific conditions, or (b) the biodegradation products of the chemical(s) of concern may cause taste and odor problems. Second, the concentration of the organic chemical may have to be lowered by another treatment technology before biodegradation can commence. Third, the organic chemical of concern may not be degraded to a concentration sufficiently low to eliminate an adverse effect; as a result, an additional treatment process(es) may be needed in order to form a treatment train that achieves the desired level of removal of the chemical from groundwater. Fourth, this technology may not work well in soil-groundwater systems with low hydraulic conductivities due to inadequate migration and circulation rates of nutrients and micro-organisms; in general, the less permeable the aquifer, the less applicable this technology will be. Fifth, if a long time period is needed for complete treatment, operation and maintenance costs may make this technology less cost-competitive with other technologies.

Another commonly-cited disadvantage of in-situ bioreclamation is its inability to handle organic chemicals in the unsaturated zone. One approach to diminish this disadvantage is to establish a "surface infiltration gallery" by discharging treated producing-well water via one of several irrigation systems at the soil surface. Treated water will percolate through the unsaturated-zone soil and supply nutrients and oxygen to stimulate the metabolism of organisms in the unsaturated zone. Also, the percolating water will aid the flushing of organic chemicals from the unsaturated zone into groundwater which is eventually circulated through the bioreclamation system.

Composting. Composting is the controlled biodegradation of the organic fraction of a solid waste pile. It has been practiced for centuries in the Orient and has been utilized in the developed countries to degrade many different types of waste (see Table 9.18). In principle, any industrial or municipal waste containing metabolizable organic chemicals is treatable by composting.

Compost systems are classified on the bases of oxygen usage, temperature, and technological approach. Compost systems can be aerobic or anaerobic. With regard to temperature, mesotrophic composting systems operate in the 15° to 25° Centigrade range, while thermophilic composting systems operate in the 45° to 65° Centigrade range. With regard to technical approach, composting systems can be operated as (a) open, or (b) mechanical systems,

TABLE 9.18　Chemicals and Wastes Degraded by Composting.

brewery wastes
cannery wastes
cotton gin wastes
crude oil
emulsions
municipal solid waste
oily sludge
oily soil
oily wastes
PCBs
pesticides
pharmaceutical wastes
polynuclear aromatic hydrocarbons
pulp & paper mill sludges
refinery sludge
RDX (an explosive)
sewage sludge
solvents
tannery waste
textile waste
TNT (an explosive)
tomato processing waste
waste oil
wood industry waste

also known as enclosed systems or in-vessel systems. In the open system, the composting mixture is either (a) stacked in a low, narrow, elongated heap known as a "windrow," or (b) stacked in a low-lying, rectangular-based pile known as a static pile. In the mechanical or closed system, the compost mixture resides in an enclosed digester.

Figure 9.4 is a general composting process flow diagram. During the first process step, the waste is mixed with a bulking agent and fertilizer. Because aerobic composting requires oxygen levels between five and fifteen percent, a bulking agent is sometimes added to provide sufficient pore space in the waste to maintain oxygen flow. Also, the bulking agent may help maintain moisture levels in the mixture between 40 to 70 percent.

Nutrients are needed to insure that metabolic activity is occurring at desirable levels. One of the more important nutrient considerations is the carbon-nitrogen ratio (C/N) ratio. The optimum C/N ratio range for most wastes is 20 to 25. The degradation rate decreases proportionately as the ratio deviates from the optimum C/N ratio range.

In some cases, organic material may be added to the waste in order to stimulate microorganisms. In an active compost mixture, the priming effect aids the degradation of many chemicals and wastes. The priming effect refers to a phenomenon in which chemicals and waste resistant to biodegradation are degraded to a greater extent when mixed with and when in close contact with chemicals and wastes which are more easily biodegraded.

In aerobic systems, aeration occurs in two ways, depending upon the type of open system utilized. In windrow systems, the compost mixture is aerated by mechanically turning over the windrow using a machine such as a front-end loader. In static pile systems, the aeration system consists of a series of perforated pipes underneath each pile which are connected to a pump that draws or blows air through the pipes.

Waste degradation commences after mixing and generally requires three to four weeks to complete for the majority of wastes. Afterwards, the compost can be cured for about 30 days during which further decomposition, stabilization, pathogen destruction, and degassing occurs. For wet wastes, a drying stage lasting from several days to several months may be needed.

The end product is a dry, inert product known as compost. It can be utilized as the bulking agent in static pile composting if it consists of (a) stable aggregates, (b) a metabolically low-energy material that contributes only a slight metabolic heat output in order to reserve the ventilation system's heat removal capacity for fresh waste/chemicals, and (c) a relatively dry material that can adsorb water from wet wastes and chemicals to improve porosity.

Compost can be sold to a number of users (see Table 9.19). Compost applications to soil, under controlled conditions, can enrich and replenish the soil with organic matter, trace minerals, and nutrients and can renovate the soil's physical, chemical, and biological properties.

It is most unfortunate that composting is not utilized to a much greater extent to treat hazardous and nonhazardous wastes and soil. It is a proven technology that has received significantly greater acceptance and use in European countries compared to the United States.

Land Treatment.   This treatment technology is also commonly known as land farming, land cultivation, land spreading, land disposal, land application, refuse farming, and sludge farming. The primary objective of land treatment is the degradation of a chemical or waste residing in topsoil by soil microorganisms. In principle, any industrial or municipal chemical or waste containing metabolizable organic chemicals is amenable to land treatment. The chemical or waste can be liquid, slurry, or solid in form. The chemical or waste of concern is either (a) spread on the soil surface, (b) injected or incorporated into the upper few centimeters of soil by mechanical manipulation such as plowing and tilling, or (c) sprayed, sprinkled, or deposited

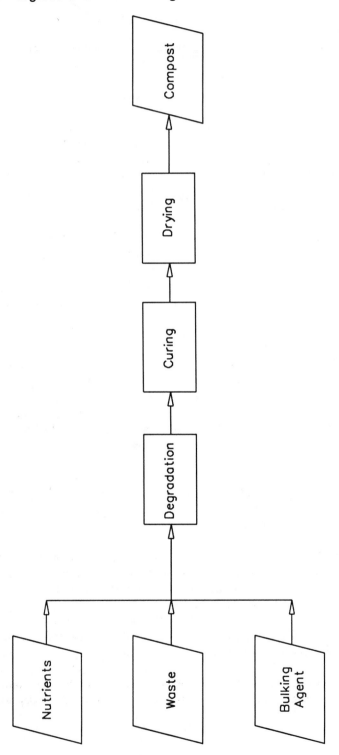

FIGURE 9.4 General composting process flow diagram.

TABLE 9.19 Major Users and Uses of Compost.

cemeteries
farmland
fertilizer companies
florists
fruit trees/orchards
gardens
golf courses
greenhouses
industrial park grounds
landfill cover
land reclamation (stip mines, sand &
                          gravel pits, etc.)
landscape contractors
military installations
nurseries
playgrounds
public parks & grounds
roadsides & median strips

by overland flow onto the soil surface.

The soil loading rate of the chemical or waste in a land treatment system depends upon (a) the physical/chemical properties of the system, and (c) the rate at which the chemical or waste can be degraded. The biodegradation and disappearance rates for various organic chemicals and wastes measured in several land treatment systems are reported in Table 9.20. It is most important to recognize that many of the chemicals and wastes listed in Table 9.20 are complex mixtures containing many different organic chemicals; for example, creosote is a derivative of coal tar which contains over 500 organic chemicals (see Table 9.21).

As with composting, nutrients are sometimes added in order to support the growth and metabolism of soil microorganisms and vegetation. Some wastes contain sufficient amounts of nutrients to support microorganisms and vegetation while others do not. It is most important to recognize that some wastes, such as municipal sludges (see Table 9.22), may contain very high amounts of nutrients and metals that may be toxic to soil microorganisms and vegetation. Also, the application of sludges with high amounts of metals to soil may result in the accumulation of these metals by horticultural and agricultural crops in amounts that may be harmful to man. Therefore, the design of land treatment systems should take into account the effect of metals

TABLE 9.20   Biodegradation/Disappearance Rates for Organic Chemicals and Wastes in Land Treatment Systems.

Chemical/Waste	Application Rate	Degradation/ Disappearance Rate	Reference
API oil-water separator sludge	8060 barrrels of sludge/acre/yr with 13.7% oil	880 barrels oil/- acre/yr	116
API separator sludge	16.5% ave. waste in soil over 16 mo.	aromatics: 0.007/day asphaltenes: 0.022/day polar compounds: 0.009/day saturates: 0.014/day	117
API separator sludge & tank bottoms & slop oil	5-10% waste in soil	degradation in 25 mo.: 30-58% for total oil, 44-71% for paraffins, 29-58% for aromatics, 9-37% for resins & asphaltenes.	118
asphalt	0.08% oil in soil	0.00038/day	119
creosote in soil	3-10% creosote	40% in 4 mo.	120
crude oil:			
Coastal	2.5% oil in soil	0.002/day	121
Heavy Arabian	2.9% oil in soil	0.0034/day	121
Saudi Arabian	10% oil in soil	60% in 60 days	122
Redwater, Alberta	6.5% oil in soil (1974, Ellerslie)	0.0019/day	123
	4.3% oil in soil (1975, Ellerslie)	0.00056/day	
	5.3% oil in soil (1974, Redwater)	0.0017/day	
	3.6% oil in soil (1975, Redwater)	0.00032/day	
crude oil spill	saturated soil	0.00014-0.00040/day for unfertilized soil; 0.00060-0.0037/day for fertilized soil.	124
home heating oil No. 2	1.9% oil in soil	0.006/day	121

TABLE 9.20 Biodegradation/Disappearance Rates for Organic Chemicals and Wastes in Land Treatment Systems. (cont.)

Chemical/Waste	Application Rate	Degradation/ Disappearance Rate	Reference
hydrocarbon residues from crude oil storage tanks	14.5g/kg soil	50.4% in 833d	125
kerosene spill	saturated soil	100% in 2 yrs.	126
oil refinery sludge (22% extractable oil)	3-6% waste in soil	100% in 1 mo; slower in proceeding months with reapplication	127
oil refinery waste	0.6-1.5 mg waste/- gram soil	73-85% in 4-6 wks.	128
oil refinery waste (28.7% oil)	>25% oil in soil over 10 yrs	degradation over 10 yrs: 64% for total waste, 99.1% for benzene, 30.4% for chrysene, 99.6% for ethylbenzene, 83.6% for 1-methyl- naphthalene, 95.2% for naphthalene, 75.3% for phenanthrene, 80.9% for pyrene, 99.9% for toluene, 99.6% for m-xylene, 99.9% for o-xylene, 99.9% for p-xylene.	129
oil-water emulsions (35-45% oil)	340,000 gal./acre	78% in 6 mo.	130
oily centrifuge sludge	6990 barrels/- acre/yr with 18% oil	>1000 barrels oil/- acre/year	116
oily sludge	120 to 200 cubic meters/hectare (8-10% oil in soil)	>cubic meters/- hectare/yr	131
POTW sludge	18 dry tons/acre	>99% in 8 yrs	132
residual fuel oil No. 2	3.5% oil in soil	0.0015/day	121

TABLE 9.20   Biodegradation/Disappearance Rates for Organic Chemicals and Wastes in Land Treatment Systems. (cont.)

Chemical/Waste	Application Rate	Degradation/ Disappearance Rate	Reference
tank bottoms (75% water, 2% solids, 23% hydrocarbons)	1.45% oil in soil	0.00083/day	133
tank bottoms (crude oil)	14% oil in soil 21% oil in soil 23% oil in soil	0.0028/day 0.0026/day 0.0018/day	134
used crankcase oil: car diesel	3.8% oil in soil 3.9% oil in soil	0.0047/day 0.004/day	121 121
waste cooling oil- water emulsions (0.5% hydrocarbons)	0.5 in./acre/wk via spray irriga- tion	>97% over 5 yrs	135
waste industrial oil	approx. 3% oil in soil	50% in 2 years; slower rates during next 3 yrs.	136
waste vacuum pump oil	9.9% oil in soil (first application: 0-50 days), 3.8% oil in soil (se- cond application: 51-116 days)	0.011/day for total C loss, first appli- cation; 0.040/day for oil loss, first appli- cation; 0.004/day for total C loss, second application; 0.034/day for oil loss, second application.	137
waxy cake: covered with black plastic uncovered	0.04% oil in soil 0.04% oil in soil	0.014/day 0.0055/day	119 119

on crops, soils, and the food chain. As with any other treatment technology, land treatment systems must be properly designed, constructed, operated, and maintained in order to produce successful results.

Other factors affecting biodegradation rates in land treatment systems include those factors discussed earlier in this chapter. Because most soil microorganisms are mesophiles, they exhibit maximum growth and activity in the 20 to 30°C range; therefore, they function best in warmer weather.

TABLE 9.21   Organic Chemicals Identified as Components of Coal Tar.

acenaphthene
1-acenaphthol
acenaphthylene
acephenanthrene
acephenanthrylene
acetaldehyde
acetic acid
acetone
acetonitrile
acetophenone
acridine
adipate esters
aliphatic esters $>C-9$
alkanes $>C-6$
alkyl acenaphthols
alkyl acridines
alkyl alcohols $>C-6$
1-aminonaphthalene
2-aminonaphthalene
aminotetralin
aminotoluene
aniline
anisoles
anthanthrene
anthracene
9,10-anthraquinone
aza-acenaphthylene
azabenzofluorene
1-azacarbazole
1-azafluoranthene
7-azafluoranthene
4-azafluorene
7-azaindole
4-azapyrene
azulene

benz(e)acephenanthrylene
benz(a)acridine
benz(c)acridine
benzanilide
benz(a)anthracene
benzathrone
benzene
benzidine
benzindole

TABLE 9.21   Organic Chemicals Identified as Components of Coal Tar. (cont.)

benzo(a)carbazole
11H-benzo(a)carbazole
benzo(b)carbazole
5H-benzo(b)carbazole
7H-benzo(c)carbazole
4H-benzo(def)carbazole
benzo(b)chrysene
benzo(b)fluoranthene
benzo(ghi)fluoranthene
benzo(j)fluoranthene
benzo(k)fluoranthene
benzo(a)fluorene
11H-benzo(a)fluorene
benzo(b)fluorene
11H-benzo(b)fluorene
benzo(c)fluorene
7H-benzo(c)fluorene
benzofluoreneamine
2,3-benzofuran
7-benzo(b)furanol
benzoic acid
benzo(a)naphthacene
benzo(b)naphtho(2,1-d)furan
benzo(b)naphtho(2,3-d)furan
benzo(b)naphtho(1,2-d)thiophene
benzo(b)naphtho(2,1-d)thiophene
benzo(b)naphtho(2,3-d)thiophene
benzonitrile
benzo(a)pentacene
benzo(rst)pentaphene
benzo(ghi)perylene
benzo(c)phenanthrene
benzo(c)picene
benzo(a)pyrene
benzo(e)pyrene
2,3-benzopyridine
3,4-benzopyridine
benzo(f)quinoline
benzo(h)quinoline
benzo(b)thiophene
benzo(k)xanthene
bicyclohexyl
2,2'-binaphthalene
2,2'-binaphthyl
biphenyl
3-(biphenyl)-4H-4-pyranone
1,2-butadiene

TABLE 9.21   Organic Chemicals Identified as Components of Coal Tar. (cont.)

1,3-butadiene
butane
butanoic acid
2-butanone
1-butene
2-butene
t-butylhydroquinone
3-n-butylphenol
4-n-butylphenol
4-sec-butylphenol
1-butyne
2-butyne

carbon disulfide
carbazole
9H-carbazole
chrysene
coronene
m-cresol
o-cresol
p-cresol
cyano-11H-benzo(b)fluorene
cyanofluorene
1-cyanonaphthalene
2-cyanonaphthalene
2-cyanotoluene
3-cyanotoluene
4-cyantoluene
4H-cyclopenteno(d,e,f)phenanthrene
1,3-cyclohexadiene
cyclohexane
cyclohexene
cyclopentadiene
cyclopentane
4H-cyclopenta(d,e,f)phenanthrene
cyclopentene

n-decane
dibenz(ac)anthracene
dibenz(ah)anthracene
dibenz(aj)anthracene
dibenzo(b,def)chrysene
13H-dibenzo(a,h)fluorene
dibenzo(b,d)furan

TABLE 9.21   Organic Chemicals Identified as Components of Coal Tar. (cont.)

dibenzo(b,d)-2-furanol
dibenzo(b,d)-3-furanol
dibenzo(b,d)thiophene
2,6-di-t-butyl-4-methylphenol
dicyclopentadiene
1,2-diethylbenzene
3,5-diethylphenol
diethylsulfide
1,2-dihydroacenaphthylene
9,10-dihydroanthracene
9,10-dihydroacridine
2,3-dihydro-1H-benz(e)indene
2,3-dihydro-1H-idene
5,12-dihydronaphthacene
1,4-dihydronaphthalene
2,3-dihydro-1H-phenalene
9,10-dihydrophenanthrene
6,7-dihydro-2-methyl-5H-1-pyridine
6,7-dihydro-3-methyl-5H-1-pyridine
6,7-dihydro-5H-1-pyridine
6,7-dihydro-5H-2-pyridine
1,2-dihydroxybenzene
1,3-dihydroxybenzene
2,2'-dihydroxybiphenyl
2,4-dihydroxytoluene
2,6-dihydroxytoluene
3,4-dihydroxytoluene
di-isopropylhydroquinone
2,3-dimethylaniline
2,4-dimethylaniline
2,5-dimethylaniline
2,6-dimethylaniline
3,4-dimethylaniline
3,5-dimethylaniline
2,3-dimethylanthracene
2,6-dimethylanthracene
2,7-dimethylanthracene
9,10-dimethylanthracene
4,7-dimethylbenzo(b)furan
5,6-dimethylbenzo(b)furan
5,7-dimethylbenzo(b)furan
3,3'-dimethylbiphenyl
3,4'-dimethylbiphenyl
3,5-dimethylbiphenyl
4,4'-dimethylbiphenyl
2,3-dimethyl-1,3-butadiene
2,2-dimethyl-3-butyne
1,3-dimethylcyclohexane

TABLE 9.21   Organic Chemicals Identified as Components of Coal Tar. (cont.)

1,4-dimethylcyclohexane
1,1-dimethylcyclopentane
4,6-dimethyldibenzo(b,d)furan
4,6-dimethyldibenzo(b,d)thiophene
2,3-dimethyl-6-ethylpryidine
dimethyl-4-indanol
4,5-dimethylindene
4,6-dimethylindene
4,7-dimethylindene
5,6-dimethylindene
5,7-dimethylindene
6,7-dimethylindene
1,3-dimethylisoquinoline
1,2-dimethylnaphthalene
1,3-dimethylnaphthalene
1,4-dimethylnaphthalene
1,5-dimethylnaphthalene
1,6-dimethylnaphthalene
1,7-dimethylnaphthalene
2,3-dimethylnaphthalene
2,6-dimethylnaphthalene
2,7-dimethylnaphthalene
3,6-dimethylphenanthrene
2,3-dimethylphenol
2,4-dimethylphenol
2,5-dimethylphenol
2,6-dimethylphenol
3,4-dimethylphenol
3,5-dimethylphenol
4,11-dimethylpicene
2,7-dimethylpyrene
2,3-dimethylpyridine
2,4-dimethylpyridine
2,5-dimethylpyridine
2,6-dimethylpyridine
3,4-dimethylpyridine
3,5-dimethylpyridine
2,3-dimethylquinoline
2,4-dimethylquinoline
2,6-dimethylquinoline
2,7-dimethylquinoline
2,8-dimethylquinoline
4,6-dimethylquinoline
4,7-dimethylquinoline
5,8-dimethylquinoline
6,8-dimethylquinoline

TABLE 9.21 Organic Chemicals Identified as Components of Coal Tar. (cont.)

dimethylsulfide
2,5-dimethylthiophene
3,5-dimethylxanthene
dinaphthofuran
dinaphtho(2,1-b:1',2'-d)thiophene

eicosane
ethanethiol
ethanol
ethylbenzene
ethylene glycol
ethylisocyanide
2-ethylnaphthalene
3-ethylpentane
2-ethylphenol
3-ethylphenol
4-ethylphenol
2-ethylpyridine
3-ethylpyridine
4-ethylpyridine
2-ethyltoluene
3-ethyltoluene
4-ethyltoluene

fluoranthene
fluorene
9H-fluorene
9H-fluorene-2-carbonitrile
2-fluorenol
heptadecane
n-heptane
1-heptene
hexadecane
hexadecanoic acid
hexamethylbiphenyl
n-hexane
1-hexene
1-hexyne
2-hexyne
3-hexyne
hydroxyanthracene
hydroxybenzofluorene
2-hydroxybenzoic acid
3-hydroxybenzoic acid

TABLE 9.21   Organic Chemicals Identified as Components of Coal Tar. (cont.)

4-hydroxybenzoic acid
2-hydroxybiphenyl
9-hydroxy-4-methylfluorene
5-hydroxy-1-methyl-3-isopropylbenzene
2-hydroxy-7-methylquinoline
2-hydroxyphenanthrene
4-hydroxyphenanthrene
2-hydroxypropanoic acid
4-hydroxythionaphthene
6-hydroxythionaphthene

indan
indano(2,1-a)indan
4-indanol
5-indanol
indene
indenofluoranthene
indenole
indeno(1,2,3-cd)pyrene
11H-indeno(1,2-b)quinoline
indole
6H-indolo(2,3-b)quinoline
2-isopropylphenol
3-isopropylphenol
4-isopropylphenol
isoquinoline

methanethiol
4-methoxybenzophenone
methylacenaphthene
methylacenaphthylene
methylacenaphthol
methylacetylene
2-methylacridine
methylaminoacenaphthylene
2-methylaniline
3-methylaniline
4-methylaniline
1-methylanthracene
2-methylanthracene
9-methylanthracene
methylbenzindole
1-methylbenzo(a)anthracene
6-methylbenzo(a)anthracene
9-methylbenzo(a)anthracene

TABLE 9.21 Organic Chemicals Identified as Components of Coal Tar. (cont.)

11-methylbenzo(a)anthracene
methylbenzofluoranthene
methylbenzofluoreneamine
2-methylbenzofuran
4-methylbenzo(b)furan
5-methylbenzo(b)furan
6-methylbenzo(b)furan
7-methylbenzo(b)furan
methylbenzo(n)naphtho(2,3-d)furan
3-methylbenzo(f)quinoline
2-methylbenzo(b)thiophene
3-methylbenzo(b)thiophene
2-methylbiphenyl
3-methylbiphenyl
4-methylbiphenyl
2-methyl-1,3-butadiene
2-methyl-2,3-butadiene
2-methyl-3-butyne
N-methylcarbazole
2-methylcarbazole
3-methylcarbazole
1-methylchrysene
6-methylchrysene
methylcyclohexane
methylcyclopentane
methylcyclopenteno(de)cinnoline
2-methyldibenzo(b,d)furan
3-methyldibenzo(b,d)furan
4-methyldibenzo(b,d)furan
2-methyldibenzothiophene
4-methyldibenzothiophene
1,1'-methylenebisbenzene
1-methyl-4-ethylbenzene
3-methyl-3-ethylpentane
2-methyl-3-ethylphenol
2-methyl-4-ethylphenol
2-methyl-5-ethylphenol
3-methyl-4-ethylphenol
3-methyl-5-ethylphenol
3-methyl-6-ethylphenol
4-methyl-2-ethylphenol
4-methyl-3-ethylphenol
1-methyl-1-ethyl-2-phenyl-2-tolylethylene
2-methyl-4-ethylpyridine
1-methylfluorene
2-methylfluorene

TABLE 9.21   Organic Chemicals Identified as Components of Coal Tar. (cont.)

3-methylfluorene
4-methylfluorene
9-methylfluorene
2-methyl-9H-carbazole
1-methyl-9H-fluorene
2-methyl-9H-fluorene
9-methyl-9H-fluorene
2-methylheptane
3-methylheptane
3-methylhexane
4-methylindan
1-methyl-5-indanol
3-methyl-4-indanol
3-methyl-5-indanol
4-methyl-5-indanol
5-methyl-4-indanol
6-methyl-4-indanol
6-methyl-5-indanol
7-methyl-4-indanol
7-methyl-5-indanol
4-methylindene
2-methylindole
3-methylindole
4-methylindole
5-methylindole
7-methylindole
methylisocyanide
1-methyl-2-isopropenylbenzene
1-methyl-4-isopropylbenzene
1-methylisoquinoline
3-methylisoquinoline
5-methylisoquinoline
6-methylisoquinoline
7-methylisoquinoline
8-methylisoquinoline
1-methylnaphthalene
2-methylnaphthalene
3-methyl-2-naphthol
3-methylnonane
4-methylnonane
3-methyloctane
2-methylpentane
methylphenanthracene
1-methylphenanthrene
2-methylphenanthrene
3-methylphenanthrene
9-methylphenanthrene
methylphenylnaphthalene

TABLE 9.21   Organic Chemicals Identified as Components of Coal Tar. (cont.)

2-methyl-1-phenylpropane
2-methyl-4-propylphenol
1-methylpyrene
2-methylpyrene
4-methylpyrene
2-methylpyridine
3-methylpyridine
4-methylpryidine
2-methylquinoline
3-methylquinoline
4-methylquinoline
5-methylquinoline
6-methylquinoline
7-methylquinoline
8-methylquinoline
2-methyltetrahydrothiophene
2-methylthionaphthene
2-methylthiophene
3-methylthiophene
methyltriphenylene
1-methyl-2-vinylbenzene
1-methyl-4-vinylbenzene

naphthacene
naphthalene
naphtho(1,2-b)chrysene
2',1':1,2-naphthofluorene
naphtho(1,2-b)furan
naphtho(2,1-b)furan
naphthoindole
1-naphthol
2-naphthol
naphtho(1,2-a)pyrene
naphthopyrrole
naphtho(1,2-b)thiophene
naphtho(2,1-b)thiophene
naphtho(2,3-b)thiophene
naphtho(2,1,8,7-klmn)xanthene
o-(2-naphthyl)-phenol
nonadecane
nonane

TABLE 9.21   Organic Chemicals Identified as Components of Coal Tar. (cont.)

octadecane
octadecanoic acid
cis-9-octadecenoic acid
octahydroanthracene
n-octane
5-oxo-4,5-dihydro-4-azapyrene

pentacene
1,3-pentadiene
n-pentane
pentanoic acid
pentaphene
1-pentene
2-pentene
1-pentyne
2-pentyne
perylene
phenanthrene
phenanthridine
5H-phenanthridin-6-one
phenanthrobenzpyrene
phenanthro(4,5-bcd)thiophene
phenol
1-phenylbenzanthrone
2-phenylbutane
1-phenylnaphthalene
2-phenylnaphthalene
2-phenylphenanthrene
1-phenylpropane
2-phenylpropane
1-phenyl-1-propanone
2-phenylpyridine
4-phenylpyridine
phthalate esters
phthalic esters
phytane
picene
pristane
propadiene
propanoic acid
propionamide
2-n-propylphenol
3-n-propylphenol
4-n-propylphenol
pyrene

TABLE 9.21 Organic Chemicals Identified as Components of Coal Tar. (cont.)

pyridine
2-pyridinecarboxylic acid
pyrrole
pyrrolo(def)phenanthrene

quinoline

styrene

tetradecane
1,2,3,4-tetrahydroacridine
1,2,3,4-tetrahydrofluoranthene
1,2,3,4-tetrahydronaphthalene
1,2,3,4-tetrahydroquinoline
1,2,3,4-tetramethylbenzene
1,2,3,5-tetramethylbenzene
1,2,4,5-tetramethylbenzene
tetramethylbiphenyl
1,2,3,5-tetramethylcyclohexane
2,3,6,7-tetramethylnaphthalene
2,3,5,6-tetramethylphenol
2,3,4,5-tetramethylpyridine
2,3,4,6-tetramethylpyridine
2,3,5,6-tetramethylpyridine
thiophene
thiophenol
thiopheno(def)phenanthrene
toluene
tri-n-butylamine
1,2,3-trihydroxybenzene
1,2,4-trihydroxynaphthalene
1,4,5-trihydroxynaphthalene
1,2,3-trihydroxypropane
1,2,3-trimethylbenzene
1,2,4-trimethylbenzene
1,3,5-trimethylbenzene
1,2,3-trimethylcyclohexane
1,2,4-trimethylcyclohexane
1,3,5-trimethylcyclohexane
1-trans-2-cis-trimethylcyclopentane
1,3,7-trimethylnaphthalene
1,6,7-trimethylnaphthalene
2,3,6-trimethylnaphthalene
2,2,4-trimethylpentane

1,2,8-trimethylphenanthrene
2,3,4-trimethylphenol
2,3,5-trimethylphenol
2,4,5-trimethylphenol
2,4,6-trimethylphenol
3,4,5-trimethylphenol
2,3,4-trimethylpyridine
2,3,5-trimethylpyridine
2,3,6-trimethylpyridine
2,4,5-trimethylpyridine
2,4,6-trimethylpyridine
3,4,5-trimethylpyridine
2,4,6-trimethylquinoline
2,4,8-trimethylquinoline
2,6,8-trimethylquinoline
trimethylthiophenes
triphenylene

xanthene
9H-xanthene
m-xylene
o-xylene
p-xylene
xylenol

Compiled from data in references 138, 139, 140, 141, & 142.

TABLE 9.22   Concentrations of Metals in Municipal Sludges.

Metal	Concentration (mg/kg)		
	Range	Mean	No. Samples
Al	300 - 110,000	24,435	397
Ag	0.13 - 150	11.4	234
As	0.007 - 109	10.7	272
B	<1 - 1400	105	266
Ba	2.25 - 2797	619	41
Be	<1 - 23	1.5	226
Cd	0.04 - 1200	57	404
Co	<1 - 250	20	396
Cr	0.36 - 15,167	596	406
Cu	0.03 - 13,380	998	406
Fe	52 - 116,000	23,838	402
Hg	0.006 - 1690	15	274
Mn	0.6 - 3800	519	398
Mo	6.0 - 3700	60	225
Ni	0.18 - 1300	138	405
Pb	2.1 - 7627	596	406
Sb	<1 - 1303	18	270
Se	0.07 - 390	12	44
Sn	25.4 - 492	205	17
Tl	0.13 - 89	26	4
Zn	4.4 - 24,000	2,243	406
pH	3.1 - 13.0	7.05	355

Compiled from data from References 143, 144, and 145.

Table 9.23   Organic Chemicals and Waste Types That Are
Degraded by Land Treatment Systems.

acid steel rolling mill sludge
agricultural wastes
API separator sludge
biosludge from organic chemical
   manufacturing processes
clay fines
coker blow down sludge
coke industry wastes
cooling oil
cooling tower sludge
crude oil
dairy processing wastes
dissolved air flotation froth
domestic wastewater
explosive manufacturing wastes
FCCU catalysts
fermentation wastes
food processing waste, wastewater, and sludges
fiber, hard, insulated, mineral,
   paper, and straw board waste
geothermal energy production
   wastes
heat exchanger bundle
   cleaning sludge
induced air flotation sludge
Kraft paper mill sludge
leather finishing industry
   wet scrubber sludges
lime sludges
municipal sludges and wastewater
munition waste
neutralized pickling liquor sludges
nitrogenous fertilizer
   manufacturing wastes
non-API separator sludge
oil-contaminated soil
oilfield brines
oilfield drilling muds and cuttings
oil-water emulsions
organic chemical manufacturing
   wastewater treatment sludges
paints and allied product wastes
paper mill waste

pesticide manufacturing wastes
pharmaceutical manufacturing
   wastes
plastics manufacturing wastewater treatment sludges
polycatalysts
POTW sludges
pulp mill waste
refinery oily wastes
refinery scale
septic tank sludge
slop oil emulsion solids
sodium cation exchange resins
spent acid sludge
spent grain liquors
steel rolling mill sludges
surfactants
synthetic fiber manufacturing
   byproducts
tank bottoms
tannery wastes
textile industry waste and
   wastewater
wood preserving industry waste
wool preserving and scouring
   waste

Soil microorganisms can function at optimum levels if soil pH in the land treatment system is in the range of 6 to 8. Because land treatment systems utilize soil at the land surface, aeration and oxygen supply are at most favorable levels to support microorganism metabolism and growth.

REFERENCES

1. Buchanan, R. E. and Gibbons, N. E. Bergey's Manual of Determinative Bacteriology. Baltimore, MD: Williams & Wilkins Co. (1974).
2. Bryan, H. H., Bryan, C. H., and Bryan, C. G. Bacteriology: Principles and Practice. New York: Barnes and Noble (1968).
3. Dorland's Illustrated Medical Dictionary. Philadelphia, PA: W. B. Saunders Co. (1974).
4. Brock, T. D. Biology of Microorganisms. Englewood Cliffs, N.J.: Prentice-Hall Inc. (1974).
5. Sax, N.I. Dangerous Properties of Industrial Materials. New York: Van Nostrand Reinhold Co. (1984).
6. Casalicchio, G. and Rossi, N. Ricerche sulla costituzione della frazione lipidica del suolo. Agrochimica **14**:505-515 (1970).
7. Stevenson, F. J. Organic Acids in Soil. *In* McLaren, A. D. and Peterson, G. H. Soil Biochemistry. New York: Marcel Dekker (1967).
8. Stevenson, F. J. and Ardakani, M. S. Organic Matter Reactions Involving Micronutrients in Soils. *In* Mortvedt, J. J., Giordano, P. M., and Lindsay, W. L. Micronutrients in Agriculture. Madison, WI: Soil Science Society of America (1972).
9. Moucawi, J., Fustec, E., and Jambu, P. Biooxidation of Added and Natural Hydrocarbons in Soils: Effect of Iron. Soil Biol. Biochem. **13**:335-342 (1981).
10. Wang, T.S.C., Hwang, P-T and Chen, C-Y. Soil Lipids Under Various Crops. Soil Science Society of America Proceedings **35**:584-587 (1971).
11. Eglinton, G. and Murphy, M. T. J. Organic Geochemistry. New York: Springer-Verlag (1969).
12. Jalal, M. A. F. and Read, D. J. The Organic Acid Composition of Calluna Heathland Soil with Special Reference to Phyto- and Fungitoxicity. Plant and Soil **70**:273-286 (1983).
13. Flaig, W. Organic Compounds in Soil. Soil Science **111**:19-33 (1971).
14. Stevenson, F. J. Humus Chemistry. New York: John Wiley (1982).
15. Sims, R. C. and Overcash, M. R. Fate of Polynuclear Aromatic Compounds (PNAs) in Soil-Plant Systems. Residue Reviews **88**:1-67 (1983).
16. Simonart, P. and Batistic, L. Aromatic Hydrocarbons in Soil. Nature **212**:1461-1462 (1966).
17. Khan, S. U. and Schnitzer, M. Sephadex Gel Filtration of Fulvic Acid: the Identification of Major Components in Two Low- Molecular Weight

Fractions. Soil Science **112**:231-238 (1971).

18. Wang, T. S. C. and Chuang, T-T. Soil Alcohols, Their Dynamics and Their Effect Upon Plant Growth. Soil Science **104**:40-45 (1967).

19. Whitehead, D. C., Dibb, H., and Hartley, R. D. Extractant pH and the Release of Phenolic Compounds from Soils, Plant Roots, and Leaf Litter. Soil Biol. Biochem. **13**:343-348 (1981).

20. Huang, P. M. and Schnitzer, M. Interactions of Soil Minerals with Natural Organics and Microbes. Madison, WI: Soil Science Society of America (1986).

21. Lambert, E. N., Seaforth, C. E., and Ahmad, N. The Occurrence of 2-Methoxy-1,4-Naphthoquinone in Caribbean Vertisols. Soil Science Society of America Proceedings **35**:463-464 (1971).

22. Morita, H. Phenolic Esters in Peat. Geoderma **13**:163-165 (1975).

23. Whitehead, D. C. Identification of p-Hydroxybenzoic, Vanillic, p-Coumaric and Ferulic Acids in Soils. Nature **202**:417-418 (1964).

24. Cheshire, M. V. Nature and Origin of Carbohydrates in Soils. New York: Academic Press (1979).

25. Brown, K. W., Donnelly, K. C., Thomas, J. C., and Davol, P. Mutagenicity of Three Agricultural Soils. The Science of the Total Environment **41**:173-186 (1985).

26. Sax, N. I. Dangerous Properties of Industrial Materials. Sixth Edition. New York: Van Nostrand Reinhold Co. (1984).

27. Dragun, J. Microbial Degradation of Petroleum Products in Soil. *In* Proceedings of a Conference on Environmental and Public Health Effects of Soils Contaminated with Petroleum Products, October 30-31, 1985, University of Massachusetts. New York: John Wiley & Sons (1988).

28. Sufita, J. M. and Bollag, J-M. Polymerization of Phenolic Compounds by a Soil-Enzyme Complex. Soil Science Society of America Journal **45**:297-302 (1981).

29. Liu, S-Y and Bollag, J. M. Enzymatic Binding of the Pollutant 2,6-Xylenol to a Humus Constituent. Water, Air, and Soil Pollution **45**:97-106 (1985).

30. Bumpus, J. A., Tien, M., Wright, D. S., and Aust, S. D. Biodegradation of Environmental Pollutants by the White Rot Fungus *Phanerochaete chrysporium*. *In* Incineration and Treatment of Hazardous Waste. Proceedings of the Eleventh Annual Research Symposium, Cincinnati, OH, April 29-May 1, 1985. EPA/600/9-85/028. Cincinnati, OH: U.S. Environmental Protection Agency (1985).

31. Pemberton, J. M. Genetic Engineering and Biological Detoxification of Environmental Pollutants. Residue Reviews **78**:1-11 (1981).

32. Kellogg, S. T., Chattersee, D. K., and Chakrabarty, A. M. Plasmid-

Assisted Molecular Breeding: New Technique for Enhanced Biodegradation of Persistent Toxic Chemicals. Science **214**: 1133-1135 (1981).

33. Barles, R. W., Daughton, C. G., and Hsieh, P. H. Accelerated Parathion Degradation in Soil Inoculated with Acclimated Bacteria under Field Conditions. Arch. Environ. Contam. Toxicol. **8**:647-660 (1979).

34. Brink, R. H. Biodegradation of Organic Chemicals in the Environment. *In* McKinney, J. D. (ed). Environmental Health Chemistry. Ann Arbor, MI: Ann Arbor Science (1981).

35. Hutzinger, O. The Handbook of Environmental Chemistry. Volume 2. Part A. New York: Springer-Verlag (1980).

36. Alexander, M. Biodegradation of Chemicals of Environmental Concern. Science **211**:132-138 (1981).

37. Goring, C. A. I. and Hamaker, J. W. Organic Chemicals in the Soil Environment. Volumes 1 & 2. New York: Marcel Dekker (1972).

38. Tabak, H. H., Quave, S. A., Mashni, C. I., and Barth, E. F. Biodegradability Studies with Organic Priority Pollutant Compounds. Journal Water Pollution Control Federation **53**:1503-1518 (1981).

39. Coover, M. P. and Sims, R. C. The Effect of Temperature on Polycyclic Aromatic Hydrocarbon Persistence in an Unacclimated Agricultural Soil. Hazardous Waste & Hazardous Materials **4**:69-82 (1987).

40. Wilson, J. T., McNabb, J. F., Cochran, J. W., Wang, T. H., Tomson, M. B., and Bedient, P. B. Influence of Microbial Adaptation on the Fate of Organic Pollutants in Groundwater. Environmental Toxicology and Chemistry **4**:721-726 (1985).

41. Bossert, I. D. and Bartha, R. Structure-Biodegradability Relationships of Polycyclic Aromatic-Hydrocarbons in Soil. Bull. Environ. Contam. Toxicol. **37**:490-495 (1986).

42. Pitter, P. Determination of Biological Degradability of Organic Substances. Water Research **10**:231-235 (1976).

43. Zoeteman, B. C. J., De Greef, E., and Brinkmann, F. J. J. Persistency of Organic Contaminants in Groundwater, Lessons from Soil Pollution Incidents in the Netherlands. The Science of the Total Environment **21**:187-202 (1981).

44. Ward, T. E. Characterizing the Aerobic and Anaerobic Microbial Activities in Surface and Subsurface Soils. Environmental Toxicology and Chemistry **4**:727-737 (1985).

45. Ventullo, R. M. and Larson, R. J. Metabolic Diversity and Activity of Heterotrophic Bacteria in Groundwater. Environmental Toxicology and Chemistry **4**:759-771 (1985).

46. Sims, G. K. and Sommers, L. E. Degradation of Pyridine Derivatives in Soil. Journal of Environmental Quality **14**:580-584 (1985).

47. Barker, J. F. and Patrick, G. C. Natural Attenuation of Aromatic Hydrocarbons in a Shallow Sand Aquifer. *In* Proceedings of the NWWA/API Conference on Petroleum Hydrocarbons and Organic Chemicals in Groundwater - Prevention, Detection, and Restoration, November 13-15, 1985, Houston, TX. Dublin, OH: National Water Well Association (1985).

48. Barker, J. F., Patrick, G. C., and Major, D. Natural Attenuation of Aromatic Hydrocarbons in a Shallow Sand Aquifer. Ground Water Monitoring Review 7:64-71 (1987).

49. Wilson, B. H., Smith, G. B., and Rees, J. F. Biotransformations of Selected Alkylbenzenes and Halogenated Aliphatic Hydrocarbons in Methanogenic Aquifer Material: A Microcosm Study. Environ. Sci. Tech. **20**:997-1002 (1986).

50. Wilson, J. T., McNabb, J. F., Balkwill, D. L., and Ghiorse, W. C. Enumeration and Characterization of Bacteria Indigenous to a Shallow Water-Table Aquifer. Ground Water **21**:134-142 (1983).

51. Bouwer, E. J. and McCarty, P. L. Transformations of 1- and 2-Carbon Halogenated Aliphatic Organic Compounds Under Methanogenic Conditions. Applied and Environmental Microbiology **45**:1286-1294 (1983).

52. Kappeler, Th. and Wuhrmann, K. Microbial Degradation of the Water-Soluble Fraction of Gas-Oil - II. Bioassays with Pure Strains. Water Research **12**:335-342 (1978).

53. Kappeler, Th. and Wuhrmann, K. Microbial Degradation of the Water-Soluble Fraction of Gas Oil - I. Water Research **12**:327-333 (1978).

54. Wilson, J. T., McNabb, J. F., Wilson, B. H., and Noonan, M. J. Biotransformation of Selected Organic Pollutants in Groundwater. Developments in Industrial Microbiology **24**:225-233 (1982).

55. Pignatello, J. J. Ethylene Dibromide Mineralization in Soils Under Anaerobic Conditions. Applied and Environmental Microbiology **51**:588-592 (1986).

56. Fogel, M. M., Taddeo, A. R., and Fogel, S. Biodegradation of Chlorinated Ethenes by a Methane-Utilizing Mixed Culture. Applied and Environmental Microbiology **51**:720-724 (1986).

57. Barrio-Lage, G., Parsons, F. Z., Nassar, R. J., and Lorenzo, P. A. Sequential Dehalogenation of Chlorinated Ethenes. Environ. Sci. Technol. **20**:96-99 (1986).

58. Moucawi, J., Fustec, E. and Jambu, P. Biooxidation of Added and Natural Hydrocarbons in Soils: Effect of Iron. Soil Biol. Biochem. **13**:335-342 (1981).

59. Roberts, P. V., Schreiner, J. E., and Hopkins, G. D. Field Study of Organic Water Quality Changes During Groundwater Recharge in the Palo Alto Baylands. Water Res. **16**:1025-1035 (1982).

60. Vogel, T. M. and McCarty, P. L. Biotransformation of Tetrachloro-

ethylene to Trichloroethylene, Dichloroethylene, Vinyl Chloride, and Carbon Dioxide Under Methanogenic Conditions. Applied and Environmental Microbiology **49**:1080-1083 (1985).

61. Parsons, F., Wood, P. R., and DeMarco, J. Transformations of Tetrachloroethene and Trichloroethene in Microcosms and Groundwater. Journal AWWA **76**:56-59 (1984).

62. Flathman, P. E. and Dahlgran, J. R. Correspondence on Anaerobic Degradation of Halogenated 1- and 2-Carbon Organic Compounds. Environ. Sci. Technol. **16**:130 (1982).

63. Strand, S. E. and Shippert, L. Oxidation of Chloroform in an Aerobic Soil Exposed to Natural Gas. Applied and Environmental Microbiology **52**:203-205 (1986).

64. Pinholt, Y., Struwe, S., and Kjoller, A. Microbial Changes During Oil Decomposition in Soil. Holartic Ecol. **2**:195-200 (1979).

65. Odu, C. T. I. Microbiology of Soils Contaminated with Petroleum Hydrocarbons. I. Extent of Contamination and Some Soil and Microbial Properties After Contamination. J. Inst. Petrol. **58**:201-208 (1972).

66. Fairbridge, E. A. and Finkl, C. W. Jr. (eds). The Encyclopedia of Soil Science. Part 1. Stroudsburg, PA: Dowden, Hutchinson, and Ross (1979).

67. Dalton, H. and Stirling, D. I. Co-Metabolism. Phil. Trans. R. Soc. Lond. B **297**:481-496 (1982).

68. Subba-Rao, R. V. and Alexander, M. Bacterial and Fungal Cometabolism of 1,1,1-Trichloro-2,2-bis(4-Chlorophenyl)ethane (DDT) and its Breakdown Products. Applied and Environmental Microbiology **49**:509-516 (1985).

69. Perry, J. J. Microbial Cooxidations Involving Hydrocarbons. Microbiological Reviews **43**:59-72 (1979).

70. Liu, D., Carry, J., and Thomson, K. Fulvic Acid Enhanced Biodegradation of Aquatic Contaminants. Bull. Environ. Contam. Toxicol. **31**:203-207 (1983).

71. Liu, D. Enhancement of PCBs Biodegradation by Sodium Ligninsulfonate. Water Research **14**:1467-1475 (1980).

72. Shimp, R. J. and Pfaender, F. K. Influence of Easily Degradable Naturally Occurring Carbon Substrates on Biodegradation of Monosubstituted Phenols by Aquatic Bacteria. Applied and Environmental Microbiology **49**:394-401 (1985).

73. You, I-S and Bartha, R. Stimulation of 3,4-Dichloroaniline Mineralization by Aniline. Applied and Environmental Microbiology **44**:678-681 (1982).

74. Brunner, W., Sutherland, F. H., and Focht, D. D. Enhanced Biodegradation of Polychlorinated Biphenyls in Soil by Analog Enrichment and Bacterial Inoculation. Journal of Environmental Quality **14**:324-328 (1985).

75. Shimp, R. and Pfaender, F. K. Influence of Naturally Occurring Humic Acids on Biodegradation of Monosubstituted Phenols by Aquatic Bacteria. Applied and Environmental Microbiology **49**:402-407 (1985).

76. Meyer, J. S., Marcus, M. D., and Bergman, H. L. Inhibitory Interactions of Aromatic Organics During Microbial Degradation. Environmental Toxicology and Chemistry **3**:583-587 (1984).

77. Schmidt, S. K. and Alexander, M. Effects of Dissolved Organic Carbon and Second Substrates on the Biodegradation of Organic Compounds at Low Concentrations. Applied and Environmental Microbiology **49**:822-827 (1985).

78. Bewley, R. J. F. and Stotzky, G. Effects of Cadmium and Simulated Acid Rain on Ammonification and Nitrification in Soil. Archives of Environmental Contamination and Toxicology **12**:285-291 (1983).

79. Bitton, G., Volk, B. G., Graetz, D. A., Bossart, J. M., Boylan, R. A., and Byers, G. E. Effect of Acid Precipitation on Soil Microbial Activity: II. Field Studies. Journal of Environmental Quality **14**:69-71 (1985).

80. Bewley, R. J. F. and Stozky, G. Simulated Acid Rain (H_2SO_4) and Microbial Activity in Soil. Soil Biol. Biochem **15**:425-429 (1983).

81. Bitton, G. and Boylan, R. A. Effect of Acid Precipitation on Soil Microbial Activity: I. Soil Core Studies. Journal of Environmental Quality **14**:66-69 (1985).

82. Atlas, R. M. Microbial Degradation of Petroleum Hydrocarbons: An Environmental Perspective. Microbiological Reviews **45**:180-209 (1981).

83. U.S. Environmental Protection Agency. Protection of Public Water Supplies From Groundwater Contamination. EPA/625/4-85/016. Cincinnati, OH: U.S. Environmental Protection Agency (1985).

84. Schnurer, J., Clarholm, M., Bostrum, S., and Rosswall, T. Effects of Moisture on Soil Microorganisms and Nematodes: A Field Experiment. Microbial Ecology **12**:217-230 (1986).

85. Orchard, V. A. and Cook, F. J. Relationship Between Soil Respiration and Soil Moisture. Soil Biol. Biochem. **15**:447-453 (1983).

86. Brown, K. W. and Donnelly, K. C. Influence of Soil Environment on Biodegradation of a Refinery and a Petrochemical Sludge. Environmental Pollution (Series B) **6**:119-132 (1983).

87. Cornfield, A. H. Effects of Addition of 12 Metals on Carbon Dioxide Release During Incubation of an Acid Sandy Soil. Geoderma **19**:199-203 (1977).

88. Bollag, J-M and Barabasz, W. Effect of Heavy Metals on the Denitrification Process in Soil. J. Environ. Qual. **8**:196-201 (1979).

89. Bewley, R. J. F. and Stotzky, G. Effects of Cadmium and Simulated Acid Rain on Ammonification and Nitrification in Soil. Arch. Environ.

Contam. Toxicol. **12**:285-291 (1983).

90. Debosz, K., Babich, H., and Stotzky, G. Toxicity of Lead to Soil Respiration: Mediation by Clay Minerals, Humic Acids, and Compost. Bull. Environ. Contam. Toxicol. **35**:517-524 (1985).

91. Bhuiya, M. R. H. and Cornfield, A. H. Incubation Study on Effect of pH on Nitrogen Mineralization and Nitrification in Soils Treated with 1,000 ppm Lead and Zinc as Oxides. Environ. Pollut. **7**:161-164 (1974).

92. Wood, J. M. and Wang, H-K. Microbial Resistance to Heavy Metals. Environmental Science & Technology **17**:582A-590A (1983).

93. Sanborn, J. R. Francis, B. M., and Metcalf, R. L. The Degradation of Selected Pesticides in Soil: A Review of the Published Literature. EPA-600/9-77-022. Washington, D.C.: U.S. Environmental Protection Agency (1977).

94. Eno, C. F. and Everett, P. H. Effects of Soil Applications of 10 Chlorinated Hydrocarbon Insecticides on Soil Microorganisms and the Growth of Stringless Black Valentine Beans. Soil Science Society of America Proceedings **22**:235-238 (1958).

95. Jones, L. W. Effects of Some Pesticides on Microbial Activities of the Soil. Bulletin 390. Utah State Agricultural College, Division of Agricultural Sciences, Agricultural Experiment Station (1956).

96. Shaw, W. M. and Robinson, B. Pesticide Effects in Soils on Nitrification and Plant Growth. Soil Science **90**:320-323 (1960).

97. Parr, J. F. and Smith, S. Degradation of Toxaphene in Selected Anaerobic Soil Environments. Soil Science **121**:52-57 (1976).

98. U.S. Environmental Protection Agency. Microbial Degradation of Pollutants in Marine Environments. EPA-600/9-79-012. Gulf Breeze, FL: U.S. Environmental Protection Agency (1979).

99. Wilson, J. T., Leach, L. E., Henson, M., and Jones, J. N. In Situ Biorestoration as a Ground Water Remediation Technique. Ground Water Monitoring Review. **6(4)**:56-64 (1986).

100. Brown, R. A. and Norris, R. D. Oxygen Transport in Contaminated Aquifers with Hydrogen Peroxide. *In* Proceedings of the NWWA/API Conference on Petroleum Hydrocarbons and Organic Chemicals in Groundwater—Prevention, Detection, and Restoration, Houston, Texas, November 5-7, 1984. Worthington, OH: National Water Well Association (1984).

101. Michelsen, D. L., Wallis, D. A., and Lavinder, S. R. In Situ Biodegradation of Dispersed Organics Using a Microdispersion of Air in Water. *In* Proceedings of the 6th National Conference on Management of Uncontrolled Hazardous Waste Sites, Washington, D.C., November 4-6, 1985. Silver Spring, MD: Hazardous Materials Control Research Institute (1985).

102. Wetzel, R. S., Davidson, D. H., Durst, C. M., and Sarno, D. J. Field Demonstration of In Situ Biological Treatment of Contaminated Groundwater and Soils. *In* Land Disposal, Remedial Action, Incineration, and Treatment of Hazardous Waste. Proceedings of the Twelfth Annual Research Symposium, April 21-23, 1986, at Cincinnati, OH. EPA/600/9-86/022. Cincinnati, OH: U.S. Environmental Protection Agency (1986).

103. Griffith, S. M. and Schnitzer, M. Organic Compounds Formed by the Hydrogen Peroxide Oxidation of Soils. Canadian Journal of Soil Science **57**:223-231 (1977).

104. U.S. Environmental Protection Agency. Office of Underground Storage Tanks. Fate and Transport of Substances Leaking From Underground Storage Tanks. Interim Report. 998-TS6-RT-CDZN-1. Washington, D.C.: U.S. Environmental Protection Agency (1986).

105. American Petroleum Institute. Literature Survey: Hydrocarbon Solubilities and Attenuation Mechanisms. Washington, D.C.: Health and Environmental Sciences Department, American Petroleum Institute (1985).

106. Goodman, D. R. and Harbison, R. D. Toxicity of the Major Constituents and Additives of Gasoline, Kerosene, and No. 2 Fuel Oil. Little Rock, AR: Division of Interdisciplinary Toxicology, University of Arkansas for Medical Sciences (1980).

107. Whitehead, W. L. and Breger, I. A. Geochemistry of Petroleum. *In* Breger, I.A. (ed). Organic Chemistry. Monograph No. 16. Earth Science Series. New York: Macmillan (1963).

108. Bruell, C. J. and Hoag, G. E. Capillary and Packed-Column Gas Chromatography of Gasoline Hydrocarbons and EDB. *In* Proceedings of the NWWA/API Conference on Petroleum Hydrocarbons and Organic Chemicals in Groundwater—Prevention, Detection, and Restoration, November 5-7, 1984, Houston, TX. Worthington, OH: National Water Well Association (1984).

109. Jhaveria, V. and Mazzacca, A. J. Bioreclamation of Ground and Groundwater by the GDS Process. Waldwick, N.J.: Groundwater Decontamination Systems, Inc. (1982).

110. Jhaveria, V. and Mazzacca, A. J. Bioreclamation of Ground and Groundwater by In Situ Biodegradation: Case History. *In* Proceedings of the 6th National Conference on Management of Uncontrolled Hazardous Waste Sites, November 4-6, 1985, Washington, D.C. Silver Spring, MD: Hazardous Materials Control Research Institute (1985).

111. Jamison, V. W., Raymond, R. L., and Hudson, J. O. Jr. Biodegradation of High-octane Gasoline in Groundwater. Developments in Industrial Microbiology **16**:305-311 (1975).

112. Jamison, V. W., Raymond, R. L., and Hudson, J. O. Jr. Biodegrada-

tion of High-octane Gasoline. *In* Proceedings of the Third International Biodegradation Symposium. London: Applied Science Publishers (1976).

113. Raymond, R. L., Jamison, V. W., and Hudson, J. O. Jr. Final Report on the Beneficial Stimulation of Bacterial Activity in Groundwaters Containing Petroleum Products. Committee on Environmental Affairs, API. Washington, D.C.: American Petroleum Institute (1975).

114. Boyer, J. D., Kosson, D. C., and Ahlert, R. C. Microbial Remediation of Soil Contaminated with 1,1,1-Trichloroethane. *In* 1986 Hazardous Material Spills Conference Proceedings, May 5-8, 1986, St. Louis, Missouri. Rockville, MD: Government Institutes, Inc. (1986).

115. Nagel, G., Kuehn, W., Werner, P., and Sontheimer, H. Sanitation of Groundwater by Infiltration of Ozone Treated Water. GWF-Wasser/Abwasser **123**:399-407 (1982).

116. Weldon, R. A. Biodisposal Farming of Refinery Oily Wastes. *In* Proceedings of the 1979 Oil Spill Conference, March 19-22, 1979, Los Angeles. Washington, D.C.: American Petroleum Institute (1979).

117. Bagawandoss, K. M., Streebin, L. E., Robertson, J. M., and Bowen, P. T. Degradation of Petroleum Fractions from Oil Refinery Wastes—A Land Treatment Study. *In* Proceedings of the NWWA/API Conference on Petroleum Hydrocarbons and Organic Chemicals in Groundwater, November 5-7, 1984, Houston, TX. Worthington, OH: National Water Well Association (1984).

118. Meyers, J. D. and Huddleston, R. L. Treatment of Oily Refinery Wastes by Land Farming. *In* Proceedings of the 34th Purdue Industrial Waste Conference, May 8-10, 1979. Ann Arbor, MI: Ann Arbor Science (1980).

119. Gudin, C. and Syratt, W. J. Biological Aspects of Land Rehabilitation Following Hydrocarbon Contamination. Environ. Pollut. **8**:107-112 (1975).

120. Ryan, J. R. and Smith, J. Land Treatment of Wood Preserving Wastes. *In* Proceedings of the National Conference on Hazardous Wastes and Hazardous Materials, March 4-6, 1986, Atlanta, GA. Silver Spring, MD: Hazardous Material Control Research Institute (1986).

121. Raymond, R. L., Hudson, J. O., and Jamison, V. W. Oil Degradation in Soil. Appl. Environ. Microbiol. **31**:522-535 (1976).

122. Yang, F-A, Chao, A., and Smallwood, C. Jr. Land Treatment of Oily Waters—Reduction of Crude Oil in Soils. *In* Proceedings of the 36th Purdue Industrial Waste Conference, May 12-14, 1981. Ann Arbor, MI: Ann Arbor Science (1982).

123. Toogood, J. A. (ed): Reclamation of Agricultural Soils After Oil Spills. Part 1: Research. Alberta Institute of Pedology Publ. M-77-11. Edmon-

ton, Alberta, Canada: University of Alberta (1977).

124. McGill, W. B. and Nyborg, M. Reclamation of Wet Forested Soils Subjected to Oil Spills. Alberta Institute of Pedology Publ. G-75-1. Edmonton, Alberta, Canada: University of Alberta (1975).

125. Cansfield, P. E. and Racz, G. J. Degradation of Hydrocarbon Sludges in the Soil. Canadian Journal of Soil Science **58**:339-345 (1978).

126. Dibble, J. T. and Bartha, R. Rehabilitation of Oil-Inundated Agricultural Land: A Case History. Soil Science **128**:56-60 (1979).

127. Pfeffer, F. M., Myers, G., Loehr, R. C., and Kincannon, D. F. Small-scale Field Evaluations of Land Treatment on an Oily Hazardous Waste. *In* Proceedings of the 39th Purdue Industrial Waste Conference, May 8-10, 1984. Boston: Butterworth Publishers (1985).

128. Loehr, R. C., Neuhauser, E. F., and Martin, J. H. Jr. Land Treatment of an Industrial Waste: Degradation and Impact on Soil Biota. *In* Proceedings of the Sixteenth Mid-Atlantic Industrial Waste Conference. Lancaster, PA: Technomic Publishing Co. (1984).

129. Olsen, R. L., Fuller, P. R., Hinzel, E. J., and Smith, P. Demonstration of Land Treatment of Hazardous Waste. *In* Proceedings of the 7th National Conference on Management of Uncontrolled Hazardous Waste Sites, December 1-3, 1986, Washington, D.C. Silver Spring, MD: Hazardous Materials Control Research Institute (1986).

130. Rice, G. B., Nichols, F. D., Snyder, H. J., and Shujins, J. Case History—Cleanup of Little Mountain Salvage Yard Acid Sludge and Oil Lagoon. *In* Proceedings of the 1976 National Conference on Control of Hazardous Material Spills, April 25-28, 1976, New Orleans, LA. Silver Spring, MD: Hazardous Materials Control Research Institute (1976).

131. Norris, D. J. Landspreading of Oily and Biological Sludges in Canada. *In* Proceedings of the 35th Purdue Industrial Waste Conference, May 13-15, 1980. Ann Arbor, MI: Ann Arbor Science (1981).

132. Demirjian, Y. A., Westman, T. R., Joshi, A. M., Rop, D. J., Buhl, R. V., and Clark, W. R. Land Treatment of Contaminated Sludge with Wastewater Irrigation. Journal of the Water Pollution Control Federation **56**:370-377 (1984).

133. Cansfield, P. E. and Racz, G. J. Degradation of Hydrocarbon Sludges in the Soil. Canadian Journal of Soil Science **58**: 339-345 (1978).

134. U.S. Environmental Protection Agency. Oily Waste Disposal by Soil Cultivation Process. Envir. Prot. Tech. Ser. EPA-R2-72-110. Washington, D.C.: U.S. Environmental Protection Agency (1972).

135. Baker, G. G. and Moe, P. G. Soil-Mediated Disposal of an Industrial Wastewater—A Case History. *In* Proceedings of the 33rd Industrial Waste Conference, May 9-11, 1978. Ann Arbor, MI: Ann Arbor Science

(1979).

136. Watts, J. R., Corey, J. C., and McLeod, K. W. Land Application Studies of Industrial Waste Oils. Environmental Pollution (Series A) **28**:165-175 (1982).

137. Francke, H. C. and Clark, F. E. Disposal of Oily Wastes by Microbial Assimilation. Report Y-1934. Washington, D.C.: U.S. Atomic Energy Commission (1974).

138. Borwitzky, H. and Schomburg, G. Separation and Identification of Polynuclear Aromatic Compounds in Coal Tar by Using Glass Capillary Chromatography Including Combined Gas Chromatography/Mass Spectrometry. Journal of Chromatography **170**:99-124 (1979).

139. Lang, K. F. and Eigen, I. Im Steinkohlenteer Nachgewiesene Organische Verbindungen. Fortschr. Chem. Forsch. **8**:91-170 (1967).

140. Novotny, M., Strand, J. W., Smith, S. L., Wiesler, D., and Schwende, F. J. Compositional Studies of Coal Tar by Capillary Gas Chromatography/Mass Spectrometry. Fuel **60**:213-220 (1980).

141. Rostad, C. E., Pereira, W. E., and Hult, M. F. Partitioning Studies of Coal-Tar Constitutents in a Two-Phase Contaminated Groundwater System. Chemosphere **14**:1023-1036 (1985).

142. U.S. Environmental Protection Agency. Environmental Hazard Rankings of Pollutants Generated in Coal Gasification Processes. EPA-600/S7-81-101. Research Triangle Park, NC: U.S. Environmental Protection Agency, Industrial Environmental Research Laboratory (1981).

143. Spiegel, S. J., Farmer, V. K., and Garvere, S. R. Heavy Metal Concentrations in Municipal Wastewater Treatment Plant Sludge. Bull. Environ. Contam. Toxicol. **35**:38-43 (1985).

144. Jacobs, L. W., Phillips, J. H., and Zabik, M. J. Toxic Organic Chemicals and Elements in Michigan Sewage Sludges—Significance for Application to Cropland. *In* Proceedings of the Fourth Annual Madison Conference of Applied Research and Practice on Municipal and Industrial Waste, September 28-30, 1981. Madison, WI: University of Wisconsin-Extension (1981).

145. Doty, W. T., Baker, D. E., and Shipp, R. F. Chemical Monitoring of Sewage Sludge in Pennsylvania. Journal of Environmental Quality **6**:421-426 (1977).

INDEX

NOTES

NOTES

NOTES